group IV IV 族	group V V 族	group VI VI 族	group VII VII 族	group VIII VIII 族
				2 Helium 氦 He 1.01
6 Carbon 碳 C 12.01	7 Nitrogen 氮 N 14.01	8 Oxygen 氧 O 16.00	9 Fluorine 氟 F 19.00	10 Neon 氖 Ne 20.18
14 Silicon 矽 Si（硅） 28.09	15 Phosphorus 磷 P 30.97	16 Sulphur 硫 S 32.06	17 Chlorine 氯 Cl 35.45	18 Argon 氬 Ar 39.45
32 Germanium 鍺 Ge 72.59	33 Arsenic 砷 As 74.92	34 Selenium 硒 Se 78.96	35 Bromine 溴 Br 79.90	36 Krypton 氪 Kr 83.80
50 Tin 錫 Sn 118.69	51 Antimony 銻 Sb 121.75	52 Tellurium 碲 Tl 127.60	53 Iodine 碘 I 126.90	54 Xenon 氙 Xe 131.30
82 Lead 鉛 Pb 207.2	83 Bismuth 鉍 Bi 208.98	84 Polonium 釙 Po (209)	85 Astatine 砹 At（砈） (210)	86 Radon 氡 Rn (222)

元素週期表
PERIODIC TABLE OF ELEMENTS

26 Iron 鐵 Fe 55.85	27 Cobalt 鈷 Co 58.99	28 Nickel 鎳 Ni 58.70	29 Copper 銅 Cu 63.55	30 Zinc 鋅 Zn 65.38
44 Ruthenium 釕 Ru 101.07	45 Rhodium 銠 Rh 102.91	46 Palladium 鈀 Pd 106.4	47 Silver 銀 Ag 107.87	48 Cadmium 鎘 Cd 112.41
76 Osmium 鋨 Os 190.2	77 Iridium 銥 Ir 192.22	78 Platinum 鉑 Pt 195.09	79 Gold 金 Au 196.97	80 Mercury 汞 Hg 200.59

前言

　　朗文科學系列圖解詞典包括生物、化學、科學、植物和地質五冊。這是一套內容既有聯繫而又各自獨立成冊的系列詞典，《朗文英漢科學圖解詞典》為其中的一冊。

　　本書收入 1500 多個基本的科學用詞，分成物理學、化學、生物學三個部分，也收入其他科學領域用的相關詞，包括地質學、氣象學及衛生保健的用詞。

　　這些詞按詞義分科目編排，英漢雙解對照，釋義簡明，概念準確。每一科目的上、下各個詞條，內容互有關聯。釋義深入淺出，易於理解；其中又標出頁碼和箭嘴號，引導讀者查找相關詞條，提供更多資料作比較，以加深理解，掌握更多詞彙。

　　書中全部使用國際制單位 (SI) 及國際理論化學與應用化學聯合會 (IUPAC) 的化學品命名法。

　　詞典中收入近 500 幅印刷精美的彩色插圖和圖表，直觀地顯示所闡釋的題目和原理，以助讀者更好理解釋義，但釋義並不依賴這些插圖。

　　詞典後部附索引，按英文字母順序排列並標注 K.K. 音標，自成一個英漢科學詞彙表，方便讀者檢索。

　　本詞典適合具高中至大學一、二年級程度的學生以及需要深入了解科學術語的非科學專業的讀者查閱使用。

朗文出版（遠東）有限公司
一九九一年十一月

朗文英漢科學圖解詞典
LONGMAN ENGLISH-CHINESE
ILLUSTRATED DICTIONARY OF
SCIENCE

Longman 朗文   YORK PRESS

Longman Group (Far East) Ltd

18/F., Cornwall House

Tong Chong Street

Quarry Bay

Hong Kong

Tel: 811 8168

Fax: 565 7440

Telex: 73051 LGHK HX

First edition published 1984

This edition published 1992

ISBN 962 359 590 5 (Hong Kong Edition)

ISBN 962 359 654 5 (Taiwan Edition)

Produced by Longman Group (Far East) Ltd.

Printed in Hong Kong

朗文出版（遠東）有限公司

香港鰂魚涌糖廠街

康和大廈十八樓

電話：811 8168

圖文傳真：565 7440

電傳：73051 LGHK HX

一九八四年初版

一九九二年第二版

國際書號 962 359 590 5（香港版）

國際書號 962 359 654 5（台灣版）

出版：朗文出版（遠東）有限公司

印刷：香港

Contents 目錄

Biology 生物學 137

The dictionary contains over 1500 basic scientific words divided into three main groups: physics, chemistry and biology. Some words may appear in two of the groups, e.g. motor[1] in physics, motor[2] in biology. Related words from other areas of science are also included: the physics part contains words from meteorology, chemistry contains words from geology and health and hygiene are found in biology. In addition the entries are arranged within the three main groups according to meaning, to help the reader to obtain a broad understanding of a particular area of science.

The main groups are further divided into subjects and parts of subjects. At the top of each page the subject is shown in bold type and the part of the subject in lighter type, for example on pages 118 and 119:

118 · CHEMICAL REACTIONS/ACIDS, BASES, SALTS

GEOLOGY/MINERALS · 119

本詞典共收 1,500 多個基本的科學用詞，分為物理學、化學、生物學三大部分編排，其中有些詞編入兩個部分中，例如 motor 這個詞編入物理學（記為 motor[1]）及生物學（記為 motor[2]）部分中。本詞典也收入其他科學領域用的相關詞，例如在物理學部分中收入氣象學用詞；化學部分中收入地質學用詞；生物學部分中收入衛生保健用詞。各詞條按照其詞義編排入上述三大部分中，旨在幫助讀者對所查閱的某一特定的科學領域獲得概括的理解。

每一個部分都分成科目，科目之下又作分段。每一頁的上方用黑體字印出有關科目名稱，並以秀麗體印出該科目下的分段名稱。例如

118 · 化學反應 / 酸、鹼、鹽
地質學 / 礦物 · 119

1. To find the meaning of a word

Look for the word in the alphabetical index at the end of the book, then turn to the page number listed.

The description of the word may contain some related words with arrows or page numbers in brackets after them. This is an indication that these words with arrows are defined nearby, and looking at that word may help in understanding the word being looked up.

(↑) the related word appears above or on the opposite page

(↓) the related word appears below or on the opposite page

(p.21) the related word appears on page 21

1. 查明詞的意義

先於詞典末尾的字母順序索引中找出欲查的詞，然後翻到該詞旁註明的頁碼。

詞的釋義中遇有一些相關詞後面帶有箭號或頁碼括在括弧中，表示該帶箭號詞的解釋就在附近，查閱這個詞的意義，有助於更好理解原先所查那個詞的意義。

(↑) 表示相關的詞出現在本詞條之前或對面頁上

(↓) 表示相關的詞出現在本詞條之後或對面頁上

(p.21) 表示相關的詞出現在第 21 頁

2. To find related words

Look for the word you are starting from in the index and turn to the page number shown. Because this dictionary is arranged by ideas, all the related words are found in a set on that or the opposite page. The illustrations will also help, as it is often difficult to give a clear description using only words.

For example, the important reproductive parts of a flower are all defined on page 211. They include *stamen*, *anther*, *stigma*, *style*, *ovary*, and *carpel*. *Pollen* and *pollination* are also described on that page, as are all the words necessary to understand pollination.

2. 查找相關的詞

先從索引中查出作為起頭的那個詞，然後翻到所註明的頁碼。由於本詞典是按照概念編排，因此可在同一頁或前、後的一頁上查出一組相關的詞。只靠文字往往難以解釋清楚，因此配以插圖幫助闡明詞義。

例如：在第 211 頁上對花的各個重要的生殖部分都作了解釋，這些部分包括："雄蕊"、"花藥"、"柱頭"、"花柱"、"子房"及"心皮"。由於這些詞對理解授粉這個詞的意義都是必要的，因此將"花粉"、"授粉"兩個詞一併放在此頁解釋。

3. To revise a subject

There are two methods. Firstly, you may wish to see if you know all the words used in that subject. Secondly, you may wish to revise your knowledge of a subject.

(a) To revise the words used in the subject of *chemical solutions*. Look up 'SOLUTION' in the alphabetical index. Turn to the page indicated, page 89. There you find the following words: *dissolve, solution, solute, solvent, saturated, unsaturated*.

Turn over to page 90, and you will find more words of related meaning to 'SOLUTION'. They are: *insoluble, solubility, concentration, dilute*. On page 91 you will see the words: *miscible, immiscible, separate, precipitation*.

All these words are relevant when you wish to write about *chemical solutions*.

(b) To revise your knowledge about a subject, for example *winds*. The only word that you can remember well is 'SEA-BREEZE'.

(i) Look up the alphabetical index and find 'SEA-BREEZE'. The page reference is 46.

(ii) Now turn to page 46. There you will find the words: *thermal, sea-breeze, land-breeze, trade wind, monsoon*.

The different kinds of winds are described. The explanation of each kind of wind depends on knowing the cause of a *thermal*, the first word on the list. After understanding this word, helped by a diagram, the other words become easier to understand. On looking at page 47 these words are seen: *whirlwind, typhoon, wind scale, cloud cirrus, cumulus, nimbus, stratus*.

These words complete the subject of *winds*, and then follows information about *clouds*. So, by using pages 46 and 47, the subject of *winds* can be revised and made clearer still by referring to the illustrations.

4. To find a word to fit a required meaning

It is almost impossible to find a word to fit a meaning in most dictionaries. It is easy with this book. For example, if you have forgotten the name of the part of the eye that is receptive to light, all you have to do is find the page reference for eye in the index and turn to that page. There is a diagram of the eye so you can see the relation of the parts, and the parts of the eye are described there, as well. You can then pick out *retina* from the others.

3. 複習科目

有兩種用法：第一，檢查您是否掌握該科目用的全部詞；第二，複習有關某一科目的知識。

(a) 複習"化學溶液"科目用的詞。先在字母順序索引中查出 SOLUTION (溶液) 這個詞，接着翻到註明的第 89 頁，即可查出"溶解"、"溶液"、"溶質"、"溶劑"、"飽和的"、"不飽和的"這些詞。

翻到第 90 頁，即可查出更多個與"溶液"這個詞的意義相關的詞，即："可溶混的"、"不溶混的"、"分離"、"分開"、"沉澱作用"這些詞。

當您想寫有關"化學溶液"這一題目的文章時，所有這些詞都是有關的詞。

(b) 複習有關某一科目，例如"風"的知識。假定您能够記起的唯一一個詞是"SEA-BREEZE"(海風) 這個詞。

(i) 查索字母順序索引，查出"SEA-BREEZE"這個詞，其頁碼是第 46 頁。

(ii) 翻到第 46 頁，可查得"上升暖氣流"、"海風"、"陸風"、"貿易風"、"季候風"這些詞。

對不同類的風都作了闡述，對各類風的解釋都有賴於理解"上升暖氣流"(thermal 這個詞排在首位) 是怎樣形成的。參照插圖閱讀，理解了這個詞的含義之後，就比較容易理解其他詞的含義。再翻到第 47 頁，可以查出"旋風"、"颱風"、"風級"、"雲"、"捲雲"、"積雲"、"雨雲"、"層雲"這些詞。

這些詞對於複習和風有關的科目已足够用了，隨後即可複習有關雲的知識。閱讀過第 46 頁和第 47 頁，既複習了有關雲的科目，也通過參照插圖閱讀獲得更明確的概念。

4. 查找適當的詞，以表達確切的意義

在大多數詞典中，想查找一個能確切表達意義的詞，幾近不可能。但使用本詞典就不難做到。例如，您忘了表達眼睛感受光線部分的名稱，您只需從索引中查出眼睛 (eye) 這個詞所在的頁碼，翻到該頁即可看到一幅插圖表示出眼睛各部分的相互關係，並有對眼睛各部分的闡述，因此您可從各個詞中選出"視網膜"(retina) 這個詞。

Useful information in science 科學上有用的資料

THE SI (SYSTEM INTERNATIONAL) UNITS 國際單位制

PREFIXES 詞頭

PREFIX 詞頭		FACTOR 因數	SIGN 符號	PREFIX 詞頭		FACTOR 因數	SIGN 符號
milli-	毫	$\times 10^{-3}$	m	kilo-	千	$\times 10^{3}$	k
micro-	微	$\times 10^{-6}$	μ	mega-	兆	$\times 10^{6}$	M
nano-	納(毫微)	$\times 10^{-9}$	n	giga-	吉	$\times 10^{9}$	G
pico-	皮(微微)	$\times 10^{-12}$	p	tera-	太	$\times 10^{12}$	T

BASIC UNITS 基本單位

UNIT	單位	SYMBOL 符號	MEASUREMENT	度量
metre	米	m	length	長度
kilogramme	千克、公斤	kg	mass	質量
second	秒	s	time	時間
ampere	安培	A	electric current	電流
kelvin	開爾文	K	temperature	溫度
mole	摩爾	mol	amount of substance	物質的量

DERIVED UNITS 導出單位

UNIT	單 位	SYMBOL 符號	MEASUREMENT	度量
newton	牛 頓	N	force	力
joule	焦 耳	J	energy, work	能、功
hertz	赫 茲	Hz	frequency	頻率
pascal	帕斯卡	Pa	pressure	壓強、壓力
coulomb	庫 倫	C	quantity of electric charge	電量
volt	伏 特	V	electrical potential	電勢
ohm	歐 姆	Ω	electrical resistance	電阻

COMMON ABBREVIATIONS 常用縮寫

a.c.	alternating current	交流電
b.p.	boiling point	沸點
d.c.	direct current	直流電
e.m.f.	electromotive force	電動勢
f.p.	freezing point	冰點、凝固點
i.r.	infra red	紅外線
m.p.	melting point	熔點
p.d.	potential difference	電勢差、電位差
r.h.	relative humidity	相對濕度
s.t.p.	standard temperature and pressure	標準溫度與壓強
u.v.	ultra violet	紫外線
v.p.	vapour pressure	蒸氣壓

COMMON CONSTANTS 常用恆量

s.t.p.   standard temperature and pressure, expressed as 1.00 atm or 760 mm Hg or 101 kPa ($=kNm^{-2}$) (Pa = pascal)   標準溫度與壓強，以 1.00 atm 或 760 毫米水銀柱高表示 或 101 kPa ($=kNm^{-2}$) 表示 (Pa = pascal 帕斯卡)

Standard volume of a mole of gas at s.t.p.,   在標準溫壓下每摩爾氣體的標準體積爲 22.4 升

The Faraday constant 法拉第常數，F, 96 500 C mol^{-1}

The Avogadro constant 亞佛加德羅常數，L, 6.02 × 10^{23} mol^{-1}

Speed of light 光速，c, 3.00 × 10^{8} ms^{-1}

1 calorie 卡路里 = 4.18 J

Specific heat capacity of water 水的比熱容量 4.18 Jg^{-1} K^{-1}

$\pi = \frac{22}{7}$ 或 3.1416

Speed of sound in air 空氣中的聲速，3.3 × 10^{2} ms^{-1}

Acceleration due to gravity 重力加速度，G, 9.81 ms^{-2}

MATHEMATICAL SYMBOLS 數學符號

$\equiv$	identically equal to	全等
$\approx$	approximately equal to	近似
$\propto$	varies directly as	與……成正比、隨……而變
∞	infinity	無窮大、無限大
ab, a·b, a × b	a multiplied by b	a 乘以 b
a/b, $\frac{a}{b}$, ab^{-1}	a divided by b	a 除以 b
sine (sin) cosine (cos) tangent (tan)	these are ratios which measure angles; they can be read from tables of angles	正弦 (sin) 餘弦 (cos) 正切 (tan) 均爲計量角度的比值，可在角度表中查得。

Physics 物 理 學

device (*n*) an object made for a special purpose to help us in our work.

tool (*n*) a device which is used with our hands to make furniture, machines, to build houses, to dig the ground.

instrument (*n*) a device which is used to do skilful work, or to measure accurately (↓).

dimension (*n*) the dimensions of a solid are its length, its breadth, and its height. Liquids and gases do not have dimensions. A solid has three dimensions; a flat surface has two dimensions (length and breadth) and a line has one dimension (length).

circumference (*n*) the line which forms a circle or is the limit of any area.

radius (*n*) the distance between the centre of a circle and its circumference or the centre of a sphere and its surface, *see diagram*.

diameter (*n*) a straight line passing from side to side through the centre of a circle or sphere, or the length of that line.

裝置(名)　爲幫助人們工作而專門製造的物件。

工具(名)　供手工製造家具、機器，建房、挖地的器件。

儀器(名)　用於技術性工作或作準確(↓)測量的器件。

空間量度(名)　固體物質的空間量度即其長度、寬度和高度。液體和氣體物質沒有空間量度。固體物質有三度；平面物質有二度(長度和寬度)；而線只有一度(長度)。

圓周線；周界(名)　形成一個圓的線或任何面積的界限。

半徑(名)　圓心至圓周之間的距離，或球心至球表面的距離。(見圖)

直徑(名)　通過圓心或球心，而兩端爲圓周或球面所截的直線或該直線的長度。

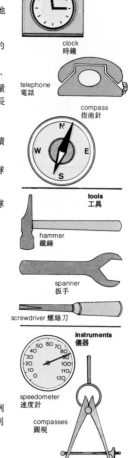

devices
裝置

clock
時鐘

telephone
電話

compass
指南針

tools
工具

hammer
鐵鎚

spanner
扳手

screwdriver 螺絲刀

instruments
儀器

speedometer
速度計

compasses
圓規

dimensions 尺度

height 高度

length 長度

breadth 寬度

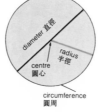

diameter 直徑

radius 半徑

centre 圓心

circumference 圓周

volume (*n*) the amount of space taken up by an object in three dimensions.

measurement (*n*) the result of measuring or the action of measuring, e.g. (1) if a rope is measured and it is 2.5 m long, then its measurement is 2.5 m; (2) a ruler was used for the measurement of the rope.

accurate (*adj*) describes a measurement that is free from mistakes, e.g. using one instrument, an object's length is 26 mm. Using another instrument, the object's length is 26.2 mm; this is a more accurate measurement. **accuracy** (*n*).

體積(名)　具長、寬、高的物體所佔據的空間。

測量；量度(名)　指量測結果或測量的動作。例如：(1)測量一條繩子，其長度爲 2.5 米，則其量度爲 2.5 米；(2)測量繩子的尺子。

準確的(形)　形容一無差錯的量度。例如用一架儀器測得一物體的長度爲 26 毫米，而用另一架儀器測得該物體爲 26.2 毫米，則後者是更加準確的測量。(名詞爲 accuracy)

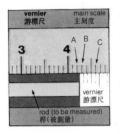

| vernier 游標尺 | main scale 主刻度 |

The 0 mark of the vernier is between lines A and B. At C a unit of the main scale meets the unit 4 of the vernier scale exactly showing that the rod is 4.24 cm or 42.4 mm.
游標尺之 0 刻度，介於線 A 與線 B 之間。在 C 點，主刻度尺的一個單位與游標尺刻度的單位會合，正確顯示桿的尺度爲 4.24cm 或 42.4 mm

scale (*n*) a set of marks, with numbers, rising from a low value to a high value.

vernier (*n*) part of an instrument used for measuring accurately, e.g. length to 0.1 mm. The vernier moves along a main scale. In the diagram the vernier reads 4.24 cm or 42.4 mm.

travel (*v*) to go from one place to another place; to go along a path, e.g. the Earth travels round the sun; a motor-car travels from one town to another town.

motion (*n*) movement; the action of travelling; going from one place to another place, e.g. the Earth is in motion round the sun; a motor-car is in motion when it is travelling. Motion is either in a straight line or in a curved line.

刻度尺；標度(名) 一組有數字的刻度，其值從低值到高值。

游標尺(名) 用於準確測量(例如測量長度短至 0.1 毫米)的儀器的部件。游標尺沿千分尺的主刻度移動。如圖，游標尺讀數 4.24 厘米或 42.4 毫米處。

運行(動) 從一處移動到另一處；沿軌道運行。例如地球環繞太陽運行；汽車從一城鎮駛到另一城鎮。

運動(名) 移動；行走的動作；從一處至另一處。例如地球繞着太陽運動；汽車行駛就是處於運動中。運動不是以直線進行就是以曲線進行。

straight line 直線

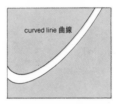

curved line 曲線

speed (*n*) the distance travelled divided by the time taken to travel the distance. If a motor-car travels 35 km in half an hour, its speed is 70 km per hour. The motion of the motor-car can be along a straight line or along a curved line.

velocity (*n*) speed in a certain direction. A motor-car travelling along a straight road with a speed of 70 km/hour also has a velocity of 70 km/hour. A motor-car travelling at 70 km/hour round a bend in a road has a velocity that is continuously changing as the road is not in a straight line.

速度(名) 行走的距離除以走過該距離所需的時間。如果一輛汽車半小時行駛 35 公里，則其速度爲每小時 70 公里。汽車可沿直線，也可沿曲線運行。

矢量速度；速率(名) 有一定方向的速度。汽車沿筆直公路以 70 公里 / 小時的速度行駛，其矢量速度也爲 70 公里 / 小時。汽車以 70 公里 / 小時沿公路上彎道行駛時，因爲公路不是筆直的，故其矢量速度也是不斷變化。

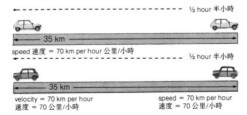

½ hour 半小時
— 35 km —
speed 速度 = 70 km per hour 公里/小時
½ hour 半小時
— 35 km —
velocity = 70 km per hour
速度 = 70 公里/小時
speed = 70 km per hour
速度 = 70 公里/小時

acceleration (*n*) the increase in velocity per unit time, i.e. the increase in velocity divided by the time taken to increase it. If the velocity of a motor-car increases from 20 metres per second (m/s) to 30 m/s in 5 seconds, then the acceleration = (increase in velocity) ÷ (time) = $(30 - 20)$ m/s ÷ 5 s = 10 m/s ÷ 5 s = 2 m/s^2 (metres per second per second) **accelerator** (*n*), **accelerate** (*v*), **accelerated, accelerating** (*adj*).

加速度(名) 每單位時間內增加的速度,即增加的速度除以增加速度所需的時間。假設一輛汽車的速度在 5 秒鐘內以每秒 20 米 (m/s) 增加到 30 m/s,那麼加速度 =(增加的速度)÷(時間)= $(30 - 20)$ m/s ÷ 5 s = 10 m/s ÷ 5 s = 2 m/s^2(米每秒每秒)。(名詞爲 accelerator,動詞爲 accelerate,形容詞爲 accelerated, accelerating)

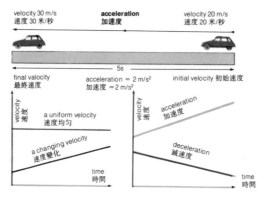

velocity 30 m/s 速度 30 米/秒 | **acceleration** 加速度 | velocity 20 m/s 速度 20 米/秒

5s

final valocity 最終速度 | acceleration = 2 m/s^2 加速度 = 2 m/s^2 | initial velocity 初始速度

velocity 速度 — a uniform velocity 速度均勻 — a changing velocity 速度變化 — time 時間

velocity 速度 — acceleration 加速度 — deceleration 減速度 — time 時間

deceleration (*n*) the rate of decrease in velocity. Deceleration is negative acceleration. **decelerate** (*v*).

rest (*n*) a state of not being in motion, e.g. a motor-car is either at rest, or it is travelling (in motion).

initial (*adj*) describes a speed, a velocity, an acceleration, or any measurement that is the first to be considered, e.g. the initial length of a spring is 10 cm. The spring is stretched and its final length is 12 cm.

uniform (*adj*) describes a speed, or a velocity, or an acceleration, that does not change with time; or also any shape, or colour, or appearance, or measurement that is the same over the whole of an object, e.g. a water-pipe of uniform diameter (p.10) has the same diameter along the whole of its length.

減速度(名) 速度減低率。減速即負加速度。(動詞爲 decelerate)

靜止(名) 處於不運動狀態。例如:一輛汽車處於靜止狀態或正在行駛(處於運動狀態)。

初始的(形) 描述最初考慮到的速度、矢量速度、加速度或任何量度。例如:一根彈簧的最初長度爲 10 cm,彈簧被拉後其最終長度爲 12 cm。

均勻的(形) 描述不隨時間而變的速度、矢量速度或加速度;任何形狀、顏色、外觀或者量度。例如:一條直徑(第 10 頁)均勻的水管,沿其整段長度的直徑都相同。

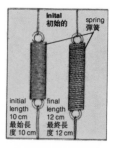

inital 初始的 | spring 彈簧

initial length 10 cm 最始長度 10 cm | final length 12 cm 最終長度 12 cm

quantity (n) (1) An indefinite amount of a material when no measurement is given, e.g. take some milk and add the same quantity of water: the quantity can be measured by volume or by weight A *quantity* of sand. (2) The different kinds of measurements made in science. Examples of scientific quantities are: length, weight, volume, velocity, density, force.

variable (adj) describes a quantity which can change, or can be changed, e.g. (1) the speed of a motor-car is variable – a person can drive it at different speeds, (2) the velocity of the wind is variable for it changes by itself. **vary** (v).

average (n) an average is the sum of variable quantities divided by the number of the quantities, e.g. the average of 10 m, 16 m, 8 m, 12 m, is: (10 m + 16 m + 8 m + 12 m) ÷ 4 = 46 m ÷ 4 = 11.5 m. **average** (adj).

量；數量(名)　(1)無指明其量度的、不確定量的物質。例如：取些牛奶並加入等量的水；其量可按容量或按重量來表示。又如沙的量。(2)科學上有不同種類的量度。例如：長度、重量、體積、速率、密度、力。

可變的(形)　描述一個可改變或可被改變的量。例如：(1)汽車的速度是可變的，因爲人可以不同的速度駕駛汽車；(2)風速是易變的，因爲它本身經常變化。(動詞爲 vary)

平均數(名)　平均數是可變量的總和除以量的個數。例如 10 m、16 m、8 m、12 m 的平均數是：(10 m + 16 m + 8 m + 12 m) ÷ 4 = 46 m ÷ 4 = 11.5 m。(形容詞爲 average)

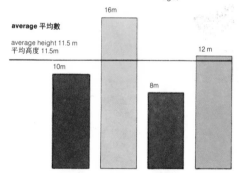

average 平均數

average height 11.5 m
平均高度 11.5m

16m

10m

8m

12 m

standard (n) a measurement that is agreed by all people. Either it is clearly described, or it can be used to see if a measuring instrument is accurate, e.g. (1) the metre is the standard of length, it has an accurate description; (2) a kilogram of metal is the standard of mass (p.14) and it is used to see if a weighing instrument is accurate. **standard** (adj).

unit (n) a standard measurement, e.g. (1) a length of 12 m has a number: 12, and a standard unit: metres; (2) a volume of 0.4 m³ has a number: 0.4, and a standard unit: cubic metres.

標準(名)　公認的量度。標準可以清楚地描述，或用來查看儀器是否準確。例如：(1)米爲長度的標準，它有一個準確的描述；(2)一千克金屬是質量(第 14 頁)的標準，它用來檢查稱重儀器是否準確。(形容詞爲 standard)

單位(名)　標準的量度。例如：(1)12 米的長度有一個數字：12，和一個標準單位：米；(2)0.4 m³ 的體積只有一個數字：0.4，和一個標準單位：立方米(m³)。

mass (*n*) a measure of the amount of a material. All materials possess mass. Masses are compared by weighing which is comparing the forces of gravitation (p.17) acting on them. It is measured in kilograms. The mass of an object never changes.

質量(名)　物質的一種量度。一切物質都具有質量。質量通過稱重作比較，而重量相當於作用於它們的地心引力(第 17 頁)。質量的計量單位為千克。任何物體的質量永遠不改變。

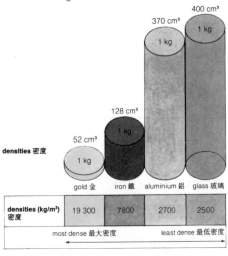

densities 密度

gold 金	iron 鐵	aluminium 鋁	glass 玻璃
52 cm³	128 cm³	370 cm³	400 cm³
1 kg	1 kg	1 kg	1 kg

densities (kg/m³) 密度	19 300	7800	2700	2500

most dense 最大密度　　　　　　least dense 最低密度

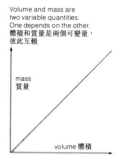

Volume and mass are two variable quantities. One depends on the other.
體積和質量是兩個可變量，彼此互賴

mass
質量

volume 體積

density (*n*) mass per unit volume, i.e. the mass of a material divided by its volume, e.g. 128.2 cm³ of iron has a mass of 1 kg: density of iron = (mass) ÷ (volume) = 1 kg ÷ 128.2 cm³ = 7.8 g/cm³ = 7800 kg/m³ (kilograms per cubic metre). Each material has a particular density, e.g. the density of water is 1 g/cm³, so density is important in identification (p.93) of materials. **dense** (*adj*).

密度(名)　每單位體積的質量，即物質的質量除以其體積。例如：128.2 cm³ 的鐵具有 1 kg 的質量：鐵的密度＝(質量)÷(體積)＝1 kg ÷ 128.2 cm³ = 7.8 g/cm³ = 7800 kg/m³，每種物質都有特定的密度。例如，水的密度是 1 g/cm³，所以密度對鑒別(第 93 頁)物質很重要。(形容詞爲 dense)

relative density relative density = (density of a material) ÷ (density of water). For iron, density is 7800 kg/m³, for water, density is 1000 kg/m³. The relative density of iron = 7800 kg/m³ ÷ 1000 kg/m³ = 7.8. Relative density is only a number, it has no units.

相對密度　相對密度＝(物質的密度)÷(水的密度)。鐵的密度是 7800 kg/m³，水的密度是 1000 kg/m³。鐵的相對密度 = 7800 kg/m³ ÷ 1000 kg/m³ = 7.8。相對密度只是一個數值，沒有單位。

specific gravity another name for relative density.

比重　相對密度的別稱。

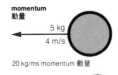

momentum
動量

5 kg
4 m/s

20 kg/ms momentum 動量

2 kg
10 m/s

20 kg m/s momentum 動量

momentum (*n*) momentum is the product of the mass and the velocity of an object. For example, an object with mass (↑) of 5 kg moving with velocity (p.11) of 4 m/s has a momentum of 20 kg m/s. A solid, or a fluid (p.39), when in motion possesses momentum.

force (*n*) a push or pull which causes (1) acceleration, *or* (2) a change in the shape of an object, *or* (3) a reaction (p.16). A force is measured by the change in momentum produced in 1 second. A force cannot be seen, only its effects can be seen. A force can also be measured by (1) the amount it stretches a spring; (2) the acceleration it gives to a mass. Force = (mass) × (acceleration). In symbols, $F = ma$. Examples of common forces are shown:

動量(名)　動量是物體的質量與其速度的乘積。例如質量(↑)爲 5 kg 的物體以 4 m/s 的速度（第 11 頁）運動，則其動量爲 20 kgm/s。固體或流體（第 39 頁）運動時都具有動量。

力(名)　引起：(1)加速度；或(2)改變物體形狀，或(3)反作用（第 16 頁）的推力或拉力。力是以 1 秒鐘內產生的動量的變化來測量的。力雖然無形，但其效果可察見。力亦可藉下列的方法測量：(1)拉伸一個彈簧的長度；(2)力使某一質量物體獲得的加速度。即力＝(質量)×(加速度)。用符號表示爲：$F = ma$。力的常見例子，如下列各圖所示：

common forces
常見的力

wind
風力

water
水力

muscles
肌力

gravity
重力

magnetism
磁化力

tendency (*n*) trying to continue an action, or to obtain a result or to maintain a certain direction, or situation. **tend** (*v*).

inertia (*n*) the tendency (↑) of an object to maintain its state of rest or uniform motion in a straight line. A force is needed to oppose inertia. **inertial** (*adj*).

傾向；趨向(名)　力求繼續一種動作，獲得一種結果，或保持一定的方向或位置。(動詞爲 tend)

慣性(名)　物體保持靜止狀態或勻速直線運動的趨向(↑)。力爲克服慣性所需。(形容詞爲 inertial)

newton (*n*) the unit of measurement for a force. A force of 1 newton gives an acceleration of 1 m/s² to a mass of 1 kg. The symbol is N.

牛頓(名) 力的量度單位。一牛頓的力使 1 kg 的物質獲得 1 m/s² 的加速度。符號爲 N。

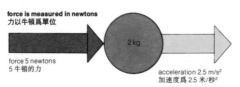

force is measured in newtons
力以牛頓爲單位

force 5 newtons
5 牛頓的力

2 kg

acceleration 2.5 m/s²
加速度爲 2.5 米/秒²

action (*n*) the effect produced by a force, e.g. a hammer (p.10) hits a piece of metal, the force of the hammer acts on the metal; the action of the force is the effect: the metal changes shape. **act** (*v*).

作用;作用力(名) 力產生的效果,例如用錘(第 10 頁)捶打一塊金屬;錘的力作用於金屬;力作用的效果是使金屬變形。(動詞爲 act)

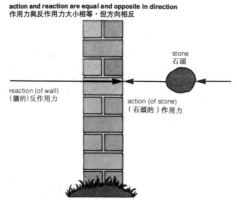

action and reaction are equal and opposite in direction
作用力與反作用力大小相等,但方向相反

stone
石頭

reaction (of wall)
(牆的)反作用力

action (of stone)
(石頭的)作用力

reaction (*n*) the opposite effect to an action, e.g. when a stone is thrown to hit a wall, the stone has an action on the wall. The wall does not move, nor is its shape changed. Instead the wall pushes back with a reaction which is equal to the action. A reaction is equal to an action, but is in the opposite direction.

反作用力(名) 對作用力的相反作用。例如拋出一塊石頭碰在牆上時,石頭對牆有作用力。牆不移動,亦無變形,反而以一個和作用力相等的反作用力把石頭反彈回來。反作用力與作用力大小相等,但方向相反。

diagram (*n*) a simple drawing using mainly lines. For example, a diagram is used to explain how a device works, or how an instrument is used. Words, called labels, help the description.

簡圖;示意圖(名) 主要用線條繪出簡單圖形。例如示意圖用來說明一個裝置如何工作。或一台儀器如何使用。標出的字有助於說明。

attraction (n) a force which tries to draw two objects together, e.g. a magnet has an attraction for a compass needle. **attract** (v).

gravity (n) the attraction of the Earth for all solids, liquids, and gases, e.g. if dropped a book falls to the floor because of gravity, that is, the Earth has an attraction for the book. Gravity depends on the masses of two objects and their distance apart. The greater the mass, the greater the attraction; the greater the distance, the less the attraction. **gravitational** (adj).

weight (n) the force of gravity on a mass which attracts the mass towards the Earth. The mass of an object does not vary. It is measured in kilograms. The weight of an object is measured in newtons (↑) and depends on the place on the Earth where it is measured. A mass of 1 kg has a weight of 9.78 N at the Equator and a weight of 9.83 N at the North or South Poles. **weigh** (v) to measure the mass or the weight of an object.

引力（名）　力圖將兩物體吸引在一起的力，例如：磁鐵對羅盤指針有吸引力。（動詞爲 attract）

重力（名）　地球對固體、液體和氣體的吸引力，例如：下落的一本書掉向地板是由於重力，即地球對書有吸引力所致。重力的大小取決於兩個物體的質量和它們的距離。質量愈大，引力就愈大；距離愈大，引力就愈小。（形容詞爲 gravitational）

重量（名）　作用於物質的地心引力，它將物質吸向地球。物體的質量是不變的，它以千克爲計量單位。物體的重量以牛頓（↑）爲計量單位，並依其在地球上被測量的位置而定。1 kg 物質的重量在赤道上爲 9.78 N，而在南極或北極則爲 9.83 N。（動詞 weigh 意爲稱量測量物體的質量或重量）

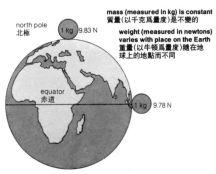

mass (measured in kg) is constant
質量(以千克爲量度)是不變的

weight (measured in newtons) varies with place on the Earth
重量(以牛頓爲量度)隨在地球上的地點而不同

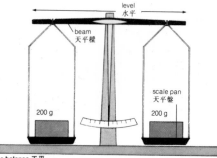

level
水平

beam
天平樑

scale pan
天平盤

200 g

200 g

a balance 天平

balance (*n*) an instrument (p.10) for measuring mass. When two equal masses are placed in the scale pans, *see diagram*, the arm of the balance is level. A balance weighs masses accurately anywhere on the Earth. **balance** (*v*).

balance (*v*) of forces in opposition, to be equal in magnitude (↓). **balance** (*n*).

spring balance an instrument for measuring weight. The weight of an object stretches the spring and the weight is read on the scale.

approximate (*adj*) describes a measurement, or a number, which is not accurate, but is near enough in magnitude (↓) to use in a calculation, e.g. the accurate speed of light is 2.997925×10^5 km/s; the approximate speed (good enough for most calculations) is 3×10^5 km/s. **approximation** (*n*), **approximate** (*v*).

significant (*adj*) describes figures which have determining value, e.g. a piece of brass has a mass of 792 g and a volume of 94 cm³; calculation of its density is $792\,g \div 94\,cm^3 = 8.425532\,g/cm^3$; 8.4 g/cm³ is accurate to 2 significant figures because there cannot be more figures in the answer than in the original measurements; the remaining figures in the answer are usually not important.

constant (*adj*) describes a quantity or a measurement which never changes in magnitude, e.g. an object falling to Earth has an acceleration which is constant at any one place; the acceleration always has the same magnitude. **constant** (*n*).

天平(名)　測量質量的儀器(第 10 頁)。當兩個質量相等的物件放在天平的兩個盤中時(見圖)，天平的樑呈水平。在地球上任何地方天平都準確地稱出質量。(動詞為 balance)

保持平衡(動)　指方向相反，量值(↓)相等的力。(名詞為 balance)

彈簧秤　一種測重儀器。物體重量使彈簧伸長，重量顯示在刻度表上。

近似的(形)　描述不準確的，但足可用於計算的量值(↓)上十分接近的量度或數值。例如，光的精確速度為 2.997925×10^5 km/s，光的近似速度(對於大多數計算來說已足夠準確)為 3×10^5 km/s。(名詞為 approximation，動詞為 approximate)

有效的(形)　描述一個具有決定性值的數字。例如，一塊黃銅的質量為 792 g，體積為 94 cm³，計算出黃銅的密度為 $792\,g \div 94\,cm^3 = 8.425532\,g/cm^3$，8.4 是準確到二位的有效數字。因為答數中不可能有比原來計算的更多數字；答數中的其餘數字通常可忽略。

恆定的(形)　描述一個量值永遠不變的量或量度，例如：物體落向地面的加速度在任何一處均恆定不變；加速度總是量值相同。(名詞為 constant)

spring balance (measures weight)
彈簧秤
(測量重量)

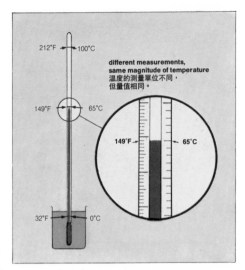

different measurements,
same magnitude of temperature
溫度的測量單位不同，
但量值相同。

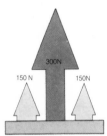

resultant of 300 N has same
effect as two forces of 150 N
300 N 的合力，
效果與兩個各 150
N 的力相同

magnitude (*n*) the size of a measurement, e.g. 2 miles and 3.22 km are different measurements of the same distance. The two measurements have the same magnitude.

vector[1] (*n*) a quantity that has a magnitude and a direction, e.g. velocity (p.11) and weight are vectors. The direction of a straight line shows the direction of the vector, its length shows the magnitude and its origin shows the point of action. A line is said to *represent* the vector.

scalar (*n*) a quantity that has magnitude but no direction, e.g. mass, density.

resultant (*n*) the single force which has the same effect as two or more forces acting on an object.

component (*n*) a force acting on an object can be considered as two, or more, forces called components, which produce the same effect.

量值；數量級（名） 指量度的大小。例如，2 英里和 3.22 公里是同一距離的不同測量單位。這兩種量度的量值相同。

矢量（名） 一個具有量值和方向的量。例如，矢量速度（第 11 頁）和重量都是矢量。直線的方向表示矢量的方向，直線長度表示矢量的量值，而直線原點表示作用點。一條線代表矢量。

無向量（名） 有量值而無方向的量。例如質量、密度。

合力（名） 指一個單一的力，其效果等於作用於一個物體上的兩個或多個力。

分力（名） 作用於一個物體上的力，可以看成產生相同效果的兩個或多個力（稱爲分力）所組成。

pushes roller with force F
合力 F 推動輪子

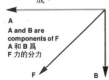

A
A and B are
components of F
A 和 B 爲
F 力的分力

F   B

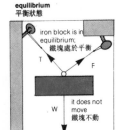

equilibrium
平衡狀態

iron block is in
equilibrium;
鐵塊處於平衡

T   F

W

it does not
move
鐵塊不動

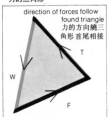

triangle of forces
力的三角形

direction of forces follow
found triangle
力的方向繞三
角形首尾相接

T

W

F

first position
原先位置

moves slightly
略微移動

returns to position
回復原先位置

stable equilibrium
穩定平衡狀態

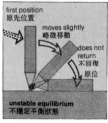

first position
原先位置

moves slightly
略微移動

does not
return
不回復
原位

unstable equilibrium
不穩定平衡狀態

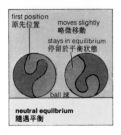

first position
原先位置

moves slightly
略微移動

stays in equilibrium
停留於平衡狀態

ball 球

neutral equilibrium
隨遇平衡

equilibrium (*equilibria*) (*n*) a state of balance (p.18) between opposing forces or effects. Forces which balance each other are *in equilibrium*. An object is in equilibrium if the forces acting on it are in equilibrium.

triangle of forces a triangle whose sides are parallel and proportional (p.23) to the three forces in equilibrium at a point. The directions of the forces must follow each other round the triangle.

stable (*adj*) the state of an object if it returns to its first position after being moved slightly, e.g. a cup tilted on a table.

unstable (*equilibrium*) (*adj*) the state of an object if on being moved slightly it moves further away and the equilibrium is upset, e.g. a pencil balanced on its point.

neutral (*equilibrium*) (*adj*) the state of an object if it stays in its new position after being moved slightly, e.g. a ball on a level floor.

work (*n*) work is done when a force moves an object. The amount of work is measured by: (strength of force) × (distance moved). The distance is measured in the direction in which the force acts. The symbol for work is *w*. $w = f \times s$. The unit of work is the joule.

平衡狀態（名）　相反的力或作用之間的平衡（第18頁）狀態。互相平衡的力處於平衡狀態，如果作用於一個物體的幾個力處於平衡狀態，那麼該物體即處於平衡狀態。

力的三角形　三角形的邊與平衡於一點的三個力平行並成正比（第23頁）。力的方向必定繞三角形首尾相接。

穩定平衡（形）　物體被略微移動後又回復原先位置的狀態。例如杯子斜置在桌子上。

不穩定平衡（形）　物體被略微移動後，如果進一步移動，將破壞平衡的狀態。例如鉛筆倒立着時的狀態。

隨遇平衡（形）　物體略被移動後停留於新位置上的狀態。例如球在平坦的地板上。

功（名）　用力移動物體時做功。功的大小可用下法求出：（力）×（移動的距離）。距離可沿力作用的方向測量出。功的符號爲 *w*。功的單位爲焦耳。$w = f \times s$

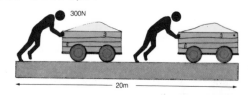

300N

20m

workdone 做的功 = 300N × 20 m = 6000j = 6kj

energy (*n*) the ability to do work. There are different forms of energy: potential energy (stored energy); kinetic energy (energy from motion, p.11); heat energy; light energy; electrical energy; chemical energy; nuclear energy. One form of energy can be *transformed* into another form. The unit of energy is the joule.

joule (*n*) one joule of work is done when a force of 1 newton moves an object a distance of 1 metre in the direction of the force. The symbol for joule is J. $1\,J = 1\,N \times 1\,m$.

power (*n*) the rate (p.23) of doing work, i.e. the amount of work done divided by the time taken to do the work. If a force does 20 J of work in 5 seconds, then the power is: $(20\,J \div 5\,s) = 4\,J/s$. The symbol for power is *P*. The unit of power is the watt.

watt (*n*) a power of 1 watt is produced when 1 joule of work is done in 1 second. The symbol for watt is W. $1\,W = 1\,J \div 1\,s$.

instantaneous (*adj*) happening in an instant, that is no time seems to be taken between a cause and its effect, e.g. when a bell is struck, no time is taken between the striking and the ringing of the bell.

能 (名) 做功的能力。能有勢能(貯藏的能)、動能(運動產生的能(第 11 頁))、熱能、光能、電能、化學能、核能等不同形式。一種形式的能可轉換爲另一種形式的能。能的單位爲焦耳。

焦耳 (名) 一焦耳的功指一牛頓的力使物體在力的方向上位移一米所做的功。焦耳的符號爲 $J \circ 1\,J = 1\,N \times 1\,m$

功率 (名) 做功的比率(第 23 頁),即所做功的大小除以做功所費的時間。設一個力在 5 秒鐘內做 20 J 的功,則功率是:$(20\,J \div 5\,s) = 4\,J/s$。功率的符號是 P。功率的單位是瓦特。

瓦特 (名) 一秒鐘內做一焦耳功所產生的功率爲一瓦特。瓦特的符號爲 W。
$$1\,W = 1\,J \div 1\,s$$

瞬時的 (形) 發生在瞬間的,即因果之間似乎無時間間隔。例如敲鐘時,敲鐘和鐘響之間無時間間隔。

instantaneous
瞬時的

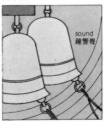

simultaneous
同時的

simultaneous (*adj*) (of *two* events) happening at the same time, e.g. two stones, held in one hand, drop simultaneously, when released.

potential energy stored energy is potential energy. Energy can be stored: by raising a weight through a height; in a stretched spring.

同時的;同步的 (形) (指兩件事情)在同一時間發生的。例如,手握兩塊石頭,手一鬆開時,石頭同時下落。

勢能;位能 貯能是勢能。能可以貯藏,提高重物至某一高度以及拉長的彈簧皆貯有能。

resistance¹ (*n*) a force which opposes a change in motion or a change in shape, e.g. a fluid offers resistance to an object moving through it.

friction (*n*) the resistance to motion between two surfaces moving over each other. Friction only appears when the surfaces try to move. If the moving force is not strong enough, friction prevents motion. **frictional** (*adj*).

overcome (*v*) to be too strong for an opposition. Friction opposes the motion of an object. If a force is strong enough it overcomes friction and the object moves.

lubricant (*n*) a material put on a surface to lessen friction. Oil is used as a liquid lubricant, graphite as a solid lubricant. **lubricate** (*v*).

阻力（名） 一種對抗運動發生變化或形狀發生變化的力。例如，流體對在其中運動的物體有阻力。

摩擦力（名） 兩個接觸面作相對運動的阻力。只有接觸面試圖運動時，才會有摩擦力。如果動力不夠大，摩擦力就會阻止運動的發生。（形容詞為 frictional）

克服（動） 強過所受到之對抗。摩擦力阻止物體運動。如果力大，足以克服摩擦力，則物體發生運動。

潤滑劑（名） 塗於物體表面以減少摩擦力的物質。油用作液體潤滑劑，石墨用作固體潤滑劑。（動詞為 lubricate）

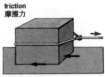

friction
摩擦力

friction prevents motion
摩擦力阻止運動

lubricant 潤滑劑

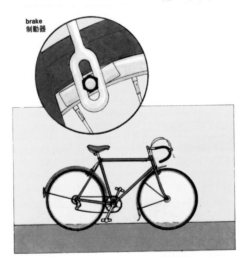

brake
制動器

brake (*n*) a device (p.10) which uses friction to stop a wheel from turning. **brake** (*v*) to use a brake to make a motor-car, cart, or engine, travel more slowly, or stop.

lever (*n*) a bar used to lift or move heavy weights. The bar turns about a point when lifting the weight. There are three types of levers: simple lever, crowbar lever and tongs lever.

制動器（名） 利用摩擦力阻止輪子轉動的一種裝置（第 10 頁）。動詞 brake 意為制動 用制動器使汽車、手推車、機車減速或停車。

槓桿（名） 用來提舉或搬動重物的桿件。提舉重物時，槓桿繞一個支點運動。槓桿有三種：簡單槓桿、橇棍槓桿和鉗式槓桿。

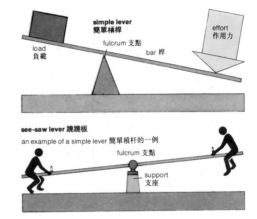

simple lever
簡單槓桿

fulcrum 支點

effort
作用力

load
負載

bar 桿

see-saw lever 蹺蹺板

an example of a simple lever 簡單槓桿的一例

fulcrum 支點

support
支座

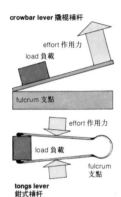

crowbar lever 撬棍槓桿

effort 作用力

load 負載

fulcrum 支點

effort 作用力

load 負載

fulcrum
支點

tongs lever
鉗式槓桿

fulcrum (*n*) the point about which a lever turns, or the support about which a lever turns.

load (*n*) (1) the weight lifted or moved by a lever, or any other machine (p.24). (2) the weight supported by a pillar; the weight on a bridge.

effort (*n*) the force (p.15) applied to a lever to lift a load; the force applied to any machine to raise a load.

exert (*v*) to bring or put into action, e.g. the earth exerts an attractive force on the moon.

input (*n*) the total amount of energy put into a device.

output (*n*) the amount of useful energy given out by a device.

ratio (*n*) the relation between two numbers or two measurements, usually with the same unit, e.g. the ratio of 7 m to 1 m is written as 7:1 or 7/1.

rate (*n*) the relation between two measurements with different units, e.g. the rate of change of distance with time (i.e. speed, measured in metres/second).

proportion (*n*) the condition of a set of ratios being equal, e.g. 1 mm/7 N and 3 mm/21 N are in proportion. The lengths measured in millimetres are proportional to the forces measured in newtons. **proportional** (*adj*).

支點(名) 槓桿繞着轉動的點或槓桿繞着轉動的支座。

負載；負荷(名) (1)槓桿或其他任何機械(第24頁)提舉或搬運的重物；(2)柱子支撐的重量；橋的載重。

作用力(名) 施加於槓桿以提起重物的力(第15頁)。施加於任何機械以舉起重物的力。

施加力(動) 使發生作用。例如，地球對月球施加吸引力。

輸入(名) 進入一裝置的總能量。

輸出(名) 一裝置放出的有效能量。

比率；比(名) 兩個數或兩個度量之間的關係，通常使用同一單位。例如，7 m 與 1 m 之比寫成 7:1 或 7/1。

率；速率(名) 不同單位的兩個度量之間的關係，例如距離隨時間的變化率(即速度是以米/秒作爲測量單位)。

比例(名) 一組比例相等的情況，例如：1 mm/7 N 和 3 mm/21 N 是成比例的。以毫米計的長度與以牛頓計的力成比例。(形容詞爲 proportional)

machine (*n*) a device (p.10) used to help in doing work; in it a force (the effort), applied at one part of the machine, overcomes another force (the load), acting at another part of the machine; or the machine changes the direction of application of a force. A lever is a simple machine.

mechanical advantage the number of times the load on a machine is greater than the effort, e.g. a load of 600 N is raised by an effort of 150 N; the mechanical advantage of the machine is: (600 N ÷ 150 N) = 4.

velocity ratio the ratio (p.23) of the distance moved by the effort to the distance moved by the load.

inclined plane a simple machine which is a sloping plane surface. Friction between the load and the inclined plane lessens the mechanical advantage.

機器；機械（名）用以幫助作功的裝置（第 10 頁）；施於機器一個部件的力（作用力）克服作用於機器另一部件的另一個力（負載）；或者機器的改變施力的方向。槓桿是一種簡單的機械。

機械效益 機器上的負載大於作用力的倍數，例如 150 N 的作用力舉起 600 N 的機械效益爲：(600 N ÷ 150 N) = 4

速比 作用力移動距離與負載移動距離之比（第 23 頁）。

斜面 傾斜的平面是一種簡單的機械。負載和斜面間的摩擦力降低機械的效益。

machines
機械

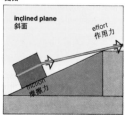

pulley (*n*) a wheel over which a rope, or chain, passes; it is used to lift weights, or to pull objects. The wheel turns on a pin fixed in a frame.
pulley-block several pulleys inside a frame.

滑輪（名） 有繩索或鏈條在其上繞過的輪子；用以提起重物或牽引物體。輪子繞固定於機架的軸釘轉動。
滑輪組 一個機架內有幾個滑輪。

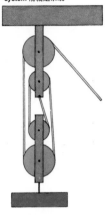

block and tackle pulley system 滑輪組系統

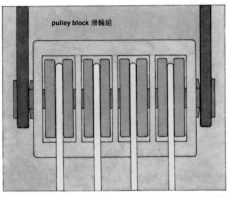

pulley block 滑輪組

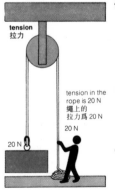

tension 拉力

tension in the rope is 20 N 繩上的拉力爲 20 N

20 N

20 N

pulley system an arrangement ot pulleys to form a machine.
tension (*n*) (1) the force exerted on or by a stretched object, e.g. a rope, a chain, a spring. The rope, chain, or spring that is stretched is *under tension*. (2) the condition of a rope etc., that is being stretched. The tension in a rope, chain or spring, is the same force along the whole of its length. **tensile** (*adj*).
revolution (*n*) (1) the motion of an object round a point outside itself, e.g. the revolution of the Earth round the Sun. (2) one complete turn of a wheel is one revolution; the circumference (p.10) is considered to revolve. **revolve** (*v*).
efficiency (*n*) the ratio (usually given as a percentage) of the energy output (p.23) to the energy input (p.23). If 200 J of work is done by the effort and 160 J of work is done on the load, then the efficiency = (160 J ÷ 200 J) × 100% = 80%. Some work is always lost because of overcoming friction in the moving parts of a machine.
perfect (*adj*) a machine which has 100% efficiency is perfect; no practical machine is perfect.

滑輪系統 構成一種機械的滑輪裝置。

拉力；拉緊狀態（名） （1）施於一件物體的力或由一件受拉伸體。例如繩子、鏈條、彈簧所施加的力。拉緊的繩子、鏈條或彈簧處於拉力下；（2）被拉緊的繩子等的狀態。繩子、鏈條或彈簧內的拉力在其整個長度內大小相同。（形容詞爲 tensile）

旋轉；轉數（名） （1）物體繞着其外側的一點運動，例如：地球繞太陽轉動；（2）輪子轉一周是一整轉；其圓周線（第 10 頁）被認爲是能旋轉的。（動詞為 revolve）

效率（名） 能量輸出（第 23 頁）與能量輸入（第 23 頁）之比（通常以百分比表示）。如果作用力做了 200 J 的功，而在重物上做的功是 160 J，那麼效率 =（160 J ÷ 200 J）× 100% = 80%。機器的運動機件總是要克服摩擦力。

完美的（形） 具有 100% 效率的機器是完美的機器。實際使用的機器並不完美。

particle (*n*) a small piece of material, such as a grain (of sand), a molecule (↓) or an atom (p.103).

molecule[1] (*n*) the smallest particle (↑) of any material that has the chemical properties (p.93) of that material. Materials are considered to be made of molecules held together by attractive (p.17) forces. **molecular** (*adj*).

cohesion (*n*) the force of attraction (p.17) between molecules of the same solid, or liquid, material, e.g. the cohesion between molecules of glass. **cohesive** (*adj*), **cohere** (*v*).

adhesion (*n*) the force of attraction (p.17) between molecules of different materials, e.g. the adhesion between molecules of water and molecules of glass. The adhesion of water to glass is stronger than the cohesion of water. **adhesive** (*adj*), **adhere** (*v*).

微粒；粒子（名） 物質的細小粒。例如：沙粒、一個分子（↓）或一個原子（第 103 頁）。

分子 任何物質中能保持該物質的一切化學性質（第 93 頁）的最小微粒（↑）。一切物質都是由其分子靠吸引（第 17 頁）力保持在一起而構成的。（形容詞爲 molecular）

內聚力（名） 在同類固體或液體物質的分子間的引（第 17 頁）力。例如玻璃分子間的內聚力。（形容詞爲 cohesive，動詞爲 cohere）

黏附力（名） 不同類物質分子間的吸引（第 17 頁）力。例如水分子和玻璃分子之間的黏附力。水黏附玻璃的力強於水的內聚力。（形容詞爲 adhesive，動詞爲 adhere）

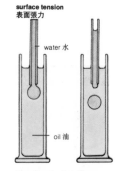

surface tension
表面張力

water 水

oil 油

formation of a drop of water in oil, ball-shape formed
在油中的一滴水形成球形

needle floating on water held up by surface tension
表面張力令針保持在水面上

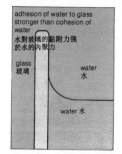

adhesion of water to glass stronger than cohesion of water
水對玻璃的黏附力強於水的內聚力
glass 玻璃
water 水
water 水

glass 玻璃
mercury 水銀
cohesion of mercury stronger than adhesion of mercury to glass
水銀的內聚力強於水銀對玻璃的黏附力

adhesion/cohesion 黏附的/內聚力

surface tension the tendency (p.15) of the surface of a liquid to behave as though covered with a skin. This is due to the cohesive forces on molecules at and near the surface not being balanced; such molecules are attracted more towards the centre of the liquid. The effect of surface tension is to make a drop of liquid take up the smallest space, usually in the shape of a ball.

capillarity (*n*) the rise or fall of liquids in a capillary tube; a capillary tube is a pipe with a very small diameter (p.10). This effect causes the *capillary rise* of water in plants. **capillary** (*adj*).

meniscus (*n*) the curved surface of a liquid in a pipe, or in a narrow vessel.

表面張力 液體表面的趨向（第 15 頁）表現恰如覆蓋有一層皮膜。這是由於表面或接近表面處的分子內聚力不平衡所致，這些分子被吸引向更靠近液體中心。表面張力的效果是使一滴液體佔有最小的空間，通常呈球形。

毛細作用（名） 液體在毛細管中的升或降；毛細管是直徑（第 10 頁）極細的管。毛細作用使水在植物體內毛細上升。（形容詞爲 capillary）

彎月面（名） 指管子中或狹窄容器中液體的彎曲表面。

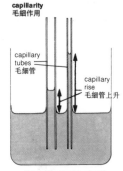

capillarity
毛細作用

capillary tubes
毛細管

capillary rise
毛細管上升

the smaller the diameter of capillary tube the greater the capillary rise
毛細管直徑越細，毛細上升越大

meniscus 彎月面

elasticity
彈性

elastic band 彈性帶

first
length
最初長度

returns
to first
length
恢復到最
初長度

extension
延伸

extended
length
延伸的長度

15N

rubber tube
first shape
橡膠管的
最初形狀

deformation
of shape
形狀的變形

returns to first shape
恢復最初形狀

iron
wire
鐵線

first shape
最初形狀

remains
deformed
變形仍存在

deformation of shape
形狀變形

kinetic (*adj*) having to do with motion (p.11). *Kinetic energy* is the energy of motion. If an object has a mass *m* and velocity *v*, and kinetic energy *E*, then $E = \frac{1}{2}mv^2$.

Brownian motion or **movement** the continuous, irregular motion of particles in a fluid on their being hit by moving molecules of the fluid. This effect can be seen in smoke.

diffusion (*n*) the spreading of molecules of a material through a solid, a liquid or a gas. A drop of ink in water slowly diffuses throughout the whole of the water. Diffusion is caused by the motion of molecules, as seen in Brownian motion (↑).

property (*n*) a description of a material, which allows it to be recognized; *physical property* a property which does not affect the chemical nature of the material; physical properties include: size; smell; colour; density (p.14); state, i.e. solid, liquid, or gas; elasticity.

elasticity (*n*) the ability of an object, or a material, to return to its first size, or shape, after being stretched by a force, and the force then taken away. **elastic** (*adj*) both rubber and a steel wire are elastic.

extension (*n*) the increase in length when an elastic solid is stretched by a force. **extend** (*v*).

deformation (*n*) an alteration in the size or shape of a solid. Within the elastic limit (p.28) an elastic solid returns to its first size or shape when the deforming force is taken away. For a plastic solid, the deformation remains when the force no longer acts. **deform** (*v*), **deformed** (*adj*).

動力學的（形） 與運動（第 11 頁）有關的。動能是運動的能。如果物體的質量爲 *m*，速度 *v*，動能爲 *E*，則 $E = \frac{1}{2}mv^2$

布朗運動 流體的微粒受流體中運動着的分子碰撞時產生連續，無規則的運動。在烟塵中可見這種效應。

擴散（名） 物質的分子散佈在固體、液體或氣體中。一滴墨水落到水中，慢慢地擴散到全部水中。正如在布朗運動（↑）中所描述的，擴散是由分子運動引致的。

性質；屬性（名） 對物質的一種描述，使人們能識別該物質，物理性質不影響物質的化學特性，物理性質包括：大小、味、色、密度（第 14 頁）、狀態（即固態、液態、或氣態）、彈性。

彈性（名） 物體或材料被外力拉伸，當外力解除之後能恢復原有大小或形狀的能力。（形容詞 elastic 意爲彈性的 橡膠和鋼絲都是彈性的）

延伸度（名） 彈性固體爲外力拉伸時所增加的長度。（動詞爲 extend）

變形（名） 固體物質尺寸或形狀的改變。在彈性極限（第 28 頁）內，彈性固體物質於變形的力消除後，可恢復到原來的大小和形狀。而可塑性的固體，在外力解除之後，仍保持其變形。（動詞爲 deform，形容詞爲 deformed）

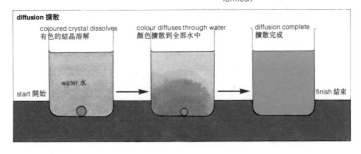

diffusion 擴散

coloured crystal dissolves
有色的結晶溶解

water 水

start 開始

colour diffuses through water
顏色擴散到全部水中

diffusion complete
擴散完成

finish 結束

stress (*n*) the force per unit area acting on a solid. The stress produces a deformation (p.27).

strain (*n*) a change in the size or shape of a solid produced by a stress. Strain is measured by: (change in size) ÷ (size before stress). If an elastic wire is 1.5 m long and a stress causes an extension of 3 mm, then the strain is 3 mm ÷ 1500 mm = 0.002. For a particular material, (stress) ÷ (strain) is a constant within the elastic limit (↓).

Hooke's law strain is proportional to stress i.e. the extension of an elastic solid is proportional to the force stretching the solid, e.g. if a wire has an extension of 3 mm under a stress of 5000 N/m², then it will have an extension of 6 mm under a stress of 10 000 N/m². 3 mm/5000 N/m² and 6 mm/10 000 N/m² are *in proportion*. There is a limit, the **elastic limit**, beyond which the solid will no longer be elastic.

limit (*n*) a line, or a point, or a point in time, or a magnitude, beyond which it is not possible to go, e.g. the limit of hearing; the limit of seeing: i.e. the greatest distance at which a person can see an object.

restoring force (*n*) the force within a material, which returns a stretched elastic solid to its first length.

rigid (*adj*) describes a solid which does not change in shape when a force acts on it, nor is it deformed by a stress, e.g. a pillar is a rigid solid.

plastic (*adj*) describes a solid which is deformed (p.27) by a stress. The stress changes the shape, or size, of the solid, and the shape remains changed when the stress is taken away e.g. hot candle wax is plastic.

應力(名)　作用於固體物質每單位面積上的力。應力引起變形(第 27 頁)。

應變；應變量(名)　應力引起的固體物質的尺寸或形狀變化。應力的測量方法如下：(尺寸的改變)÷(受應力前的尺寸)。設彈性金屬線長 1.5 m，應力使之延伸了 3 mm，則應變量爲 3 mm ÷ 1500 mm = 0.002。對於一個特定的物質來說，在彈性極限(↓)範圍內，(應力)÷(應變量)爲一個常數。

虎克定律　應變量與應力成正比，即彈性固體物質的延伸量與拉長該固體物質的力成正比。例如金屬線在 5000 N/m² 的應力下，其延伸長度爲 3 mm，那麼在 10 000 N/m² 的應力下，其延伸長度是 6 mm。3 mm/5000 N/m² 與 6 N/mm/10 000 N/m² **成比例**。但有個極限，即**彈性極限**，超越此極限，固體物質不再有彈性。

極限點(名)　不可逾越的一條線或一個點，或一個時間點，或者一個量值。例如，聽力的極限、視力的極限，即人所能見物的最大距離。

回復力(名)　物質內部的力，此力使拉長的彈性固體物質恢復到原有長度。

剛性的(形)　描述一個固體物質，當受力作用時不變形，應力亦不能使它變形。例如，柱子是剛性固體。

可塑的(形)　描述應力可使之變形(第 27 頁)的固體物質。應力改變其形狀或尺寸，且當應力解除後，形狀仍然保持不變。例如，熱蠟燭是可塑的。

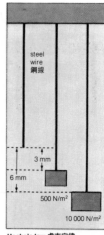

Hooke's law 虎克定律

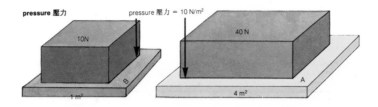

pressure (n) force per unit area, i.e. the force spread over a particular area. Force is measured in newtons and area in square metres, so pressure is measured in newtons per square metre. In the diagram opposite, a force of 40 N acts on an area of 4 m² at surface A. The pressure is 40 N/4 m² = 10 N/m². For surface B, the pressure is 10 N/1 m² = 10 N/m². The symbol for pressure is *p*. Pressure is *transmitted* by solids, and fluids in vessels, from one place to another; pressure is *applied* to a surface, a liquid, or a gas. **press** (v).

pascal (n) the unit of pressure; the symbol is Pa. 1 Pa = 1 N/m².

壓力；壓强(名) 每單位面積上所受的力，即分佈在一特定平面上的力。力的測量單位爲牛頓，面積的測量單位爲平方米，所以壓力的測量單位是牛頓／米²。在上頁圖中，一個 40 N 的力作用於平面 A 的 4 m² 面積上。壓力是 40 N/4 m² = 10 N/m²。平面 B 上的壓力爲 10 N/1 m² = 10 N/m²。壓力的符號是 *p*。壓力通過固體和容器中的流體從一處傳到另一處；壓力可施於表面、液體或氣體。(動詞爲 press)

帕斯卡(名) 壓力單位；符號是 Pa。
1 Pa = 1 N/m²。

tank
水槽

boiler
鍋爐

water gauge
水位計

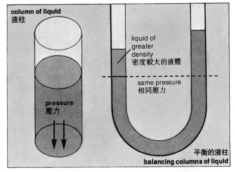

column of liquid
液柱

liquid of greater density
密度較大的液體

same pressure
相同壓力

pressure
壓力

balancing columns of liquid
平衡的液柱

column of liquid liquid contained in a pipe. The pressure on the bottom surface depends on the height of the liquid and its density. Columns of liquids can be balanced (p.18), and used to find the relative density (p.14) of liquids.

water gauge a device (p.10) for measuring the depth of water in a vessel such as a boiler.

spirit level a device containing a bubble of air, or oil, in a liquid (usually alcohol). The spirit level is used to test whether a surface is horizontal.

液柱 盛於管中的液體。底平面所受壓力取決於液體的高度和密度。液柱之間可以平衡(第18頁)，並用以求出不同液體的相對密度(第14頁)。

水位計 一種用於測量容器如鍋爐中水的深度的裝置(第10頁)。

酒精水準儀 在液體(通常是酒精)中含有氣泡或油泡的一種裝置。酒精水準儀用於測試一個平面是否水平。

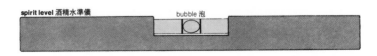

spirit level 酒精水準儀

bubble 泡

Pascal's law the pressure on a liquid in a closed vessel is transmitted (*see* pressure p.29), without becoming less, in all directions. In the diagram, the pressure at A (40 N/cm²) causes a pressure of 40 N/cm² on B.

hydraulic press a machine using hydraulic pressure. The effort (200 N) produces a pressure of 40 N/cm² on the large piston at B. Force = (pressure) × (area). The force produced at B = 40 N/cm² × 500 cm² = 20 000 N. The mechanical advantage (p.24) is 20 000 N ÷ 200 N = 100. A *hydraulic jack* and *hydraulic brakes* act in the same way as a hydraulic press.

帕斯卡定律 密封容器內的液體所受壓力向各個方面傳送(參見第 29 頁" 壓力")時,壓強並不減小。圖中 A 處的壓力(40 N/cm²)在 B 處產生 40 N/cm² 的壓力。

水壓機 一種利用水的壓力的機器。作用力(200 N) 在 B 處 的 大 活 塞 上 產 生 一 個 40 N/cm² 的壓強。力=(壓強)×(面積)。在 B 上 產 生 的 力= 40 N/cm² × 500 cm² = 20 000 N。機械效益(第 24 頁)是 20 000 N ÷ 200 N = 100。**液壓起重機和水力閘**工作的方式與水壓機相同。

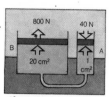

Pascal's law 帕斯卡定律

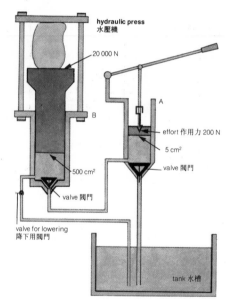

hydraulic press
水壓機

20 000 N

B

A

effort 作用力 200 N

5 cm²

valve 閥門

500 cm²

valve 閥門

valve for lowering
降下用閥門

tank 水槽

atmospheric pressure the pressure (p.29) caused by the air in the Earth's atmosphere on the surface of the Earth. The pressure varies continuously from day to day. Normal pressure (i.e. not high, not low) is taken as 101 325 Pa. The approximate value is 101 kPa. This will support a column of mercury 760 mm high.

大氣壓強;大氣壓 壓強(第 29 頁)是由地球大氣層中的空氣產生的。它時刻不停地變化。正常的氣壓(即不高,不低的氣壓)取值爲 101 325 Pa。近似值爲 101 kPa。它能支承 760 mm 高的水銀柱。

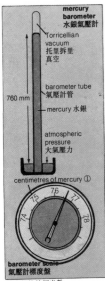

mercury
barometer
水銀氣壓計

Torricellian
vacuum
托里拆里
真空

barometer tube
氣壓計管

760 mm

mercury 水銀

atmospheric
pressure
大氣壓力

centimetres of mercury ①

barometer scale
氣壓計標度盤

①水銀柱的厘米數

barometer (*n*) an instrument for measuring atmospheric pressure. A mercury barometer, *see diagram*, uses mercury, a liquid metal, in a tube closed at the top, with the lower end in a vessel containing mercury. **barometric** (*adj*).

Torricellian vacuum the space above the column of mercury in a barometer. There is nothing in the vacuum.

vacuum (*n*) a space with nothing in it; this is an *absolute* vacuum. A space from which as much air (or other gas) as possible has been taken; this is a *high* vacuum. Taking less air produces a *low* vacuum, also called a *partial* vacuum.

aneroid barometer a thin circular metal box, from which air has been taken, (leaving a partial vacuum) measures atmospheric pressure. An increase in atmospheric pressure causes the box to bend inwards; this movement is passed on by levers so that a needle turns on the scale, *see diagram*.

氣壓計（名） 測量大氣壓的儀器。水銀氣壓計（見圖）是將水銀（一種液態金屬）裝入一根管內，管頂端封閉，下端置於盛水銀的容器內構成的。（形容詞爲 barometric）

托里拆利真空 氣壓計水銀柱上部的空間。此真空內一無所有。

真空（名） 絕對真空是不存在任何氣體的空間。高真空是抽去盡可能多的空氣（或其他氣體）的空間。抽去較少空氣形成低真空，稱爲部分真空。

無液氣壓計 測量大氣壓力用的圓形薄金屬盒，其中已抽去空氣（部分真空）。大氣壓升高使盒子向內凹進；從而帶動槓桿刻度盤上的指針擺動（見圖）。

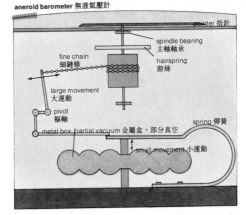

aneroid barometer 無液氣壓計

pointer 指針

spindle bearing
主軸軸承

fine chain
細鏈條

hairspring
游絲

large movement
大運動

pivot
樞軸

spring 彈簧

metal box, partial vacuum 金屬盒，部分真空

small movement 小運動

altimeter (*n*) an aneroid barometer suitably altered to measure height, by measuring the change in atmospheric pressure with height. A mercury barometer falls approximately 1 mm for a rise in height of 12 m.

高度計（名） 經適當改裝的氣壓計。它通過測量大氣壓力隨高度的變化來測量高度。每升高 12 米，水銀氣壓計下降約 1 毫米。

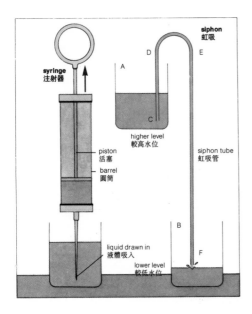

siphon
虹吸

syringe
注射器

D

A

C

E

higher level
較高水位

siphon tube
虹吸管

piston
活塞

barrel
圓筒

B

liquid drawn in
液體吸入

lower level
較低水位

F

gas 氣體

U-tube U 形管

manometer 流體壓力計

suction (*n*) the drawing in of a fluid to produce a partial vacuum which is filled by atmospheric pressure. **suck** (*v*).

syringe (*n*) a device (p.10) for sucking in liquid and then pushing it out under pressure. A syringe has a barrel in which a piston moves up and down. When the piston (*see diagram*) is raised, atmospheric pressure pushes liquid in. When the piston is lowered, the liquid is pushed out.

siphon (*n*) a device for taking liquid out of a vessel and passing it to a lower level. A curved tube, *see diagram*, full of liquid, is used. The side EF must be longer than the side CD for the tube to siphon water out of vessel A and pass it to vessel B.

manometer (*n*) an instrument for measuring small pressures of gases. A U-tube contains liquid; the pressure of the gas pushes the liquid (usually water or mercury) round the tube.

吸入(名) 形成部分真空，再藉大氣壓使流體充滿入此空間，從而吸入流體。(動詞為 suck)

注射器(名) 一種吸入液體再加壓力將之排出的裝置(第 10 頁)。注射器圓筒內的活塞可上下移動。活塞(見圖)上升時，大氣壓力將液體推入；活塞下降時，液體被排出。

虹吸管(名) 一種從容器內吸出液體送入低液位的裝置。用一根彎管(見圖)內充滿液體。管子的 EF 段必須長於 CD 段，從容器 A 吸出水送入容器 B。

流體壓力計(名) 測量氣體微小氣壓的儀器。U形管內有液體，氣體的壓力推動液體(通常是水或水銀)沿管子流動。

valve[1] (*n*) a device (p.10) which: (1) controls the flow of a liquid or a gas through a pipe, or (2) allows liquid or gas to flow through a pipe in one direction only.

pump (*n*) a machine for: (1) taking liquids or gases from one place and passing them to another place; (2) forcing gases into smaller spaces. A pump uses valves and pistons.

閥門；活門(名)　一種裝置(第 10 頁)，用於(1)控制液體或氣體通過管子；或(2)讓液體或氣體在管內只向一個方向流動。

泵(名)　用於(1)從一處抽出液體或氣體送往另一處；(2)將氣體壓入較小的空間的一種機器。泵使用閥和活塞。

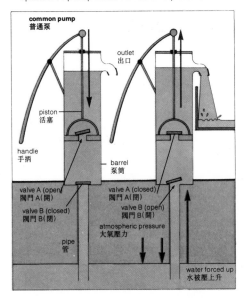

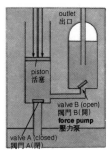

common pump (*n*) a machine used to raise water or other liquids. A piston moves up and down in the pump barrel, *see diagram*, the pump draws liquid into its barrel and pushes the liquid out through an outlet. This pump cannot raise water more than 10 m.

lift pump another name for common pump.

force pump a pump that draws liquid into its barrel (by atmospheric pressure) but forces the liquid out under pressure to a height greater than 10 m depending on the force acting on the piston.

普通泵(名)　一種用於提水或其他液體的機器。活塞在泵筒內上下移動(見圖)，泵將液體吸入筒內並將液體往排出口排出。這種泵最高只能將水升高到 10 米。

升液泵　普通泵的別稱。

壓力泵　利用大氣壓力將液體抽入泵筒內，再靠施於活塞的力加壓將液體壓升到 10 米以上的一種泵。

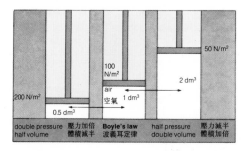

Boyle's law the law that using a particular mass of gas, at a fixed temperature: (pressure of gas) × (volume of gas) is constant. In the diagram: $(200\,N/cm^2 \times 0.5\,dm^3) = (100\,N/m^2 \times 1\,dm^3) = (50\,N/dm^3 \times 2\,dm^3)$. Boyle's law in symbols is $pV = K$ (a constant).

Bernouilli's principle when liquid flows through a pipe of varying diameter, the amount of energy per kilogram of liquid does not change. When the velocity of flow is greatest, the pressure is least. Liquid flows faster through a narrow part of the tube, and slower through a wider part.

Venturi tube a tube which shows Bernouilli's principle. Two wide diameter tubes are joined by a narrow tube. Mercury manometers (p.32) show the pressure is low in the narrow tube and higher in the wide tubes.

波義耳定律 一定質量的氣體在溫度不變時，（氣體的壓力）×（氣體的體積）＝常數。圖中 $(200\,N/cm^2 \times 0.5\,dm^3)$ ＝ $(100\,N/m^2 \times 1\,dm^3)$ ＝ $(50\,N/dm^3 \times 2\,dm^3)$。波義耳定律以符號表示爲：$pV = K$（常數）。

伯努利原理 液體流過直徑不一的管時，每千克液體的能量不變。流速最大時，壓力最小。液體流過狹管部分時流速較快，而流過較寬的部分時較慢。

文丘里管；文氏管 用以證明伯努利原理的管子。一條窄管連接兩條大直徑管。水銀壓力計（第 32 頁）顯示小管內壓力低，大管內壓力高。

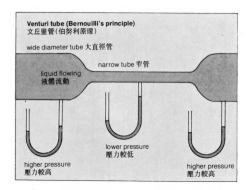

stream line flow 流線型流動

turbulent flow 湍流流動

aerofoil (aeroplane wing)
翼型面 (飛機機翼)

submerged
浸沒於水中的

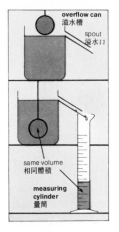

overflow can
溢水槽

spout
溢水口

same volume
相同體積

measuring
cylinder
量筒

streamline (*adj*) describes the flow of a liquid, or gas, which has no sudden change of direction.

streamlined (*adj*) describes the shape of a solid such that any liquid or gas passing it will have streamline flow.

turbulent (*adj*) describes the flow of a liquid, or gas, in which the flow changes rapidly in direction and magnitude. Turbulent is the opposite of streamline.

aerofoil (*n*) a surface, e.g. of an aeroplane which helps the aeroplane to remain in the air, or to change direction, e.g. wings are aerofoils.

suspend (*v*) (1) to hang from a support, e.g. to suspend a load by a rope from a pulley; (2) to be held or to remain in a position, e.g. dust suspended in the air. **suspended** (*adj*).

immerse (*v*) to put a solid object partially, or wholly, under the surface of a liquid, usually water, *see diagram*.

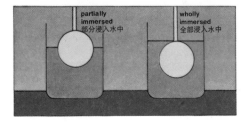

partially
immersed
部分浸入水中

wholly
immersed
全部浸入水中

submerge (*v*) to put a solid object completely into water, usually at a depth.

displacement (*n*) (1) the volume (or weight) of fluid pushed out of place by an object; (2) the difference between the first position of an object and any later position, *see diagram*. **displaced** (*adj*), **displace** (*v*).

overflow can a metal can with a spout. The can is filled with water and an object submerged in the water. The object displaces water through the spout. The volume of displaced water equals the volume of the object.

measuring cylinder a tall glass vessel which measures, in cubic centimetres, the volume of a liquid.

流線型(形) 描述一種液體或氣體的流動,其流動方向不會突然改變。

流線型的(形) 描述固體的一種形狀,任何液體或氣體流過時將呈流線型流動。

湍流的(形) 描述一種液體或氣體的流動,其流動方向和量值迅速改變。"湍流的"是"流線的"反意詞。

翼型面(名) 一種平面,例如使飛機在空中飛行的平面或改變飛行方向的平面,例如機翼是翼型面。

懸掛;懸浮(動) (1)吊在支架上,例如用繩子將一重物懸掛在滑輪上;(2)保持或停留在某一位置,例如灰塵懸浮在空氣中。(形容詞為 suspended)。

浸沒(動) 將一固體物質部分或全部置於液面之下,通常是水面下(見圖)。

浸沒水中(動) 將一固體物質完全置於水中,通常在某一深度的水中。

排液量;位移(名) (1)為一物體排出的流體體積(或重量);(2)物體初始位置和終結位置之間的距離。(見圖)。(形容詞為 displaced,動詞為 displace)

溢水槽 有溢水口的金屬槽。槽中注滿水,一物體浸入水中。該物體將水經溢水口排出。所排出水的體積等於該物體的體積。

量筒 以立方厘米為單位測量液體體積的高身玻璃容器。

upthrust (*n*) a force acting in an upward direction.
Archimedes' principle when an object is
wholly, or partially, immersed (p.35) in a liquid,
there is an apparent loss in weight. The loss in
weight is caused by an upthrust on the object.
The upthrust is equal to the weight of displaced
(p.35) liquid. Archimedes' principle can be used
for gases as well as liquids.

上推力（名）　向上的作用力。
阿基米德原理　當物體全部或部分地浸入（第35
頁）液體中時，出現一種表觀失重現象。失
重是由對物體的向上推力引起的。上推力等
於被排開（第35頁）的液體的重量。阿基米
德原理適用於氣體及液體。

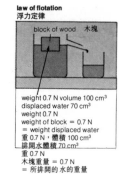

law of flotation
浮力定律

block of wood　木塊

weight 0.7 N volume 100 cm³
displaced water 70 cm³
weight 0.7 N
weight of block = 0.7 N
= weight displaced water
重 0.7 N，體積 100 cm³
排開水體積 70 cm³
木塊重量 = 0.7 N
= 所排開的水的重量

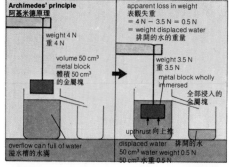

Archimedes' principle
阿基米德原理

weight 4 N
重 4 N

volume 50 cm³
metal block
體積 50 cm³
的金屬塊

apparent loss in weight
表觀失重
= 4 N – 3.5 N = 0.5 N
= weight displaced water
排開的水的重量

weight 3.5 N
重 3.5 N

metal block wholly
immersed
全部浸入的
金屬塊

upthrust 向上推

overflow can full of water
溢水槽的水滿

displaced water　排開的水
50 cm³ water weight 0.5 N
50 cm³ 水重 0.5 N

flotation (*n*) a floating object displaces its own
weight of the fluid in which it is floating. This
is the law of flotation; it is a special case of
Archimedes' principle. **float** (*v*).
buoyancy (*n*) (1) the upthrust of a fluid on an
object which is immersed (p.35) in it; (2) the
tendency (p.15) of an object to float in a fluid.
Buoyancy decides whether a solid object floats
or sinks. **buoyant** (*adj*) (1) able to float; (2)
able to keep an object floating.
balloon (*n*) a bag-like object filled with a gas
of low density (p.14). The bag is made of a soft,
but strong, material. The balloon displaces air;
if the upthrust of the displaced air is greater
than the weight of the balloon, then the balloon
rises in the air.
hydrometer (*n*) a floating instrument (p.10) used
for measuring the density (p.14) of a liquid. It has
a hollow glass body with a long stem, and a
weight at the bottom to make it float upright. The
stem shows a scale of densities; the higher
densities are at the bottom of the scale.

浮（名）　浮體在流體中排開與自身重量相等的流
體，此即浮力定律；是阿基米德原理的一個
特例。（動詞為 float）

浮力（名）　(1)流體對浸入（第35頁）於其中的物
體的向上推力；(2)物體浮在流體中的傾向
（第15頁）。浮力決定固體物體的沉浮。形
容詞 buoyant 意為：(1)能浮起的；(2)能保
持使物體浮起的。

氣球（名）　充入低密度（第14頁）氣體的袋狀
物，由柔軟而結實的材料製成。氣球排走空
氣；如果被排去空氣的向上推力大於氣球的
重量，氣球就上升到空中。

液體比重計（名）　用於量度液體密度（第14頁）
的浮式儀器（第10頁）。係一個帶長管的空
心玻璃體，底部的重物使它直立浮起。長管
上標有密度刻度尺；刻度尺的底端標示刻度
密度較大。

ballon
氣球

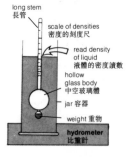

long stem
長管

scale of densities
密度的刻度尺

read density
of liquid
液體的密度讀數

hollow
glass body
中空玻璃體

jar 容器

weight 重物

hydrometer
比重計

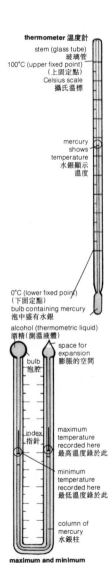

thermometer 溫度計
stem (glass tube) 玻璃管
100°C (upper fixed point) (上固定點)
Celsius scale 攝氏溫標

mercury shows temperature 水銀顯示溫度

0°C (lower fixed point) (下固定點)
bulb containing mercury 泡中盛有水銀

alcohol (thermometric liquid) 酒精(測溫液體)

space for expansion 膨脹的空間
bulb 泡腔

index 指針
maximum temperature recorded here 最高溫度錄於此

minimum temperature recorded here 最低溫度錄於此

column of mercury 水銀柱

maximum and minimum thermometer 最高和最低溫度計

temperature (*n*) a measure, using a scale, of how hot, or how cold, an object, an organism (p.147), or the atmosphere is.

heat (*n*) a form of energy which materials possess from the kinetic energy of their molecules (p.26); heat is measured in joules (p.21). The physical effects of heat are: (1) to change the temperature of a material; (2) to change the state (p.39) of a material; (3) to cause expansion (p.38). Heat is transferred (p.45) from a material at a higher temperature to one at a lower temperature. A change in heat content of an object is measured by its heat capacity (p.42) multiplied by the change in temperature. **heat** (*v*).

thermometer (*n*) any instrument for measuring temperature; it uses a physical property of a material that changes regularly with a change in temperature. The commonest kind uses the expansion of mercury in a glass tube, *see diagram*; other thermometric liquids, such as alcohol, can be used instead of mercury. **thermometric** (*adj*).

fixed point a standard temperature. The upper fixed point on thermometer scales is the boiling point (p.41) of pure water; the lower fixed point is the temperature of melting ice (p.20).

Celsius scale a temperature scale with 100 degrees between the lower fixed point (0°C) and the upper fixed point (100°C). The symbol is: °C.

kelvin (*n*) the S.I. unit of temperature (↑); the symbol is: K. 1 K = 1°C. The lower fixed point is 273 K and the upper fixed point is 373 K.

maximum and minimum thermometer a thermometer which measures the highest (maximum) and the lowest (minimum) temperatures recorded during a particular length of time, usually one day, *see diagram*. The thermometric (↑) liquid is alcohol and it pushes a column of mercury round a U-shaped thermometer. The mercury has two metal indices to record (p.41) the maximum or the minimum temperatures. The indices are reset by a magnet.

溫度(名) 利用標度尺測量物體、生物體(第147頁)或大氣冷熱程度的尺度。

熱(名) 物質具有其分子(第26頁)所含之動能，這是能的一種形式。熱的測量單位爲焦耳(第21頁)。熱的物理效應是：(1)改變物質的溫度；(2)改變物質的狀態(第39頁)；(3)引起膨脹(第38頁)。熱從溫度較高的物質傳遞(第45頁)到溫度較低的物質。物體含熱量的變化可由溫度的變化乘以熱容量(第42頁)求出。(動詞爲 heat)

溫度計(名) 測量溫度的任何儀器；它利用物質的物理性質隨溫度變化而有規律地變化的原理而設計。最普通的一種溫度計是利用玻璃管內水銀的膨脹性，(見圖)，可用其他測量液體如酒精代替水銀。(形容詞爲 thermometric)

固定點(名) 係一個標準溫度。溫度計標度上固定點是純水的沸點(第41頁)；下固定點是冰的熔化(第20頁)溫度。

攝氏溫標 溫度的一種標度，其下固定點(0°C)和上固定點(100°C)之間分爲100°。符號爲°C。

開爾文(名) 溫度(↑)的國際單位制；符號爲K。1K = 1°C。下固定點爲273K，上固定點爲373K。

最高最低溫度計 一種在特定時間內(通常爲一天)測量並記錄最高(最大)和最低(最小)溫度的溫度計，(見圖)。測溫的(↑)液體是酒精，酒精推動水銀柱繞過 U 型溫度計，水銀有兩個金屬指針，記錄(第41頁)最高和最低溫度。磁鐵使指針復位。

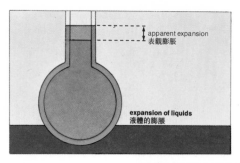

apparent expansion
表觀膨脹

expansion of liquids
液體的膨脹

expand (v) to increase in length, area, or volume of a solid, or in volume of a fluid. The increase is caused by a rise in temperature, and for gases is also caused by a decrease in pressure (Boyle's law p.34). **expansion** (n).

apparent expansion the measured expansion of a fluid in a vessel. On heating, both the vessel and the fluid expand, so the true expansion of the fluid is its apparent expansion added to the expansion of the vessel.

contract (v) to decrease in length, area, or volume of a solid, or in volume of a fluid. The decrease is caused by a fall in temperature and for gases is also caused by an increase in pressure (Boyle's law p.34), e.g. a metal rod contracts on cooling. **contraction** (n).

coefficient (n) a constant ratio which measures a particular property for a change in quantity of a material, e.g. the coefficient of linear expansion is the increase in length per metre of a solid for a rise in temperature of 1°C.

bimetallic (adj) made of two different metals, e.g. iron and brass. A bimetallic strip bends when heated, as the two metals have different coefficients of expansion.

thermostat (n) a device for keeping a fluid, or an object, at a constant temperature. The bimetallic strip of brass and iron, *see diagram*, bends when hot and straightens when cold. When cold it makes an electric contact to start a heating coil. When sufficiently hot, the contact is broken.

膨脹(動)　增加固體的長度、面積或體積，或增加流體的體積。溫度上升引起膨脹，壓力引起氣體膨脹。(波義耳定律，見第 34 頁)。(名詞爲 expansion)

表觀膨脹　流體在容器中實測的膨脹。在加熱時，容器和流體都膨脹。所以流體的真膨脹是其表觀膨脹與容器膨脹之和。

收縮(動)　減少固體的長度、面積或體積，或減少流體的體積。溫度下降引起收縮。壓力增大也引起氣體收縮。(波義耳定律，見第 34 頁)。例如，金屬在冷卻時收縮(名詞爲 contraction)

系數(名)　用於量度材料量變特性的一個常數比。例如線型膨脹系數爲每米長的固體在溫度升高 1°C 時所增加的長度。

雙金屬的(形)　由兩種不同金屬，例如鐵和黃銅製成的。加熱時，雙金屬條因兩種金屬的膨脹系數不同而彎曲。

恆溫器(名)　使流體或固體保持恆溫的裝置。黃銅和鐵製成的雙金屬條(見圖)在熱時彎曲，冷時伸直。冷時它使電觸頭啟動加熱線圈。熱到一定程度時，觸頭又斷開。

expansion of metals
金屬的膨脹

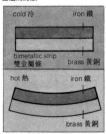

cold 冷　　　　iron 鐵

bimetallic strip
雙金屬條　　　brass 黃銅

hot 熱　　　　iron 鐵

brass 黃銅

simple thermostat
簡單恆溫器

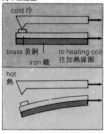

cold 冷

brass 黃銅　　to heating coil
iron 鐵　　　往加熱線圈

hot
熱

state of matter all materials are solids, liquids, or gases. These are the three states of matter.

solid (*n*) a solid possesses molecules (p.26) held together by strong forces, so solids have a definite shape and a definite volume. **solid** (*adj*), **solidify** (*v*).

liquid (*n*) a liquid possesses molecules held together by weaker forces than those in solids, so a liquid has a definite volume but no shape. **liquid** (*adj*), **liquefy** (*v*).

gas (*n*) a gas possesses molecules which are free to move about with no forces holding them together, so a gas has no definite volume and no shape. **gaseous** (*adj*).

fluid (*n*) any material that flows, i.e. a liquid or a gas. **fluid** (*adj*).

Charles' law the volume of a fixed mass of gas at a constant pressure is proportional to its temperature (measured in kelvin). In symbols: $V \propto T$.

gas law Boyle's law and Charles' law are made into one law. For a fixed mass of gas: (pressure) × (volume) is proportional to (temperature); the temperature is measured in kelvin. In symbols: $pV \propto T$.

物態 一切物質都可呈固體、或液體或氣體狀態。此即物質三態。

固體(名) 固體的分子(第 26 頁)由强大的力保持在一起,故有一定形狀和體積。(形容詞爲 solid,動詞爲 solidify)

液體(名) 液體的分子由較固體爲弱的力保持在一起,故有一定的體積,但無一定的形狀。(形容詞爲 liquid,動詞爲 liquefy)

氣體(名) 氣體的分子可自由運動。沒有力將分子保持在一起,故氣體無一定的形狀和體積。(形容詞爲 gaseous)

流體(名) 指任何流動的物質,即液體或氣體。(形容詞爲 fluid)

查理定律 在壓力不變時,一定量的氣體體積與溫度(用凱氏度作爲量度單位)成正比。用符號表示爲:$V \propto T$。

氣體定律 波義耳定律和查理定律合而成一新定律。對一定質量的氣體而言(壓强)×(體積)與溫度成正比;溫度的量度單位爲凱氏度,用符號表示爲 $pV \propto T$。

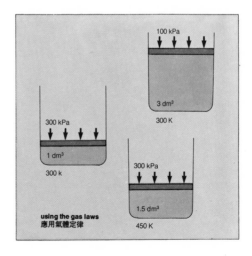

100 kPa

3 dm³

300 K

300 kPa

1 dm³

300 k

300 kPa

1.5 dm³

450 K

**using the gas laws
應用氣體定律**

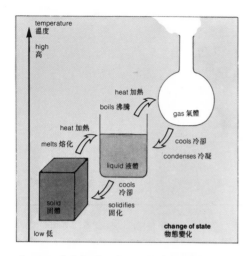

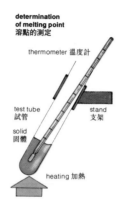

change of state the change of a material from one state to another, e.g. from a solid to a liquid, or a liquid to a gas. Such changes are mainly caused by heating or cooling.

melt (*v*) to change, when heated at a constant temperature, from a solid to a liquid, e.g. ice melts to form water. Compare dissolve (p.89).
melted (*adj*), **molten** (*adj*).

solidify (*v*) to change, when cooled, from a liquid to a solid, e.g. liquid wax, on cooling, becomes solid. Compare crystallization (p.110) and sublimation (p.92).

melting point the temperature at which a solid melts. Each solid has its own melting point, e.g. the melting point of ice is 0°C, of copper is 1083°C. It is a physical property useful for identification (p.93) of materials. Abbreviation: m.p.

freezing point the temperature at which a liquid solidifies. For a pure material, it is the same temperature as the melting point.

boil (*v*) to change, when heated at a constant temperature, from a liquid to a vapour (↓), e.g. water boils at 100°C to form steam; alcohol boils at 78°C to form alcohol vapour.

物態變化 物質從一種狀態轉化爲另一種狀態。例如從固體轉化爲液體,或從液體轉化爲氣體。這些變化主要由加熱或冷卻引致。

融化;熔化(動) 物質在恆定溫度下加熱,從固體轉化爲液體。例如冰融化成水。與"溶解"(第89頁)這個詞作比較。(形容詞爲melted,molten)。

凝固;固化(動) 物質冷卻時從液體轉化爲固體。例如液體蠟冷卻後變爲固體。與"結晶"(第110頁)和"升華"(第92頁)這兩個詞作比較。

熔點 固體熔化時的溫度。每一種固體都有一定的熔點。例如冰的熔點爲0°C,銅的熔點爲1083°C。熔點對於鑒別(第93頁)不同物質的物理性質是有用的。縮寫爲m.p.。

凝固點;冰點 液體凝固時的溫度。純淨的物質,其凝固點與熔點的溫度相同。

沸騰;煮沸(動) 物質在恆溫下加熱,從液體轉化爲蒸汽(↓)。例如水在100°C時沸騰成蒸汽;酒精在78°C時沸騰化爲酒精蒸汽。

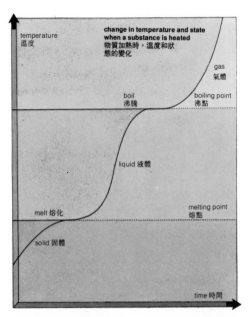

change in temperature and state
when a substance is heated
物質加熱時，溫度和狀
態的變化

temperature
溫度

gas
氣體

boil
沸騰

boiling point
沸點

liquid 液體

melt 熔化

melting point
熔點

solid 固體

time 時間

boiling point the temperature at which a liquid boils. Each liquid material has its own boiling point, e.g. the boiling point of water is 100°C; of alcohol is 78°C at normal atmospheric pressure. It is a physical property useful for identification (p.93) of materials. Abbreviation: b.p.

vapour (*n*) a vapour is a gas which can be liquefied by increasing the pressure without changing the temperature. It is the gaseous form of a material which is a solid or a liquid at room temperatures, e.g. petrol vapour; water vapour. **vaporize** (*v*).

record (*v*) (1) to write a reading taken from an instrument, or an observation (p.93), e.g. to record room temperature; to record the hours of sunshine; (2) for an instrument, to keep showing the same reading, e.g. the index of a thermometer records the maximum (p.37) temperature during a day. The record, in both cases, can always be seen. **record** (*n*).

沸點 液體沸騰時的溫度。每種液體都有自己的沸點。例如在正常氣壓下水的沸點爲100°C，酒精的沸點爲78°C。沸點是一種對鑒別（第 93 頁）材料有用的物理性質。縮寫爲：b.p.。

蒸汽；汽（名） 蒸汽是一種在溫度不變，壓力升高條件下能液化的氣體。它是室溫下爲固體或液體的材料的氣體狀態。例如石油蒸汽、水蒸汽。（動詞爲 vaporize）

記錄（動） （1）寫下儀器上的讀數或觀察（第 93 頁）得的讀數，例如記錄室溫；記錄日照的時數；（2）儀器一直顯示同一讀數，例如溫度計上的指示器記錄下一天中的最高溫度。以上兩種情況下的記錄都是可見的。（名詞爲 record）

evaporation (*n*) the change of a liquid to a vapour at a temperature below, or at, its boiling point, e.g. (a) the evaporation of rain water without the water boiling; (b) when a salt solution is boiled (p.40), the liquid evaporates leaving the salt. **evaporate** (*v*).

condensation (*n*) the change of a vapour, or gas, to a liquid when cooled, e.g. the condensation of steam on a cold surface. Compare liquefaction, the changing of a vapour to a liquid by pressure alone.

calorie[1] (*n*) a former unit of heat. One calorie is the quantity of heat needed to raise the temperature of 1 g of water by 1°C. 1 calorie = 4.18 J.

calorimeter (*n*) any apparatus (p.88) used for measuring quantities of heat, usually by finding the rise in temperature of a known mass of water. The simplest calorimeter is a metal vessel used with a thermometer. **calorimetric** (*adj*), **calorimetry** (*n*).

蒸發作用(名)　液體在溫度低於或達到沸點時變成蒸汽。例如，(a)雨水無需沸騰就化成蒸汽；(b)鹽溶液煮沸(第 40 頁)時，液體蒸發而留下鹽。(動詞爲 evaporate)

冷凝作用(名)　蒸汽或氣體冷卻時變爲液體。例如水蒸汽在冷面上冷凝。與“液化”這個詞比較，液化僅靠加壓的方法使蒸汽變爲液體。

卡路里(名)　過去用的熱量單位。一卡等於使一克水的溫度升高 1°C 所需的熱量。1 卡 = 4.18 焦耳。

量熱器(名)　指任何用於測量熱量的儀器(第 88 頁)，通常是藉求出已知量水的溫升作測量。最簡單的量熱計是一個和溫度計一起使用的金屬容器。(形容詞爲 calorimetric，名詞爲 calorimetry)

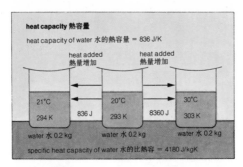

heat capacity 熱容量

heat capacity of water 水的熱容量 = 836 J/K

heat added 熱量增加　　heat added 熱量增加

21°C　294 K　836 J　20°C　293 K　8360 J　30°C　303 K

water 水 0.2 kg　water 水 0.2 kg　water 水 0.2 kg

specific heat capacity of water 水的比熱容 = 4180 J/kgK

heat capacity the number of joules (p.21) needed to raise the temperature of an object by 1 K (1°C), e.g. the heat capacity of a copper vessel is 40 J/K; 240 J are needed to raise its temperature by 6 K.

specific heat capacity the number of joules needed to raise the temperature of 1 kg of a substance by 1 K, e.g. the specific heat capacity of copper is 400 J per kg per kelvin. If a copper calorimeter has a mass of 0.1 kg, its heat capacity is 40 J/K.

熱容量　物體的溫度提高 1K (1°C) 所需的焦耳(第 21 頁)數。例如銅容器的熱容量爲 40 J/K；銅的溫度升高 6K 需要 240 J 的熱量。

比熱容　將 1 kg 物質的溫度提高 1K 所需的焦耳數。例如，銅的比熱容爲 400 J/kg · K。假設銅量熱計的質量爲 0.1 kg，則其比熱容爲 40 J/K。

latent heat the heat needed to change the state of matter (p.39) of a material. While the latent heat is given to the material, its temperature remains constant, *see diagram*.

specific latent heat the number of joules needed to cause a change of state, without change of temperature, for 1 kg of a substance; it is a physical property of a substance. Each substance can have two specific latent heats: (1) of fusion (or melting); (2) of vaporization. These two quantities are constant for a particular substance, e.g. the specific latent heat of fusion of water is 336 kJ/kg and of vaporization of water is 2260 kJ/kg.

refrigeration (*n*) the use of energy to take heat away from an object. In a **refrigerator**, vapour is compressed by a pump; liquid is formed when the vapour is cooled and passed to an evaporator. The liquid takes heat from the refrigerator to supply the latent heat of vaporization, thus cooling the food in the refrigerator, *see diagram*.

潛熱 改變物質的物態(第 39 頁)所需的熱量。當給物質提供潛熱時,其溫度保持不變。(見圖)

比潛熱 使 1 kg 物質的物態發生變化而溫度不發生變化所需之焦耳數。比潛熱是物質的物理性質。每一種物質都可有兩個比潛熱:(1) 熔融(或熔化)的比潛熱;(2) 蒸發的比潛熱。對一特定物質而言,這兩個量都是恆定的。例如水的融化比潛熱爲 336 kJ/kg,蒸發比潛熱爲 2260 kJ/kg。

致冷(名) 利用能量來排出物體的熱量。在冰箱裏,泵壓縮蒸汽,蒸汽冷卻,進入蒸發器後化爲液體。液體從冰箱中吸收熱量提供蒸發潛熱,從而使冰箱中的食物變冷,(見圖)。

evaporator
蒸發器

expansion valve
膨脹閥

condenser
pump 冷凝泵

refrigerator 冰箱

refrigeration 致冷

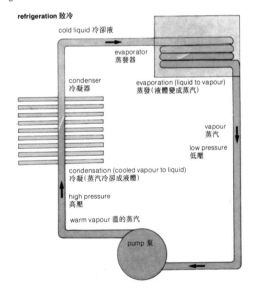

cold liquid 冷卻液

evaporator
蒸發器

condenser
冷凝器

evaporation (liquid to vapour)
蒸發(液體變成蒸汽)

vapour
蒸汽

low pressure
低壓

condensation (cooled vapour to liquid)
冷凝(蒸汽冷卻成液體)

high pressure
高壓

warm vapour 溫的蒸汽

pump 泵

engine (n) a device which uses the properties of a working fluid to transform heat and other forms of energy into mechanical energy, e.g. a steam engine (↓) uses heat for energy and uses water as the working fluid and supplies mechanical energy.

cylinder (n) part of an engine in which a vapour expands, *see diagram*. Holes in the cylinder, closed by valves, allow the vapour to enter and leave after expansion.

piston (n) a solid circular piece of metal which can move up and down in a cylinder. Expansion of a vapour forces the piston down, thus producing mechanical energy.

steam engine heat energy boils water and produces steam under pressure; the steam expands in a cylinder, forcing the piston down. The expanded steam is led away and the piston returns. This movement is repeated. The up and down motion of the piston is changed to a circular motion by a crank, *see diagram opposite.*

reciprocating engine an engine which uses a cylinder, piston and crank to produce circular motion of a wheel, e.g. a steam engine. The motion of the wheel returns the piston.

turbine (n) an engine in which a shaft, *see diagram opposite*, is turned by the force of a stream of fluid on blades fixed to the shaft. In a steam turbine, steam from pipes is directed onto the blades.

internal combustion engine a mixture of petrol vapour and air enters the cylinder of the engine and is exploded by a spark. Heat from the explosion makes the gases expand and force the piston down. A four-stroke engine has (1) down-stroke; petrol mixture sucked in; (2) up-stroke; mixture compressed; (3) down-stroke; mixture exploded; (4) up-stroke; hot gases pushed out.

conservation (n) keeping a quantity constant, i.e. unchanging; preventing loss or waste. *Conservation of mass:* materials cannot be made or destroyed, they can only be transformed into other materials. *Conservation of energy:* energy cannot be made or destroyed, it can only be transformed. **conserve** (v).

發動機；引擎(名) 利用工作流體的性質，將熱能或其他形式的能轉化爲機械能的一種裝置。例如，蒸汽機(↓)用熱作能源，水作工作流體，將水加熱產生機械能。

汽缸(名) 發動機的一個部件，蒸汽在其內膨脹，(見圖)。汽缸的汽口用滑閥封閉，可放入蒸汽，並在膨脹後排出蒸汽。

活塞(名) 可在汽缸中作上下運動的實心圓形金屬部件。蒸汽膨脹迫使活塞下移，從而產生機械能。

蒸汽機 熱能使水沸騰產生壓力蒸汽；壓力蒸汽在汽缸內膨脹迫使活塞下移。排出膨脹的蒸汽，活塞回到原位。重複此運動。活塞的上下運動通過曲軸轉變爲旋轉運動。(見下頁圖)。

往復式發動機 一種利用汽缸、活塞和曲軸使輪子產生旋轉運動的發動機，例如蒸汽機。輪子運動使活塞復位。

渦輪機(名) 一種發動機，流體的蒸汽作用於固定在發動機主軸上的葉片，從而帶動主軸旋轉，(見下頁圖)。從管道來的蒸汽噴射向葉片。

內燃機 汽油蒸汽和空氣的混合氣進入發動機汽缸，由電火花引爆。爆炸產生的熱使氣體膨脹，迫使活塞下移。發動機的四個衝程是：(1)下衝程：吸進汽油混合氣；(2)上衝程：壓縮混合氣；(3)下衝程：混合氣爆炸；(4)上衝程：排出熱氣。

守恆(名) 保持量恆定，即保持量不變；防止損耗或浪費。質量守恆：物質既不能創生也不能消滅，只能轉化爲其他物質。能量守恆：能量既不能創生，也不能消滅，只能轉化。(動詞爲 conserve)

internal combustion engine
內燃機

1 st stroke
第一衝程
petrol vapour sucked in
吸入石油蒸汽

2 nd stroke
第二衝程
vapour compressed
蒸汽壓縮

3rd stroke
第三衝程
spark explosion
點火 膨脹

working stroke
工作衝程

4 th stroke
第四衝程
hot gases pushed out
排出熱氣

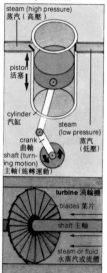

steam engine 蒸汽機

steam (high pressure)
蒸汽（高壓）

piston
活塞

cylinder
汽缸

crank
曲軸
shaft (turn-
ing motion)
主軸(施轉運動)

steam
(low pressure)
蒸汽
（低壓）

turbine 渦輪機

blades 葉片

shaft 主軸

steam or fluid
水蒸汽或流體

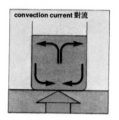

convection current 對流

radiation
輻射

lamp emits heat and light
燈泡發出光和熱

transfer (v) to move an object, or anything else, from one place to another by any means, e.g. to transfer heat from a hot liquid to a cold vessel. **transfer** (n).

conduction (n) the passing of heat through a solid. There is hardly any conduction of heat in fluids. Also the passing of an electric current (p.74) through a substance. **conduct** (v), **conductor** (n).

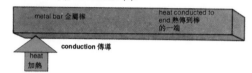

metal bar 金屬棒

heat conducted to end 熱傳到棒的一端

conduction 傳導

heat 加熱

non-conductor (n) a substance that does not conduct heat or an electric current (p.74). Most non-metals are non-conductors, while metals are good conductors.

convection (n) the transfer of heat in a fluid by the rising of hot fluid and the sinking of cold fluid to take the place of the hot fluid. A **convection current** is formed by the movement of the fluid.

ventilation (n) the use of convection currents (↑) to supply a building with fresh air or with cool air. **ventilator** (n), **ventilate** (v).

radiation¹ (n) the transfer of heat from a hot object through space to a cold object, e.g. the radiation of heat from the sun to the Earth. Such heat is called *radiant heat*; it does not need any material for the transfer of heat. Compare conduction and convection, which need a material for the transfer of heat.

emission (n) the sending out of radiant heat (↑). Also the giving out of light, sound, radio waves, other kinds of radiation (p.83), and electrons (p.106), e.g. the emission of heat from the sun; the emission of light from a lamp. **emit** (v), **emitter** (n), **emissive** (adj).

absorption (n) the taking in of energy or fluids, e.g. (a) the absorption of radiant heat (↑) by a surface; (b) the absorption of carbon dioxide gas by sodium hydroxide solution. **absorb** (v), **absorptive** (adj), **absorbent** (n).

傳遞；轉移（動） 用任何方法將物體或其他物件從一處移到另一處。例如將熱量從熱液體傳導到冷的容器。(名詞爲 transfer)

傳導（名） 熱量過固體。流體幾乎不熱傳導。電流(第 74 頁)流過物質也稱傳導。(動詞爲 conduct，名詞爲 conductor)

非導體（名） 不能導熱或導電(第 74 頁)的物質。大部分非金屬都是非導體，金屬爲良導體。

對流（名） 熱在流體中的傳導是靠熱流體上升，冷流體下降取代熱流體的方法進行的。流體運動產生對流。

通風（名） 利用對流(↑)爲建築物提供新鮮空氣或冷氣。(名詞爲 ventilator，動詞爲 ventilate)

輻射（名） 熱量從熱物體通過空間傳遞到冷物體。例如太陽的熱向地球輻射。此種熱稱爲"輻射熱"；它不需要任何物質來傳熱。傳導和對流的比較：兩者都需要有傳熱物質;。

發射；放射（名） 輻射熱(↑)的放出。光波、聲波、無線電波及其他各種輻射(第 83 頁)和電子(第 106 頁)的放出也可稱爲發射，例如太陽熱的發射。燈光的輻射。(動詞爲 emit，名詞爲 emitter，形容詞爲 emissive)

吸收（名） 接收能量或流體。例如(a)表面吸收輻射熱(↑)；(b)氫氧化鈉溶液吸收二氧化碳氣。(動詞爲 absorb，形容詞爲 absorptive，名詞形 absorbent)

thermal (*adj*) of heat, e.g. (a) the thermal efficiency of an engine is the ratio of the work done by the engine to the heat energy supplied by the fuel (p.129); (b) a thermal current is a stream of air rising rapidly from a hot surface on the earth.

sea breeze a wind produced by a convection current (p.45) during the daytime. When the sun shines, land heats up more quickly than the sea. So hot air rises over the land and cooler air from the sea takes its place, thus forming a sea breeze.

land breeze a wind produced by a convection current during the evening. The land cools more quickly than the sea when the sun goes down. So hot air rises over the sea and cool air from the land takes its place; this forms a land breeze.

trade wind a wind that blows in a fixed direction for a particular season or for the whole year.

熱的；熱力的(形)　指有關熱量的。例如：(a) 發動機的熱效應爲發動機所作功與燃料(第129頁)所提供熱能之比；(b)暖氣流是從曬熱的地球表面迅速上升的空氣。

海風　日間空氣對流(第45頁)產生的風。陽光普照時，陸地較海面熱得快，所以陸地的熱空氣上升，海上來的較冷的空氣取代其位置，從而形成海風。

陸風　晚間空氣對流產生的風。日落後，陸地較海面涼得快，所以海面的熱空氣上升，陸上來的冷空氣取代其位置，從而形成陸風。

信風；貿易風　在某一特定季節，或全年都向着一個固定方向吹的風。

monsoon (*n*) a regular wind blowing in one direction for one part of the year and in the opposite direction for the remaining part of the year. In Asia, during summer, the land around the Arabian Sea and the Indian Ocean is very hot, and the sea is cool, so a convection current, similar to a sea breeze, is formed; this is the south-west monsoon. In winter the sea is warmer than the land, so the wind blows in the opposite direction, from the north-east; this is the north-east monsoon.

季侯風(名)　一種有規律的風。一年之中的部分時間內吹向一個方向，其餘時間則吹向相反方向。夏季時亞洲阿拉伯海和印度洋周圍的陸地炎熱，而海面涼爽，所以空氣形成對流；與海風相似；這是西南季風。冬季海面較陸地溫暖，所以風從東北方向朝相反方向吹，形成東北季風。

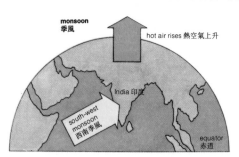

cirrus
捲雲

cumulus
積雲

nimbus
雨雲

stratus
層雲

whirlwind (n) air moving in a circular motion round a centre of low pressure. The low pressure is caused by a convection current (p.45) from a small area of very hot land, as found in deserts. A whirlwind is small in size and may suck up sand and dust as it moves over the surface of the Earth.

cyclone (n) (1) a large area of low pressure over part of the Earth. The pressure is lowest at the centre. Winds circle round and into the area of low pressure; rain usually falls; (2) a violent wind in the tropics caused by a smaller area of low pressure. **cyclonic** (adj).

wind scale a scale describing the speed of wind. The highest speed is 12, the lowest is 1.

Beaufort scale another name for wind scale.

hurricane (n) a wind with a speed of more than 120 km/hr; the strongest wind in the Beaufort scale (scale 12).

typhoon (n) a wind with a speed of more than 120 km/hr. It is the same as a hurricane (↑). The name typhoon is used in the Pacific Ocean, and the name hurricane is used in the Atlantic Ocean.

cloud (n) a large number of very small drops of water formed by water vapour condensing when warm air moves up to a higher level where the temperature is lower. The bottom of a cloud is at a height where the air temperature is that of the dew point (p.50).

cirrus (n) white clouds, shaped like feathers, at a height of 7.6 to 10 km, which are sometimes separate, and sometimes arranged regularly in groups. They are formed from ice crystals.

cumulus (n) thick clouds, looking like cotton wool, with a definite shape, and generally with a flat base. They are formed at a height of 3 to 5 km by warm air rising, becoming cooler and water vapour condensing.

nimbus (n) thick clouds, usually grey, with no definite shape, forming at a height of about 2 km. Rain or snow usually comes from nimbus clouds.

stratus (n) low clouds, formed in layers, usually covering a large area of sky. They are formed in calm weather; if they become thicker, rain may fall.

旋風（名） 圍繞一低壓中心作圓周運動的空氣。如在沙漠中所見，低壓是由酷熱小塊陸地上的對流（第 45 頁）產生的。旋風範圍小，當它移過地面時，能捲起沙和灰塵。

氣旋（名） （1）地球某處的上空的一個大面積低壓區，其中心的壓力最低。風繞低壓區旋轉並旋入低壓區，通常伴有雨；（2）在熱帶地區由小面積低壓引起的熱帶暴風。（形容詞爲 cyclonic）

風級 描述風速的等級。風速最高爲 12 級，最低爲 1 級。

蒲福風級 風級的別稱。

颶風（名） 風速大於 120 公里／小時的風，是蒲福風級中最強的風（12 級）。

颱風（名） 風速大於每小時 120 公里的風。和颶風（↑）相同等級。颱風這一名稱用於太平洋，而在大西洋則稱颶風。

雲（名） 熱空氣上升到較高空處，由於溫度較低，水蒸汽冷凝成無數細小的水滴。雲的底層所處高空，氣溫爲露點（第 50 頁）。

捲雲（名） 形似羽毛的白色雲層，處於 7.6 到 10 公里的高。捲雲時而分散時而有規律聚成團。冰晶形成捲雲。

積雲（名） 密雲，看似棉絨，有一定形狀，一般都有一平的底部。積雲是在 3 至 5 公里的高空，由上升的暖空氣冷卻，水蒸汽冷凝而成。

雨雲（名） 密雲，通常呈灰色，無一定形狀，在大約 2 公里的高空形成。雨和雪通常都是由雨雲所形成。

層雲（名） 低空雲層，一層層形成，通常遮蓋大片天空。它們在無風的天氣形成；層雲厚時，可能要下雨。

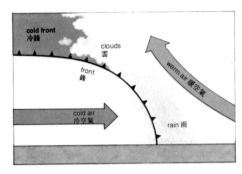

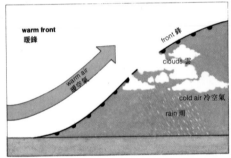

cold front a cold front is formed when cold air moves into an area of warm air. The cold air forces the warm air to rise and cumulus clouds are formed.

warm front a warm front is formed when warm air moves into an area of cold air over which it rises. Cloud is formed in layers, slowly changing from white to grey, and becoming thicker. Rain usually falls.

ridge of pressure a long, narrow area of high atmospheric pressure which comes from a greater area of high pressure.

anti-cyclone a large area of high pressure over part of the Earth; the pressure is highest at the centre. Winds circle round and out of the area of high pressure. The weather is usually clear and bright.

冷鋒 冷空氣進入暖空氣區時形成冷鋒。冷空氣迫使暖空氣上升形成積雲。

暖鋒 暖空氣進入冷空氣區時，上升到冷空氣上面形成暖鋒。雲一層層地形成，慢慢地由白色轉灰色，成爲密雲。通常會下雨。

高壓脊 由更大面積的高氣壓區來的狹長面積的高氣壓。

反氣旋 地球某處上空的大面積高氣壓區，其中心的氣壓最高。風繞高氣壓區旋轉並離開此區，通常是天氣晴朗。

humidity (*n*) a measure of the amount of water vapour in the atmosphere. Usually the *relative humidity* is measured as a percentage; 100% indicates atmospheric air saturated (p.89) with water vapour and 0% indicates completely dry air. **humid** (*adj*), **humidify** (*v*).

hygrometer (*n*) an instrument for measuring the humidity of the atmosphere, e.g. the hair hygrometer which uses the expansion and contraction of hair with changes in the humidity.

濕度(名)　大氣中水汽含量的量度。**相對濕度**通常以百分比計；100% 表示大氣中含水汽達飽和(第 89 頁)，0% 則表示完全乾燥。(形容詞爲 humid，動詞爲 humidify)

濕度表(名)　測量大氣濕度的儀器。例如毛髮濕度計，它利用毛髮隨濕度變化膨脹和收縮的現象來測量大氣濕度。

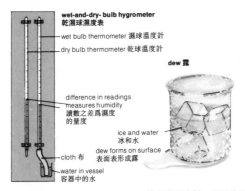

wet-and-dry- bulb hygrometer
乾濕球濕度表
wet bulb thermometer 濕球溫度計
dry bulb thermometer 乾球溫度計

difference in readings measures humidity
讀數之差爲濕度的量度

cloth 布
water in vessel
容器中的水

dew 露

ice and water
冰和水
dew forms on surface
表面表形成露

wet-and-dry bulb hygrometer this instrument has two thermometers. One thermometer has a wet cloth round its bulb, and records a low temperature because of evaporation of the water. The dry bulb records atmospheric temperature. The difference between the readings of the two thermometers is a measure of the humidity.

precipitation[1] (*n*) the forming of drops of liquid from a vapour when the temperature has fallen below a certain value and the drops are large enough to fall. The vapour is mixed with a gas, e.g. water vapour in atmospheric air where precipitation happens when the temperature falls below the dew point (p.50), and water falls as rain or snow. **precipitate** (*v*).

dew (*n*) drops of water formed on solid surfaces by condensation (p.42) of water vapour from the air, e.g. dew is formed on grass during the night.

乾濕球濕度計　這種儀器有兩支溫度計。一支的球部包有濕布，記錄水蒸發而致的低溫。乾球則記錄大氣溫度。兩支溫度計讀數之差爲濕度的量度。

降水；降水量(名)　當溫度降到一定數值以下，水汽形成液滴，水滴大到可以滴落的情形。水汽與氣體相混和。例如大氣中的水汽就是這樣。當溫度降至露點(第 50 頁)以下，大氣中的水蒸氣凝結而下雨或下雪。(動詞爲 precipitate)

露(名)　空氣中水汽冷凝(第 42 頁)在固體表面形成的水滴。例如夜間草上形成的露水。

dew point the temperature at which the air becomes saturated (p.89) with water vapour and condensation takes place, forming dew. Above the dew point, atmospheric air is unsaturated.

mist (n) rapid cooling of the air causes very small drops of water to form by condensation (p.42); this is a mist. A mist is a cloud formed at ground level. If the water condenses on dust in the air, a fog is formed.

fog (n) – see mist (↑).

rain (n) water drops formed by precipitation, falling from clouds.

rain-gauge (n) an instrument for measuring the amount of rain which falls in a certain time, e.g. in 24 hours. The rain is caught by a funnel and passed into a bottle, *see diagram*. The water is measured in a measuring cylinder, which gives a reading of the rainfall.

rainfall (n) the depth, in cm, of rain water in a given time if it did not flow away.

snow (n) small crystals (p.110) of ice falling from clouds. Water drops form ice crystals if the air temperature is below 0°C.

ice (n) formed from water at 0°C, when water becomes solid. Ice forms on rivers and lakes, less often on the sea.

露點 空氣中的水汽達到飽和(第 89 頁)並發生冷凝的溫度,即形成露水的溫度。在露點以上時,大氣是不飽和的。

薄霧;霾(名) 空氣迅速冷卻,因冷凝作用(第 42 頁)形成極小的水滴,此即薄霧。薄霧是地面上形成的雲。如果水冷凝在空氣中的塵埃上,則成霧。

霧(名) 見“薄霧”(↑)。

雨(名) 凝結所形成並從雲中降落的水滴。

雨量器(名) 測量在一定時間內(例如 24 小時內),降雨量的儀器。雨收集到漏斗裏,流入瓶中(見圖)。雨水用一標有雨量讀數的量筒來測量。

雨量(名) 假定雨水不流失,在一定的時間內的雨水深度,以 cm 表示。

雪(名) 從雲中降落的小冰晶(第 110 頁)。氣溫在 0°C 以下,水滴形成冰晶。

冰(名) 在 °C 時形成。此時水變成固體。河流、湖泊上可結冰,海上很少結冰。

snow and ice 雪和冰
snow falling 降雪
ice 冰
river 河流
snow 雪
water 水

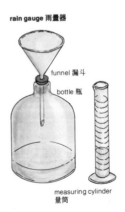

rain gauge 雨量器
funnel 漏斗
bottle 瓶
measuring cylinder 量筒

the Earth's atmosphere
地球的大氣
approximate temperature
平均溫度　height in kilometres
　　　　　　高度，公里

1200°C	400
1200°C	350
1150°C	300
1100°C	250

IONOSPHERE
電離層

900°C	200

air ionised
空氣電離
radio waves reflected
反射無綫電波

200°C	150
-75°C	100

jet stream
急流
highest balloon flight
氣球達到的最大高度

-60°C no clouds 無雲
no wind 無風
STRATOSPHERE 同溫層
-60°C TROPOPAUSE 對流層頂
TROPOSPHERE 對流層
20°C clouds 雲　wind 風

atmosphere (n) the gases around the Earth, or any other heavenly body. **atmospheric** (adj).

troposphere (n) the layer of the atmosphere nearest the Earth; it varies in height from 10 km at the Poles to 18 km at the equator. The temperature falls with height above the Earth. In this layer, clouds are formed, and there are many convectional air currents from unequal heating of the Earth's surface.

tropopause (n) a thin layer of the atmosphere, about 5 km thick, between the troposphere and the stratosphere. The temperature no longer falls with height; fast air currents, called **jet streams**, are formed in this layer.

stratosphere (n) an upper layer of the atmosphere, about 25 km thick, above the tropopause. The air temperature does not increase with height; it is about −80°C over the equator and about −40°C over the Poles. No clouds or air convection currents are formed in the layer.

ionosphere (n) an upper layer of the atmosphere, about 350 km thick, above the stratosphere, starting at an average height of 50 km above the Earth. In it the air is ionized (p.101) by ultra-violet light (p.83) from the sun, and it reflects radio waves. The ionosphere is separated from the stratosphere by an **ozone layer**.

space (n) above the ionosphere is space. It contains no gases. The sun and its planets occupy a part of space and stars, galaxies, comets and all the stellar systems are found here. Only electromagnetic waves travel through space, e.g. light waves, radio waves etc.

climate (n) the conditions of an area on the Earth's surface such as the variation in temperature, rainfall, and humidity. There are four important types of climate: tropical; sub-tropical; temperate; polar. **climatic** (adj).

forecast (v) to say what events will happen after having seen past events, e.g. to forecast the weather, after taking measurements of temperature, humidity and wind changes. Forecasting is not accurate, but it is not expected to be very far wrong.

大氣（名）　地球或其他任何天體周圍的氣體。（形容詞爲 atmospheric）

對流層（名）　離地球最近的大氣層；其高度距地球 10 公里（在兩極）到 18 公里（在赤道）之間。溫度隨距地面的高度增加而下降。在這氣層內形成雲，並且由於地球表面受熱不均勻而產生多股對流氣流。

對流層頂　一層薄的大氣層，厚約 5 公里。介於對流層和同溫層之間。溫度不再隨高度增加而下降；在這一氣層形成快速氣流，又稱**急流**。

同溫平流層（名）　在對流層頂的上面的上層大氣層，厚約 25 公里。氣溫不隨高度增加而上升；在赤道上空約爲 −80°C，在兩極上空約爲 −40°C。在這一層中既無雲也無空氣對流。

電離層（名）　在同溫層的上面的上層大氣層，厚約 350 公里，開始於距地面平均 50 公里的高空。此一層中的空氣被太陽的紫外線（第 83 頁）電離（第 101 頁），因此可反射無線電波。**臭氣層**將電離層和同溫層隔開。

太空（名）　電離層的上面是太空。太空中沒有氣體。太陽及其行星佔據部分太空，恆星、銀河系、彗星和所有的星系也都位於太空，只有電磁波，例如光波、無線電波等才能通過太空。

氣候（名）　地球表面一個地區的狀況，比如溫度、雨量、濕度的變化。氣候分四種主要類型：熱帶氣候，亞熱帶氣候，溫帶氣候和極地氣候。（形容詞爲 climatic）

預報；預測（動）　研究過去的事件後，推測未來的事件。例如在測量了溫度、濕度和風的變化後，預報天氣。預報不很準確，但也不致差得很遠。

source (*n*) the place from which light, sound, other forms of energy, or materials come, e.g. (a) a lamp is a source of light; (b) a fire is a source of heat; (c) some minerals are the source of metals.

medium (*n*) the means by which a wave motion (p.65) travels. A medium can be material, e.g. a gas, a liquid, or a solid, or it can be non-material, i.e. a vacuum. Light does not need a material medium; it travels through a vacuum. Sound needs a material medium.

propagation (*n*) the sending of a wave motion by a medium (↑), e.g. (a) the propagation of light by a vacuum; (b) the propagation of sound waves by air. **propagate** (*v*).

rectilinear (*adj*) formed of straight lines, travelling in a straight line, e.g. the rectilinear propagation of light is the sending of light waves in a straight line by any medium.

transparent (*adj*) describes any solid or liquid medium through which light can travel to form an image (p.56), e.g. the glass in a window is transparent.

translucent (*adj*) describes any solid or liquid medium through which light can travel, but no clear image can be formed, e.g. waxed paper is translucent but not transparent.

源;源頭(名) 產生光、聲及其他形式的能或物質之處。例如:(a)燈是光源;(b)火是熱源;(c)一些礦物是金屬的來源。

介質;媒質(名) 波動(第 65 頁)傳播的工具。介質可能是物質。例如:一種氣體,一種液體或一種固體;它也可能是非物質。例如真空。光不需要物質介質,它可通過真空傳播。聲音則需要物質介質。

傳播(名) 通過介質(↑)傳播波動。例如:(a)光通過真空傳播;(b)聲波通過空氣傳播。(動詞爲 propagate)

直線的(形) 由直線構成的,以直線傳播。例如光的直線傳播是光波通過任何介質所做的直線傳播。

透明的(形) 描述任何一種固體或液體的介質,光能透過此介質成像(第 56 頁)。例如窗玻璃是透明的。

半透明的(形) 描述任何一種固體或液體介質,光能透過但不能清晰成像。例如蠟紙是半透明而非透明的。

sources of light 光源

source of sound 聲源

transparent 透明的

opaque 不透明的

translucent
半透明的

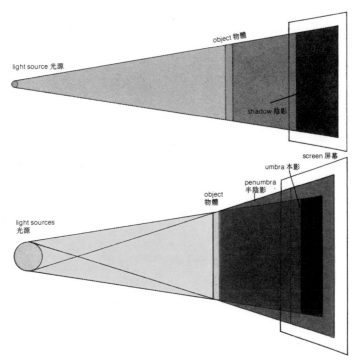

opaque (*adj*) describes any solid or liquid through which light cannot travel.

shadow (*n*) the dark area formed by an object which stops light. A shadow is formed because light travels in straight lines. The shadow has the same shape as the object when the shadow is seen on a screen (p.56). If the light source is very small, the shadow will be sharp.

umbra (*n*) the area in a shadow from which light is completely cut off.

penumbra (*n*) a lighter area between the umbra and the edge of a shadow; some light reaches it because the source is not small. A very small light source forms an umbra only; other sources form an umbra and a penumbra.

不透明的(形)　描述光不能透過的任何一種固體或液體。

陰影(名)　物體擋光而形成的陰暗區域。陰影的形成是因爲光呈直線傳播。在屏幕上(第56頁)看到的陰影，它的形狀與原物體一樣。如果光源極小，陰影很清晰。

本影(名)　光線被完全阻隔的陰影區域。

半陰影(名)　本影和陰影邊緣之間較亮的區域；有些光達到這個區域，因爲光源不小。非常小的光源僅形成本影，其他光源形成本影和半陰影。

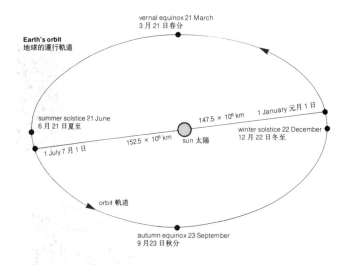

vernal equinox 21 March
3 月 21 日春分

Earth's orbit
地球的運行軌道

summer solstice 21 June
6 月 21 日夏至

147.5 × 10⁶ km

1 January 元月 1 日

152.5 × 10⁶ km

sun 太陽

winter solstice 22 December
12 月 22 日冬至

1 July 7 月 1 日

orbit 軌道

autumn equinox 23 September
9 月 23 日秋分

Earth (*n*) the third planet (↓) at an average distance of 150 000 000 km from the sun. It is almost spherical in shape, being slightly flattened at the Poles. The equatorial radius is 6378.388 km; the mass is 5.976×10^{24} kg.

orbit (*n*) the path followed by one object moving round another object, e.g. the orbit of the Earth round the sun. Gravitational (p.17) attraction between the Earth and the sun keeps the Earth in its orbit; the orbit takes 365¼ days to complete.

moon (*n*) a body in orbit round the Earth, held in orbit by gravitational attraction between it and the Earth. The distance from the Earth to the moon is 384 400 km; the mass is 7.35×10^{22} kg, and it takes 28 days to complete one orbit. As the Earth is itself in orbit, the path of the moon is as shown in the diagram. The moon has neither water nor atmosphere, is cold, and reflects (p.56) light from the sun.

sun (*n*) the source of light and heat for the Earth. It is spherical, its diameter is about 1 392 000 km, and its mass about 2×10^{30} kg. The surface is at a temperature of about 6000°C.

地球（名） 太陽系的第三個行星（↓），與太陽的平均距離是 150 000 000 km。形狀接近球體，兩極略呈扁平。赤道半徑 6378.388 km，質量是 5.976×10^{24} kg。

天體等的運行軌道（名） 一物體繞另一物體運行所循的路線，例如地球繞太陽運行的軌道。地球和太陽之間的萬有引力（第 17 頁）使地球保持在軌道上；循軌道運行一週需 365.25 天。

月球（名） 繞地球軌道運行的天體，月球和地球之間的萬有引力使月球保持在軌道上。地球到月球的距離是 384 400 km；質量爲 7.35×10^{22} kg，循軌道運行一週需 28 天。地球沿自己的軌道運行時，其路線如圖示。月球上既無水也無大氣，而且寒冷，反射（第 56 頁）太陽光。

太陽（名） 地球上光和熱的源泉。太陽爲球體。直徑約爲 1 392 000 km，質量約爲 2×10^{30} kg。表面溫度約爲 6000°C。

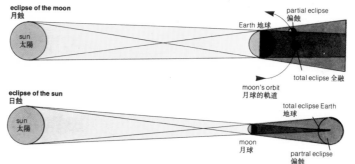

eclipse of the moon
月蝕

sun
太陽

Earth 地球

partial eclipse
偏蝕

total eclipse 全蝕

moon's orbit
月球的軌道

eclipse of the sun
日蝕

sun
太陽

total eclipse Earth
地球

moon
月球

partral eclipse
偏蝕

eclipse (*n*) the darkening of a heavenly body when it moves into the shadow of another heavenly body. An eclipse of the moon is seen when the shadow of the Earth falls on the moon; an eclipse of the sun is seen when the shadow of the moon falls on the Earth.

total eclipse the condition when a heavenly body is completely in the umbra (p.53) of another heavenly body so that no light falls on it at all.

partial eclipse only part of a heavenly body is in the umbra of another heavenly body, so part of it receives light.

phase (*n*) the state, position, or condition of an object or a quantity which passes through a number of events in order, and then repeats the same events again and again, e.g. the phases of the moon are (1) new moon (2) first quarter (3) full moon (4) third quarter (1) new moon, and so on. *See diagram.*

蝕(名) 當一天體運行進入另一天體的影子時，該天體的遮暗現象。當地球的影子落在月球上時，可看到月蝕；當月球的影子落在地球上時，可看到日蝕。

全蝕 一天體完全進入另一天體的本影內(第53頁)，以致光完全照不到它時的狀態。

偏蝕 只有一部分天體進入另一天體的本影內，所以它還有一部分可接收到光。

位相；盈虧(名) 一物體或一個量按順序通過一系列事件時的狀態、位置或情況，這一系列事件一再重複發生。例如月球的盈虧是：(1)新月，(2)上弦月，(3)滿月，(4)下弦月，(5)新月等等。(見圖)。

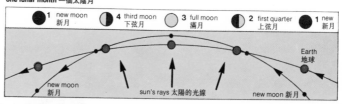

one lunar month 一個太陰月

1 new moon 新月   4 third moon 下弦月   3 full moon 滿月   2 first quarter 上弦月   1 new 新月

new moon 新月   sun's rays 太陽的光線   new moon 新月   Earth 地球

planet (*n*) a body in orbit round the sun. The Earth is a planet. Planets are not luminous (↓). (See back endpapers.) **planetary** (*adj*).

luminous (*adj*) giving out light.

行星(名) 繞太陽運行的天體。地球是行星。行星不發光(↓)。(見封底內頁)(形容詞爲planetary)

發光的(形) 發出光的。

angle (*n*) a measurement of a change in direction, measured in degrees or radians. There are 360° or 2π radians in a complete circle. **angular** (*adj*).

reflect (*v*) to change the direction of a line or path by means of a surface, e.g. the surface of water reflects light from the sun. **reflection** (*n*).

incident (*adj*) meeting, hitting, or falling on a surface, e.g. an incident ray of light. **incidence** (*n*).

mirror (*n*) a very smooth surface which reflects light, e.g. a metal plate or a glass plate with silver on the back.

plane mirror a flat mirror, the common kind of mirror.

ray (*n*) (1) any one line of a number of lines starting from the same point; (2) a line representing the direction of light; (3) a line of particles, in motion one after the other, going in a particular direction.

角度（名）　方向改變的量度，以度數或弧度測量。整圓爲 360° 或 2π 弧度。（形容詞爲 angular）

反射（動）　藉一個表面改變線路或路線的方向。例如水面反射太陽光。（名詞爲 reflection）

入射的（形）　遇到，碰到或落到一平面上，例如光的入射線。（名詞爲 incidence）

鏡（名）　極光滑，能反射光的表面。例如，一塊金屬或一塊背面塗有銀的玻璃板。

平面鏡　一種平鏡，即普通的鏡子。

射線；光線（名）　(1) 由同一點發出的許多線之一；(2) 表示光的方向的一條線；(3) 一個跟一個朝一特定方向運動的一系列粒子。

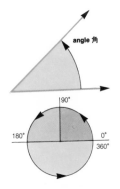

angle 角

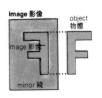

image 影像
object 物體
image 影像
mirror 鏡

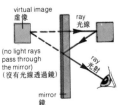

virtual image 虛像
ray 光線
(no light rays pass through the mirror)（沒有光線透過鏡）
ray 光射
mirror 鏡

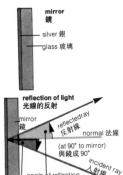

mirror 鏡
silver 銀
glass 玻璃

reflection of light 光線的反射
mirror 鏡
reflected ray 反射線
normal 法線
(at 90° to mirror) 與鏡成 90°
incident ray 入射線
angle of reflection 反射角
angle of incidence 入射角

image (*n*) a picture of an object formed by a mirror or a lens (p.58) or formed on the retina (p.205) of the eye, e.g. the image of a person when he looks at himself in a mirror. An image is formed when rays of light, starting from the same point on an object, meet at a point which is the same point on the image as on the object, *see diagram*.

screen (*n*) (1) a flat surface on which a picture can be formed by rays of light, e.g. the screen of a cinema; (2) a thin object, like a wall, which stops light, magnetism (p.69) or any undesirable effect from reaching a place. **screen** (*v*).

real image an image through which rays of light pass, and thus can be put on a screen.

virtual image an image from which rays of light appear to come. As no light rays actually pass through the image, it cannot be put on a screen.

影像（名）　由鏡子或透鏡（第 58 頁）形成的，或在眼睛視網膜（第 205 頁）上形成的物像，例如人照鏡子時出現的人像。從物體同一點發出的光線射到與物體影像上對應的一點時形成的影像。（見圖）

屏幕（名）　(1) 光線可在其面形成圖像的平面。例如，電影屏幕；(2) 像一幅牆的薄物體，可阻止光、磁力（第 69 頁）或任何不良的效應達到某一處。（動詞爲 screen）

實像　光線能通過的影像，因此它能投映在屏幕上。

虛像　看上去能發出光線的影像。因爲實際上光線不能透過此影像故不能投映在屏幕上。

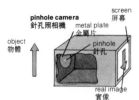

pinhole camera
針孔照相機
object 物體
screen 屏幕
metal plate 金屬片
pinhole 針孔
real image 實像

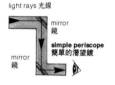

light rays 光線
mirror 鏡
simple periscope 簡單的潛望鏡
mirror 鏡

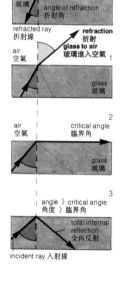

refraction
折射
angle of incidence
入射角
incident ray
入射線
air
空氣
glass
玻璃
angle of refraction
折射角
refracted ray
折射線

refraction
折射
glass to air
玻璃進入空氣
air
空氣
glass
玻璃
1

air
空氣
critical angle
臨界角
glass
玻璃
2

angle > critical angle
角度 > 臨界角
total internal reflection
全內反射
3

incident ray 入射線

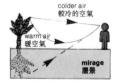

colder air
較冷的空氣
warm air
暖空氣
mirage
屬景

pinhole camera a box with a screen at one end and a very small hole in a thin metal plate at the other end. A real image is formed on the screen.

periscope (*n*) an instrument for seeing objects which are above eye-level, e.g. for looking over a wall. A simple periscope uses mirrors; prisms (p.61) can be used instead of mirrors. Lenses can be added so that the periscope also acts as a telescope (p.60).

refraction (*n*) the bending of a ray of light as it passes from one medium to another. **refract** (*v*).

Snell's law for a ray of light that is refracted, the sine of the angle of incidence divided by the sine of the angle of refraction is a constant ratio for any angle of incidence.

refractive index a measure of the ability of a medium (p.52) to refract light. For a particular medium it is equal to the speed of light in a vacuum divided by the speed of light in the medium. It is also measured by: refractive index = (sine angle of incidence) ÷ (sine angle of refraction). In symbols: $n = \sin i / \sin r$.

critical angle when a ray of light passes from one medium to another and the angle of refraction is greater than the angle of incidence (as from glass to air) then the least angle of incidence for which no refraction takes place is the critical angle. At the critical angle, the angle of refraction would be 90°, but the ray is reflected instead by the surface of the medium. For glass to air, the critical angle is 42°.

total internal reflection if the angle of incidence is greater than the critical angle of a medium, total internal reflection takes place.

mirage (*n*) the reflection of light by a layer of very warm air, which has been heated by the Earth. An object and its reflected image are seen; this gives the appearance of a water surface.

針孔照相機 一端有屏幕，另一端的薄金屬片上有一極小孔的盒子。在屏幕上形成實像。

潛望鏡(名) 一種觀看視線對平以上物體，例如越牆觀察用的儀器，簡單的潛望鏡使用平面鏡；也可用三稜鏡(第61頁)代替平面鏡；如加用透鏡也能起望遠鏡(第60頁)的作用。

折射(名) 光線從一種介質進入另一種介質時的曲折現象。(動詞為 refract)

斯涅耳折射定律 對任何入射角的折射光線，入射角正弦與折射角正弦之比為一個常數。

折射率；折光率 介質(第52頁)折光能力的尺度。對一種特定介質而言，折射率等於光在真空中的速度與光在介質中的速度之比。還可用於下列方法測量：折射率＝(入射角的正弦)÷(折射角的正弦)。用符號表示為 $\eta = \sin i / \sin r$。

臨界角 光線從一種介質進入另一種介質，而且折射角大於入射角(如從玻璃進入空氣)時，不發生折射的最小入射角即為臨界角。在臨界角時，折射角為90°，但光線都被介質表面反射回來。從玻璃進入空氣的臨界角為42°。

全內反射 如果入射角大於介質的臨界角，則產生全內反射現象。

海市蜃樓；蜃景(名) 接近地面的一層熱空氣反射光線的現象。可以看到一個物體及其反射的像；這就形成水面的幻景。

lens (*n*) (*lenses*) a piece of glass (or other transparent material) with one or both sides curved, so that it refracts light, and can form an image.

focus (*n*) (1) the point at which rays of light, which have been coming closer together, finally meet; this is a *real focus* as on a screen. (2) the point from which rays appear to come; this is a *virtual focus*. The *principal focus* of a lens is the point through which parallel rays will pass, or will appear to pass, after being refracted by a lens, *see diagram*.

focal length the distance between the centre of a lens and its principal focus (↑).

converging lens a lens which causes the rays of a parallel beam of light to come closer and pass through a point, i.e. the rays *converge*. It produces a real image at its principal focus. Such a lens is thicker at the centre than at the edge.

diverging lens a lens which causes the rays of a parallel beam of light to spread out and appear to come from a point, i.e. the rays *diverge*. It produces a virtual image at its principal focus. Such a lens is thinner at the centre than at the edge.

透鏡(名) 折射光並能成像，其一面或兩面呈曲面的一種鏡片(或其他透明材料)。

焦點(名) (1)光線愈來愈靠近最後會聚之點；即**實焦點**，如屏幕上的焦點。(2)光線似乎是從一點(即**虛點**)發出的。透鏡的"主焦點"係平行光線經透鏡折射後，行將通過的或似將通過的點(見圖)。

焦距 透鏡中心和主焦點(↑)之間的距離。

會聚透鏡 一種使平行光束愈來愈靠近並通過一點的透鏡，即能使光線"會聚"的透鏡。在其主焦點上產生實像。這種透鏡是凸形的。

發散透鏡 一種使平行光束散開並看似發自一點的透鏡，即光線"發散"的透鏡。在其主焦點上產生虛像。這種透鏡是凹形的。

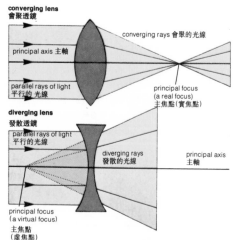

converging lens
會聚透鏡

converging rays 會聚的光線

principal axis 主軸

parallel rays of light
平行的 光線

principal focus
(a real focus)
主焦點(實焦點)

diverging lens
發散透鏡

parallel rays of light
平行的 光線

diverging rays
發散的光線

principal axis
主軸

principal focus
(a virtual focus)
主焦點
(虛焦點)

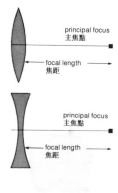

principal focus
主焦點

focal length
焦距

principal focus
主焦點

focal length
焦距

convex surface
凸面

curved mirror
曲面鏡

concave surface
凹面

convex (adj) describes a surface which curves outwards. A lens with two convex surfaces is a converging lens.

concave (adj) describes a surface which curves inwards. A lens with two concave surfaces is a diverging lens.

curved mirror a mirror with either a convex or a concave surface, capable of bringing parallel rays of light to a real image (concave mirror) or a virtual image (convex mirror).

magnify (v) to make an image larger than the object. **magnifying** (adj).

magnification (n) the ratio of the image size to the object size in one direction (one dimension).

principal axis a line passing through the centre of a lens or curved mirror and at a right angle to it.

凸的(形) 描述凸出的表面。雙凸面透鏡是一種會聚透鏡。

凹的(形) 描述凹進的表面。雙凹透鏡是一種發散透鏡。

曲面鏡 能使平行光線產生實像(凹面鏡)或虛像(凸面鏡)的凸面或凹面鏡。

放大(動) 使影像大於原物體。(形容詞爲 magnifying)

放大率(名) 影像的大小與物體的大小之比。(按一個空間方向計)

主軸 通過透鏡或曲面鏡的中心並與鏡面成直角的線。

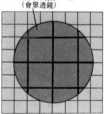

magnified image (×2)
(a converging lens)
放大的鏡像(×2)
(會聚透鏡)

magnifying glass
放大鏡

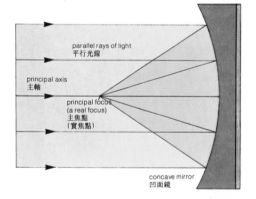

parallel rays of light
平行光線

principal axis
主軸

principal focus
(a real focus)
主焦點
(實焦點)

concave mirror
凹面鏡

crystalline lens a transparent structure, a part of the human eye, (p.203) which acts as a converging lens, focusing light onto the retina (p.205) of the eye to form an image. The lens is suspended from a circular muscle which can contract and make the lens more convex; this shortens the focal length so the eye can focus on near objects.

accommodation (n) the action of altering the focal length of a lens to see near objects. At rest the eye is focused on distant objects.

水晶體 一種透明的結構,人眼(第 203 頁)的一部分,其作用像會聚透鏡,將光線聚焦於眼的視網膜(第 205 頁)上,形成影像。水晶體懸在一塊圓形的肌肉上,這種肌肉能夠收縮,並使水晶體更外凸從而縮短焦距,使眼能看見近的物體。

調節(名) 爲了看到近的物體而改變水晶體焦距的動作。眼睛休息時聚焦於遠處的物體上。

defect (*n*) a missing part, property or characteristic, a wrong part or anything that causes a whole to be imperfect, e.g. (a) an eyeball with the wrong shape causes a defect in sight; (b) the loss of elasticity (p.27) in the crystalline lens is a defect of sight causing loss of accommodation (p.59). **defective** (*adj*).

缺陷（名） 缺欠的部分、性質或特性，有毛病的部分或使整體變成不完美的任何情況。例如：(a)眼珠變形引致視力缺陷；(b)水晶體失去彈性（第27頁）是視力喪失調節（第59頁）機能的缺陷。（形容詞爲 defective）

spectacles 眼鏡

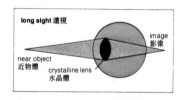

long sight 遠視

image 影像
near object 近物體
crystalline lens 水晶體

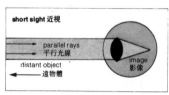

short sight 近視

parallel rays 平行光線
distant object 遠體
image 影像

long sight a defect in sight caused by the eye focusing near objects behind the retina (p.205). The defect is overcome by using converging lenses in spectacles.

遠視 由於眼睛將近物聚焦於視網膜（第205頁）後方引致的視力缺陷。這種缺陷可用會聚透鏡做眼鏡片來矯正。

short sight a defect in sight caused by the eye focusing distant objects in front of the retina. The defect is overcome by using diverging lenses in spectacles.

近視 由於眼睛將遠物聚焦於視網膜前方引致的視力缺陷。這種缺陷可用發散透鏡做眼鏡片來矯正。

camera (*n*) a device for obtaining photographs. A box has at one end a converging lens which focuses an image on a light-sensitive **film** at the other end. A **shutter**, in front of the lens, prevents light entering the camera until it is opened to take a photograph. The camera can be focused by moving the lens away from the film. A **diaphragm** controls the size of the **aperture** (↓), *see diagram*.

照相機（名） 一種拍攝相片的裝置。盒子的一端裝有一個會聚透鏡將影像聚焦在另一端的感光膠卷上。透鏡前方有一**快門**，它在被打開拍攝照片之前，阻止光線進入照相機。照相機可以通過使透鏡離開膠卷而聚焦。光圈控制光孔（↓）的大小（見圖）。

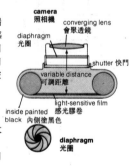

camera 照相機
converging lens 會聚透鏡
diaphragm 光圈
shutter 快門
variable distance 可調距離
light-sensitive film 感光膠卷
inside painted black 內側塗黑色
diaphragm 光圈

aperture (*n*) an opening into a space; the size of an opening put in front of a lens.

光孔（名） 通向外面的孔；透鏡前面的開孔的大小。

telescope (*n*) a device for seeing distant objects. Two or more tubes slide into each other; at one end is a converging lens (the objective) and at the other end is another converging lens (the eyepiece). These two lenses form a simple astronomical telescope.

望遠鏡（名） 一種用於看遠處物體的裝置。兩個或多個管筒互相套在一起，相互間可滑動，一端是會聚透鏡（物鏡），另一端是另一個會聚透鏡（目鏡）。這兩個透鏡構成一具簡單的天文望遠鏡。

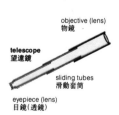

objective (lens) 物鏡
telescope 望遠鏡
sliding tubes 滑動套筒
eyepiece (lens) 目鏡（透鏡）

objective (*n*) the lens in a telescope or microscope (↓) nearest to the object.

物鏡（名） 望遠鏡或顯微鏡（↓）中離物體最近的透鏡。

eyepiece (*n*) the lens in a telescope or microscope (↓) through which a person looks.

目鏡（名） 望遠鏡或顯微鏡（↓）中，供人用眼睛觀察的透鏡。

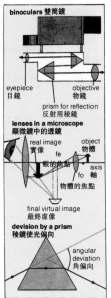

binoculars 雙筒鏡

eyepiece 目鏡
objective 物鏡

prism for reflection 反射用稜鏡

lenses in a microscope 顯微鏡中的透鏡

real image 實像
object 物體
fe
眼的焦點
axis 軸
fo
物體的焦點

final virtual image 最終虛像

devision by a prism 稜鏡使光偏向

angular deviation 角偏向

binoculars (n) a device (p.10) which uses two telescopes, one for each eye, for seeing objects at a distance. Prisms (↓), using internal reflection, are used to increase the distance between objective (↑) and eyepiece (↑).

binocular (adj) using two eyes for looking at the same object.

microscope (n) a device for magnifying (p.59) very small objects. Two converging lenses are used as shown in the diagram. **microscopic** (adj).

prism (n) a triangular prism is a piece of glass (or other transparent material) with two equal triangular faces joined by three rectangular, flat faces; it is used for refracting or internally reflecting light. **prismatic** (adj).

deviation (n) a change in direction of the path of an object or a wave motion (p.65), e.g. a deviation of a light ray by a prism. **deviate** (v).

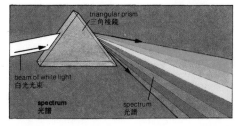

triangular prism 三角稜鏡

beam of white light 白光光束

spectrum 光譜

spectrum 光譜

rainbow 虹

rain 雨

direct vision spectroscope 直視分光鏡

crown glass prisms 冕牌玻璃稜鏡

eye sees spectrum 眼睛觀察光譜
slit 狹縫

flint glass prisms 火石玻璃稜鏡

lens 透鏡
white light 白光

spectrum (n) the result obtained when light is passed through a prism and is split into the following colours: red, orange, yellow, green, blue, indigo, violet. Light from the sun is white light, and it is made up of the colours of the spectrum. **spectral** (adj).

rainbow (n) the colours of the spectrum formed by refraction (p.57) and reflection (p.56) of light from the sun by raindrops. The effect is seen when a person has his back to the sun, and rain in front of him.

spectroscope (n) an instrument for looking at the colours of the spectrum. A simple spectroscope has five glass prisms in a tube (a direct vision spectroscope), see diagram.

雙筒望遠鏡(名) 一個裝置(第 10 頁),上有兩個望遠鏡,各供一隻眼睛觀察遠方的物體。利用稜鏡(↓)內部的反射作用來增大物鏡(↑)和目鏡(↑)之間的距離。

雙目的(形) 用兩隻眼睛觀察同一物體的。

顯微鏡(名) 用來放大(第 59 頁)極小物體的裝置。如圖所示,這種裝置使用兩個會聚透鏡。(形容詞為 microscopic)

稜鏡(名) 三角稜鏡是由三個矩形平面連接兩個全等的三角形平面構成的一塊玻璃(或其他透明物質);用於使光線折射或內部反射。(形容詞為 prismatic)

偏向;偏轉(名) 物體或波動(第 65 頁)路線方向的改變,例如使三稜鏡光線偏轉。(動詞為 deviate)

光譜(名) 光通過三稜鏡產生的結果,光線被分成以下的顏色:紅、橙、黃、綠、藍、靛青、紫。太陽光是白色光,它是由光譜的這些顏色組成的。(形容詞為 spectral)

虹(名) 陽光被雨點折射(第 57 頁)和反射(第 56 頁)所形成的光譜顏色。當一個人背對太陽面對雨時,可以看到這種現象。

分光鏡(名) 一種觀察光譜顏色的儀器。簡單的分光鏡裝有五個玻璃三稜鏡的筒管(直視分光鏡)(見圖)。

star (*n*) a luminous (p.55) heavenly body which remains in a definite place relative to other stars. Compare planet (p.55), comet and meteor (↓). The sun is a star.

light year the distance covered by light in 1 year, i.e. approx. 10^{13} km; used in measuring distances of stars from the Earth.

magnitude² (*n*) a measure of the brightness of a star. The scale is: first magnitude, second magnitude, etc., with first magnitude as the brightest. A star of one magnitude is approximately 2.51 times brighter than a star of the next lower magnitude.

constellation (*n*) a small group of stars fixed relative to each other; the group is given a name, e.g. the Plough.

nebula (*n*) (*nebulae*) a milky, luminous (p.55) spot in the sky; a great cloud of very hot gas round some stars. New star systems may be formed from nebulae. **nebular** (*adj*).

solar (*adj*) of the sun, e.g. a solar day, the time for the Earth to make one complete turn relative to the sun.

sidereal (*adj*) measured relative to the stars, e.g. in a sidereal day, the Earth makes one complete turn relative to the stars. A sidereal day is about 4 minutes shorter than a solar day.

stellar (*adj*) of a star, e.g. a constellation is a stellar system.

lunar (*adj*) of the moon, e.g. a lunar month is the time from one new moon to the next new moon.

system¹ (*n*) a group of objects or materials which depend on each other and act in agreement with scientific laws to form a whole, e.g. the solar system is a group of planets together with the sun, all obeying the law of gravity and acting on each other, thus forming a whole. **systematic** (*adj*).

galaxy (*n*) a very large group of stars, containing thousands of millions of stars, forming a stellar (↑) system. The solar system is part of a galaxy called the Milky Way. Other galaxies are placed irregularly in space. **galactic** (*adj*).

universe (*n*) the system of all the galaxies. **universal** (*adj*).

恆星（名） 它相對於其他星球保持固定的一種發光的（第 55 頁）天體。試比較行星（第 55 頁）、彗星和流星（↓）。太陽是一顆恆星。

光年 光在一年中所行經的路程，約爲 10^{13} 公里；光年用於測量恆星和地球之間的距離。

星光度；星等級（名） 恆星亮度的尺度。記數法為：第一星光度，第二星光度等等。第一星光度最亮。一顆第一星光度的恆星比下一顆較低星光度的恆星約亮 2.51 倍。

星座（名） 彼此位置相對固定的小星群；這種星群常有命名，例如大熊星座。

星雲（名） 天空中乳白發光的（第 55 頁）亮點；這是包圍某些星球的大片熾熱氣體雲。星雲可能形成新的星系。（形容詞爲 nebular）

太陽的（形） 指屬太陽或有關太陽的，例如一個太陽日是地球相對於太陽自轉一週的時間。

恆星的（形） 以恆星爲基準所測量的。例如一個恆星日是地球以恆星爲準自轉一週。一個恆星日比一個太陽日大約縮短 4 分鐘。

恆星的（形） 指屬恆星或有關恆星的。例如一個星座就是一個恆星的星系。

太陰的（形） 指屬月球或有關月球的。例如一個太陰月就是從一個新月到下一個新月的時間。

體系（名） 互相依賴並按科學定律起作用而形成一個統一整體的一組物體或物質。例如：太陽系是和太陽在一起的一羣行星，它們都遵循萬有引力定律互相起作用，形成一個統一的整體。（形容詞爲 systematic）

星系（名） 由幾十億顆恆星形成的一個極巨大的恆星（↑）系。太陽系是銀河星系的一部分。其他星系不規則地位於太空中。（形容詞爲 galactic）

宇宙（名） 由全部星系組成的一個體系。（形容詞為 universal）

constellation 星座
the Plough 北斗七星
ursa major 大熊星座

nebula 星雲

milky way 銀河系

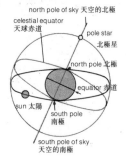

Earth and its celestial sphere 地球及其天球
north pole of sky 天空的北極
celestial equator 天球赤道
pole star 北極星
north pole 北極
equator 赤道
sun 太陽
south pole 南極
south pole of sky 天空的南極

comet
彗星

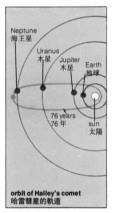

Neptune
海王星
Uranus
天王星
Jupiter
木星
Earth
地球
76 years
76 年
sun
太陽

orbit of Halley's comet
哈雷彗星的軌道

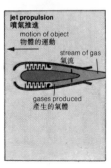

jet propulsion
噴氣推進

motion of object
物體的運動

stream of gas
氣流

gases produced
產生的氣體

comet (*n*) a luminous (p.55) heavenly body with a bright tail, which moves in an orbit (p.54) round the sun. A comet has small mass, but often has great size. The tail points away from the sun. Some comets, e.g. Halley's comet, are regularly seen from Earth, while other comets disappear from the solar system.

meteor (*n*) a solid body (either iron or stone) which enters the Earth's atmosphere and shines brightly because of the heat produced by friction (p.22) with the air; small meteors are completely burned to gases. **meteoric** (*adj*).

meteorite (*n*) a large meteor which is not completely burned to gas, but falls to the Earth.

meteor shower a large number of meteors entering the Earth's atmosphere, mainly seen when the Earth crosses the orbit of a comet.

satellite (*n*) (1) a smaller heavenly body held in orbit (p.54) round a bigger body by gravitational (p.17) attraction, e.g. the moon is a satellite of the Earth. (2) an object, made by man, and put into an orbit by a rocket (↓) is called an artificial satellite.

jet propulsion the pushing of an object in one direction by a stream of gas sent out in the opposite direction. The gas is usually produced by combustion (p.112). The lower the density of the atmosphere, the more efficient is this kind of propulsion.

reaction propulsion a name for jet propulsion.

projectile (*n*) an object pushed or thrown forward in the air by a sudden, great force, so that the object continues in motion when the force no longer acts, e.g. a bullet, fired from a gun, is a projectile. **project** (*v*), **projectile** (*adj*).

rocket (*n*) a projectile driven by jet propulsion (↑). The rocket contains its own propellants (↓) and does not need air for combustion, so it does not depend on the Earth's atmosphere, and can travel in space.

propellant (*n*) a slow explosive, e.g. gunpowder, used to apply force to a projectile. It is a substance which either burns explosively or supplies oxygen for the explosion. Rocket propellants are either solid or liquid substances.

彗星（名） 帶有明亮尾部的發光（第 55 頁）天體，它繞太陽的軌道（第 54 頁）運行。彗星的質量小，但往往體積巨大。彗星背向太陽。有些彗星，例如哈雷彗星，在地球上可以定期看見，而其他一些彗星則從太陽系消失了。

流星（名） 進入地球大氣層的實體（鐵或石）由於與空氣摩擦（第 22 頁）產生熱而發光；小的流星則完全燃燒而化成氣體。（形容詞爲 meteoric）

隕星（名） 巨大的流星，它尚未完全燃燒化成氣體而落到地球上。

隕石雨 進入地球大氣層的大量流星，主要是在地球越過彗星軌道時可以看見隕石雨。

衛星（名） （1）由於萬有引力（第 17 頁）的作用而保持在繞較大天體的軌道（第 54 頁）上的較小的天體。例如：月球是地球的衛星。（2）人造的、由火箭（↓）送入軌道的物體稱爲人造衛星。

噴氣推進 向反方向噴出氣流推動物體向前。氣體通常是由燃燒（第 112 頁）產生的。大氣的密度愈低，這種推進的效率就愈高。

反向推進 噴氣推進的別稱。

拋射體（名） 一個物體爲一個突發的巨大推力推向或拋向空中，力不再作用時，該物體仍繼續運動。例如，槍射出的子彈是一拋射體。（動詞爲 project，形容詞爲 projectile）

火箭（名） 用噴氣推進（↑）發射的拋射體。火箭本身有推進劑（↓），不需要空氣助燃，所似它不依賴地球的大氣就能在太空中飛行。

推進劑；發射火藥（名） 一種慢性炸藥。例如火藥，用於給拋射體施力。它是一種燃燒爆炸或給爆炸供氧的物質。火箭推進劑爲固體或液體物質。

vibrate (*v*), of an elastic (p.27) material, to move regularly, backwards and forwards. An elastic rod, held at one end, vibrates with the free end moving backwards and forwards. When a fluid vibrates, each molecule moves backwards and forwards about a fixed point, because of the fluid's elasticity. **vibration** (*n*), **vibrator** (*n*), **vibratory** (*adj*).

frequency (*n*) the number of times an event is regularly repeated in unit time, e.g. the number of vibrations in 1 second. The symbol for frequency is *f*. **frequent** (*adj*).

period (*n*) the time taken for one complete event in a number of events, when each event takes the same time for completion, e.g. the period of revolution of the Earth round the sun is 1 year.

periodic (*adj*) describes an event which occurs at regular periods. **periodicity** (*n*).

hertz (*n*) the S.I. unit of frequency. A periodic event has a frequency of 1 hertz if its period is 1 second. (*period*) = 1/(*frequency*) when the period is measured in seconds, and the frequency in hertz. The symbol is: Hz.

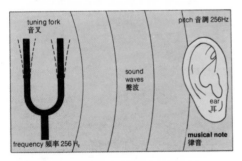

musical note a sound produced by vibrations of regular frequency of the air, e.g. the sound made by a bell and propagated (p.52) by the air.

amplitude (*n*) (1) the distance between the middle and the outer position of a vibrating body. (2) the distance between the middle and the top (or bottom) of a wave (↓).

振動 (動)　指彈性 (第 27 頁) 材料有規律地來回運動。彈性棒的一端固定，則隨自由端來回移動而產生振動。當流體振動時，其各個分子都由於流體的彈性，而在一固定點附近來回運動。(名詞爲 vibration，vibrator，形容詞爲 vibratory)

頻率 (名)　單位時間內某一事件有規律地重複的次數。例如一秒鐘內的振動次數。頻率的符號爲 *f*。(形容詞爲 frequent)

週期 (名)　在一系列事件中，當每個事件都用相同時間完成時，完成一個完整事件所需的時間。例如，地球繞太陽公轉的週期是一年。

週期的 (形)　描述以有規律地按週期發生的事件。(名詞爲 periodicity)

赫茲 (名)　頻率的國際制單位。如果週期性事件的週期爲一秒，則其頻率爲一赫茲。當週期以秒爲單位時，(週期) = 1 / (頻率)，頻率以赫茲爲單位。符號爲 Hz。

vibrating rod
振動棒

律音　空氣以規律頻率振動所產生的聲音。例如空氣傳播 (第 52 頁) 的鐘聲。

振幅 (名)　(1) 振動物體中間位置和外部位置之間的距離；(2) 波 (↓) 的中間位置和波峰 (或波谷) 之間的距離。

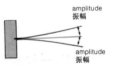

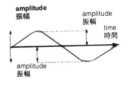

pitch (*n*) the place of a note in a musical scale; it is a measure of the frequency of vibration of the source of the note, e.g. a high frequency vibration produces a note of high pitch.

quality (*n*) a property of a musical note which makes it different from another note of the same pitch and loudness. It depends on the overtones (p.67) produced by a musical instrument and enables different instruments to be recognized.

音調（名）　音階中音符的位置；音調是音符聲源振動頻率的量度。例如，高頻振動產生高調律音。

音質；音色（名）　律音的特性，它使音調和音量相同的律音之間有所區別。律音取決於樂器產生的泛音（第 67 頁），可以靠它識別不同的樂器。

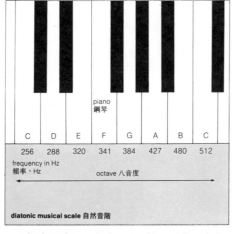

diatonic musical scale 自然音階

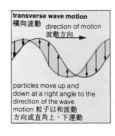

transverse wave motion
橫向波動
direction of motion
波動方向
particles move up and down at a right angle to the direction of the wave motion 粒子以和波動方向成直角上、下運動

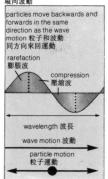

longitudinal wave motion
縱向波動
particles move backwards and forwards in the same direction as the wave motion 粒子和波動同方向來回運動
rarefaction 膨脹波
compression 壓縮波
wavelength 波長
wave motion 波動
particle motion 粒子運動

musical scale musical notes of increasing pitch at regular intervals (p.68) form a musical scale.

wave motion the sending of energy by a periodic (↑) movement in a medium (p.52) in the form of a wave, *see diagram*. In a material medium, the particles move about a central position and produce the wave motion.

transverse wave a wave in which particles of a material medium move up and down, about a central position, at a right angle to the direction of the wave motion.

longitudinal wave a wave in which particles of a material medium move backwards and forwards, about a central position, in the same direction as the wave motion. Sound waves are longitudinal waves.

音階　按規定音程（第 68 頁）逐步升高音調的音符組成音階。

波動　通過介質（第 52 頁）以波的形式進行週期性（↑）運動來放出能量（見圖）。在物質介質中，粒子在中心位置附近運動並產生波動。

橫波　物質介質的粒子在波中心位置的附近上下運動，並與波動的方向成直角的一種波。

縱波　物質介質的粒子在波中心位置附近前後運動，其方向與波相同的一種波。聲波是縱波。

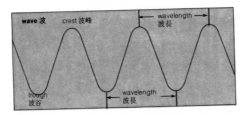

wavelength (*n*) the distance between a point in a wave and the next point at the corresponding place moving in the same direction, i.e. the distance between one crest and the next crest, *see diagram*, or between two troughs. The symbol for wavelength is λ. The wavelength is equal to the speed of the wave motion (*v*) divided by its frequency (*f*). The frequency of a wave motion is the number of crests (or troughs) that pass a point in 1 second. In symbols: $v = f\lambda$.

sound wave a longitudinal wave produced by a vibrating object. As the object vibrates, it sends out: (1) a wave of high pressure, known as a **compression**; (2) a wave of low pressure, a **rarefaction**. In between, in air, the pressure returns to atmospheric pressure. Sound needs a material medium for its propagation (p.52). A sound wave has a speed of about 330 m/s in air. The wavelength of a sound wave is the distance between one compression and the next.

natural frequency the frequency of vibration of a solid object, or a column of fluid, when free to vibrate and not acted upon by an outside force.

resonance (*n*) the cause, in an object, of very large amplitudes (p.64) of vibration which are produced when a periodic (p.64) force, with the same frequency as the natural frequency of the object, is applied to the object. The object is in resonance with the force, e.g. a loud note of the same frequency as that of a wine glass can break the glass by resonance. **resonant** (*adj*).

echo (*n*) the effect produced by the reflection of sound from a surface, e.g. an echo from the sea bed, used in an echo sounder, *see diagram*.

波長（名） 波內一點與相鄰的同向運動同相位點之間的距離，即一個波峰與相鄰波峰之間的距離（見圖）或兩個波谷之間的距離。波長符號爲 λ。波長 (λ) 等於波動速度 (*v*) 除以波動頻率 (*f*)。波動頻率是波峰（或波谷）一秒鐘內通過一點的次數。用符號表示爲 $v = f\lambda$。

聲波 由振動物體產生的縱波。當物體振動時發出：(1)高壓波，稱爲壓縮波；(2)低壓波，稱爲膨脹波。在空氣中，聲波之間，壓力恢復到大氣壓力。聲音需要物質介質傳播（第52頁）。聲波在空氣中傳播的速度約爲 330 m/s。聲波的波長是一個壓縮波和下一個壓縮波之間的距離。

自然頻率 固體或流體柱體在自由振動及不受外力影響時產生的振動頻率。

共振（名） 當週期性（第64頁）外力作用於物體，而其頻率與物體的自然頻率相同時，物體產生非常大的振幅（第64頁），引起共振。物體與外力處於共振狀態。例如，一個頻率與酒杯的頻率相同的響亮律音，可能由於共振而使酒杯破裂。（形容詞爲 resonant）

回聲（名） 聲音從一個表面反射而產生的效應。例如回聲探測器所用的來自海底的回聲（見圖）。

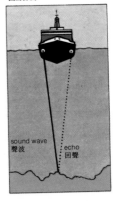

echo sounding
回聲探測

sound wave
聲波

echo
回聲

stationary wave the kind of wave formed when a stretched string vibrates, *see diagram*. The waves do not move along the string; they are caused by two waves, of the same frequency, travelling in opposite directions and opposing each other's motion. The wavelength (↑) of a stationary wave is twice the distance between two nodes (↓), or between two antinodes (↓). Stationary waves are also produced in vibrating columns of air.

駐波　拉緊的弦振動形成這種波（見圖）。駐波不沿弦運動；它們是由兩列頻率相同、傳播方向相反，運動相向的波引起的。駐波的波長（↑）爲兩個波節（↓）間或兩個波腹（↓）間距離的兩倍。空氣柱振動時也產生駐波。

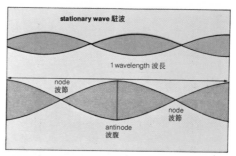

node¹ (*n*) a point in a stationary wave where there is no amplitude (p.64) of vibration. **nodal** (*adj*).

波節（名）　無振幅（第 64 頁）的駐波中的一個點。（形容詞爲 nodal）

antinode (*n*) a point in a stationary wave where there is the biggest amplitude of vibration. An antinode is midway between two nodes.

波腹（名）　駐波中振幅最大的一個點。腹點位於兩個波節的正中。

fundamental frequency the lowest frequency of vibration of a stationary wave. For a stretched string the fundamental frequency of vibration is produced when the wavelength is twice the length of the string. For a wind instrument, the fundamental frequency is produced when the wavelength is four times the length of the vibrating column of air. When vibrating with its fundamental frequency alone, a musical instrument gives out a **pure note**.

基頻　駐波的最低振幅。拉緊的弦，當波長爲弦長的兩倍時，產生振動的基頻。對管樂器而言，當波長爲振動的空氣柱長的四倍時，才產生基頻。樂器只以基頻振動時，才發出純音。

overtone (*n*) a note of higher frequency than the pure note (↑) given out by a musical instrument. A stretched string can produce overtones as shown on page 68. Most musical instruments produce overtones, and the addition of these overtones forms the quality (p.65) of the note.

泛音（名）　樂器發出的頻率高於純音（↑）的律音。如第 68 頁所示，拉緊的弦可發出泛音。大部分樂器都能發出泛音。這些泛音的加入形成律音的音質（第 65 頁）。

harmonic (*n*) an overtone (p.67) with a frequency that is equal to the fundamental frequency multiplied by a whole number, e.g. the second harmonic has a frequency twice that of the fundamental frequency, the third harmonic, three times that of the fundamental frequency. Harmonics can be added to a wave which has a single frequency to produce a **complex** wave.

諧音；諧波(名) 頻率等於基頻整數倍的泛音(第 67 頁)。例如：第二諧音的頻率爲基頻的兩倍。第三諧音的頻率爲基頻的三倍。諧音加在單頻波上可產生**複合波**。

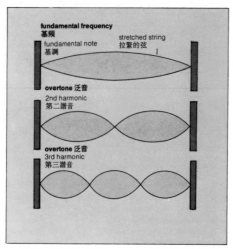

interval (*n*) (1) a space in distance between two objects, e.g. an interval of 3 metres between posts in a line, or a space in time between two events, e.g. the interval between meals. (2) the difference between two points on a scale. The interval between the pitch of two musical notes is measured by the ratio of their frequencies, e.g. two notes with pitches of 288 Hz and 256 Hz have a frequency interval of 9:8.

間隔；音程(名) (1)兩物體間距的空間。例如：一行柱子中，柱子的間隔三米；或兩個事件之間的間歇時間。例如：兩餐之間的間歇時間；(2)音階上兩點間的差異。兩個律音音調間的差異可由其頻率之比測出。例如：兩個音調爲 288 Hz 和 256 Hz 的音，其音程爲 9:8。

octave (*n*) (1) the interval between any two frequencies having a ratio of 2:1, e.g. notes with pitches of 512 Hz and 256 Hz are an octave apart *see diagram* (p.65). (2) in any wave motion, the band of frequencies between any two frequencies which are in the ratio of 2:1.

八音度；倍頻程(名) (1)比率爲 2:1 的任意兩個頻率之間的音程，例如：音調爲 512 Hz 和 256 Hz 的兩個音是一個分開的八音度(見圖)(第 65 頁)；(2)任何波動中，任意兩個比率是 2:1 的頻率間的頻帶。

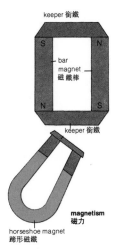

keeper 衛鐵

S N

bar
magnet
磁鐵棒

N S

keeper 衛鐵

magnetism
磁力

horseshoe magnet
蹄形磁鐵

like poles repel
同極相斥

N

S

N

S

unlike poles attract
異極相吸

magnet (n) a solid object that attracts iron and attracts or repels (↓) other magnets. When free to turn, it points in a north-south direction. It possesses the property of **magnetism** (n). **magnetic** (adj), **magnetize** (v).

permanent magnet a magnet that does not lose its magnetism; usually made of steel, or an alloy (p.103) of steel.

temporary magnet a piece of iron, or other magnetic material, that is a magnet only as long as it is influenced by magnetizing force; soft iron is often used for temporary magnets.

pole (n) points near each end of a magnet at which the magnetism appears to be strongest. One end is called a north pole because it points north when the magnet is free to turn; the other end is a south pole. **polar** (adj).

keeper (n) a piece of soft iron put against the north pole of one magnet and the south pole of the same magnet or another magnet; it prevents any tendency (p.15) of a permanent magnet to lose its magnetism.

like (adj) (1) two north poles or two south poles are called like poles. (2) two positive charges (p.71) or two negative charges are called like charges. (3) same in nature, or character.

unlike (adj) (1) a north pole and a south pole are unlike poles. (2) a positive charge and a negative charge are unlike charges.

repel (v) to make an object, or anything else, go away, e.g. like magnetic poles repel each other and unlike magnetic poles attract each other.

magnetic field the space round a magnet or an electric current in which a magnetic material experiences a magnetic force of attraction, or a magnet sets in the direction of the magnetic force from the magnet.

磁體（名） 能吸鐵以及吸引或排斥（↓）其他磁體的一種固體物件。磁鐵可以自由轉動時，它指向南北方向。磁體具有**磁力**的屬性。（形容詞爲 magnetic，動詞爲 magnetize）

永磁體 不會失去磁力的磁體。通常由鋼或合金（第 103 頁）鋼製成。

暫時磁體 鐵塊或其他磁性材料，只在受磁力影響時，才成爲磁體。軟鐵常用作暫時磁體。

極（名） 接近磁體末端磁力表現最強的兩個端點。一端稱爲北極，因爲它自由轉動時指向北方；另一端稱爲南極。（形容詞爲 polar）

衛鐵（名） 一塊軟鐵跨放在磁體的北極和南極上，或另一磁體的南極上；它防止永磁體失去磁力的傾向（第 15 頁）。

相同的（形） （1）兩個北極或兩個南極稱爲相同的極；（2）兩個正電荷（第 71 頁）或兩個負電荷稱爲同性電荷；（3）性質或特性相同。

不同的（形） （1）北極和南極是不同的極；（2）正電荷和負電荷是異性電荷。

排斥（動） 使一件物體或其他東西離開。例如，相同磁極互相排斥，而不同的磁極互相吸引。

磁場 磁性材料在磁鐵或電流周圍受磁力吸引的空間或磁體的磁力方向使另一磁體定位的範圍。

magnetic field 磁場

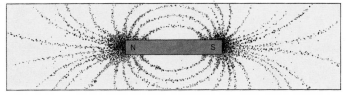

N S

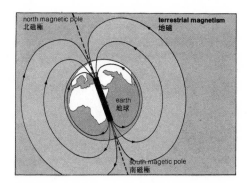

north magnetic pole
北磁極

terrestrial magnetism
地磁

earth
地球

south magnetic pole
南磁極

terrestrial magnetism the magnetism of the Earth; it exerts (p.23) a magnetic field covering the Earth, similiar to the field that a powerful magnet, at the centre of the Earth, would produce.

compass (*n*) an instrument with a magnet, free to turn on a pivot, in a case, with the directions north, south, east and west marked on a scale. In a **mariner's compass**, two or more magnetic needles are fixed to a card which floats in alcohol and water. The compass is used to find directions on the Earth.

dip circle an instrument with a magnetic needle, free to turn in a vertical plane, which is pulled down (i.e. dips) at one end by the Earth's magnetic field.

magnetic variation the angle between magnetic north and geographical north at any point on the Earth. The north pole of the Earth's magnetic field is not in the same place as the North Pole of the Earth. Magnetic variation lies between 30°W to 30°E, depending on the position of the compass on the Earth.

magnetic declination another name for magnetic variation.

magnetic dip the angle between the horizontal and the direction of the Earth's magnetic field. It varies between 0° about the equator to 90° at the Earth's magnetic poles.

magnetic inclination a name for magnetic dip.

地磁　地球的磁力；地磁施加(第 23 頁)的磁場覆蓋整個地球，就像地球中心有一個強磁體產生的磁場。

羅盤　由裝在盒子內的一根繞樞軸自由轉動的磁針和標有東西南北的刻度盤組成的儀器。**航海羅盤**中有兩個或多個磁針固定在浮於酒精或水面的方位板上。羅盤用於判別在地球上所處的方向。

磁傾儀　一種帶有磁針的儀器，能在直立平面上自由轉動。地球磁場吸引磁針的一端向下(即磁傾角)。

磁偏差　在地球表面任何地點的磁北極與地理北極間的夾角。地磁場的北極和地球的北極並不同一處。磁偏差在西 30° 到東 30° 之間，視羅盤在地球上的位置而定。

磁偏角　磁偏差的別稱。

磁傾角　水平面和地球磁場方向之間的夾角。夾角的大小從赤道附近的 0° 到地球磁極 90° 之間。

磁傾　磁傾角的別稱。

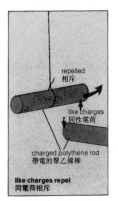

repelled
相斥

like charges
同性電荷

charged polythene rod
帶電的聚乙烯棒

like charges repel
同電荷相斥

discharge of a cloud
雲的放電

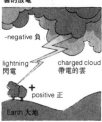

-negative 負

lightning
閃電

charged cloud
帶電的雲

+ positive 正

Earth 大地

charge (n) (1) the correct quantity of a material that is put in, or is needed, by a device, e.g. a charge of explosives used in a mine. (2) a charge of electricity cannot be described; its effects are: (a) an attraction for an unlike (p.69) charge or repulsion for a like charge; (b) the production of a spark when the charge escapes to earth, i.e. discharged; (c) the forming of an electric field. Symbol for electric charge: Q. **charged** (adj), **charge** (v), **discharge** (v), **discharge** (n).

electrostatic (adj) describes effects caused by electric charges at rest, such as an electric charge on an object.

spark (n) an instantaneous (p.21) appearance of light and sound in an electrostatic discharge of short duration.

lightning (n) a very large spark caused by the discharge of a cloud carrying an electrostatic charge.

thunder (n) the sound produced by the discharge of a charged cloud.

裝填量；電荷(名) (1)一種裝置所裝或所需要的準確物料量。例如礦用的炸藥裝填量。(2)電荷難於描述；但可見其效應：(a)異性(第69頁)電荷相吸，同性電荷相斥；(b)電荷向地面逸散時，即放電時產生電火花；(c)形成電場。電荷的符號爲 Q。(形容詞爲charged，動詞爲 charge，discharge，名詞爲 discharge)

靜電的(形) 描述電荷在靜止狀態時所引起的效應。例如物體上電荷產生的效應。

電火花(名) 短暫靜電放電瞬間出現的(第21頁)光和聲。

閃電(名) 帶靜電的雲放電產生的巨大電火花。

雷(名) 帶電雲放電發出的聲音。

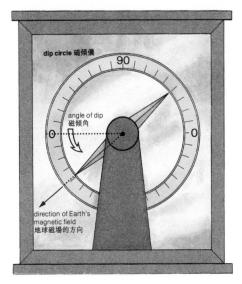

dip circle 磁傾儀

90

angle of dip
磁傾角

0

0

direction of Earth's magnetic field
地球磁場的方向

primary cell a device which produces a flow of electric charge, i.e. an electric current, by means of a chemical reaction. The simplest cell has two pieces of different metals, in acid, or alkali, contained in a vessel.

local action an effect caused by impurities (p.99) in the zinc plate of a primary cell. Bubbles of hydrogen are formed on the plate and the efficiency of the cell is made less. Rubbing the zinc with mercury prevents local action.

polarization (*n*) an effect produced when a simple primary cell produces electric current. Bubbles of hydrogen form on the copper or positive plate and the electric current quickly falls to a small value. All primary cells suffer from polarization. **polarize** (*v*).

原電池　利用化學反應產生電荷流，即電流的一種裝置。最簡單的電池由兩塊不同金屬置於酸溶液或鹼溶液的容器中的極板組成。

局部作用　原電池鋅板中的雜質（第 99 頁）引致的一種效應，結果在鋅板上產生氫氣泡，降低電池效力。用汞擦鋅可防止這種局部作用。

極化（名）　簡單原電池產生電流時的一種效應。銅陽極板上產生氫氣泡，而同時電流值很快降低。所有的原電池都會因極化而損壞。（動詞爲 polarize）

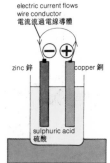

electric current flows
wire conductor
電流流過電線導體

zinc 鋅　　　copper 銅

sulphuric acid
硫酸

simple primary cell
簡單原電池

symbol for a cell 電池的符號
-primary or secondary
原電池或蓄電池

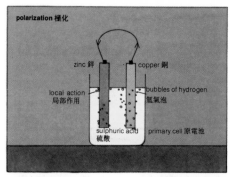

polarization 極化

zinc 鋅　　copper 銅
local action　　bubbles of hydrogen
局部作用　　氫氣泡
sulphuric acid　primary cell 原電池
硫酸

depolarization (*n*) the action to prevent polarization. A substance is added which chemically combines (p.94) with the hydrogen formed; such a substance is called a **depolarizer** (*n*). **depolarize** (*v*).

Leclanché cell a primary cell using a zinc rod and a carbon rod. The carbon rod has manganese (IV) oxide (manganese dioxide) powder around it, and is contained in a porous pot. The zinc rod and the porous pot stand in a solution (p.89) of ammonium chloride. The zinc rod is negative and the carbon rod is positive. The manganese (IV) oxide is a depolarizer (↑).

去極化作用（名）　防止極化的作用。加入一種能與氫產生化學結合（第 94 頁）的物質；這種物質稱爲去極劑。（動詞爲 depolarize）

勒克朗歇電池　一種使用鋅棒和碳棒的原電池。碳棒周圍爲氧化錳(IV)（即二氧化錳）粉末一起放入一個素燒瓶內。將鋅棒和素燒瓶置於氯化銨溶液（第 89 頁）中。鋅棒爲陰極，碳棒爲陽極。氧化錳(IV)爲去極劑(↑)。

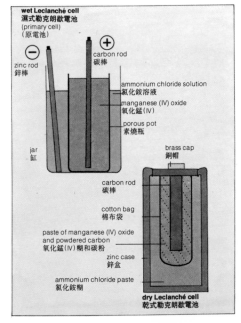

wet Leclanché cell
濕式勒克朗歇電池
(primary cell)
（原電池）

zinc rod
鋅棒

carbon rod
碳棒

ammonium chloride solution
氯化銨溶液

manganese (IV) oxide
氧化錳(IV)

porous pot
素燒瓶

jar
缸

brass cap
銅帽

carbon rod
碳棒

cotton bag
棉布袋

paste of manganese (IV) oxide
and powdered carbon
氧化錳(IV)糊和碳粉

zinc case
鋅盒

ammonium chloride paste
氯化銨糊

dry Leclanché cell
乾式勒克朗歇電池

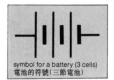

symbol for a battery (3 cells)
電池的符號（三節電池）

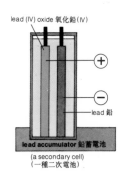

lead (IV) oxide 氧化鉛(IV)

lead 鉛

lead accumulator 鉛蓄電池

(a secondary cell)
（一種二次電池）

battery (*n*) a number of primary or secondary (↓) cells working together.

dry battery one or more dry cells working together. A dry cell is a Leclanché cell using ammonium chloride in the form of a paste and not a solution.

secondary cell an electric cell which produces an electric current by a chemical reaction. After being discharged (p.71) the cell can be recharged by passing an electric current through it in the opposite direction. A primary cell cannot be used in this way.

storage cell another name for secondary cell.

accumulator (*n*) a storage cell, or a battery of storage cells. The commonest kind is the lead-acid accumulator, with a negative plate of lead and a positive plate of lead (IV) oxide (lead dioxide) in sulphuric acid.

電池組(名)　數個聯合使用的原電池或可充電電池(↓)。

乾電池組　一個或多個一起使用的乾電池。乾電池是勒克朗歇電池，它使用糊狀氯化銨，而不使用氯化銨溶液。

二次電池　利用化學反應產生電流的電池。電池放電(第 71 頁)盡後，可以反方向通電流重新充電。原電池則不能再充電。

二次電池英文亦稱 **storage cell**。

蓄電池(名)　一個蓄電池或蓄電池組。最普通的一種是鉛酸蓄電池，其鉛陰極板和氧化鉛(IV)(二氧化鉛)陽極板置於硫酸中。

electrical (*adj*) concerning electricity, but not always carrying an electric current, e.g. electrical engineer.

electric (*adj*) describes any device which uses or produces a charge of electricity or any effect of such charge, e.g. an electric cell, an electric bell, an electric field.

current (*n*) (1) a fluid moving in a particular direction, or the motion of the fluid. (2) the flow of electric charge through a solid or a liquid. An electric current is measured in amperes. The symbol is: *I*.

conductor (*n*) a material through which an electric current or heat can flow. All metals are good conductors of both electric charge and heat. **conduct** (*v*).

insulator (*n*) a material or an object which prevents the flow of an electric current or heat. Most non-metals and all gases are insulators, e.g. sulphur and air are insulators.

insulate (*v*) to prevent electric current or heat escaping from where they are used and to prevent them from entering where they are not wanted. **insulation** (*n*) (1) the action of insulating. (2) material used for insulating.

electromotive force the force which drives an electric current round an electrical circuit (p.76). An electric cell (↑) or a generator (p.80) produce an electromotive force; for any one kind of cell, its electromotive force is constant if there is no polarization (p.72). Abbreviation: e.m.f.; symbol: E. An e.m.f. is measured in volts (↓).

電的(形)　和電有關，但未必總是載電流，例如電氣工程師。

電力的(形)　描述使用或產生電荷的任何裝置或此種電荷的任何效應。例如電池、電鈴、電場。

流；電流(名)　(1)朝特定方向運動的流體或流體的運動；(2)流過固體或液體的電荷流。電流以安培爲計量單位。符號爲 *I*。

導體(名)　電流或熱能流過的物質或材料一切金屬都是電荷和熱的良導體。(動詞爲 conduct)

絕緣體(名)　阻止電流或熱流動的材料或物體。大部分非金屬和所有的氣體都是絕緣體。例如：硫和空氣是絕緣體。

使絕緣(動)　阻止電流或熱從其所用之處逸出或阻止它們進入不需用之處。名詞 **insulation** 意爲：(1)絕緣作用；(2)絕緣材料。

電動勢　推動電流沿電路(第76頁)流動的力。電池(↑)或發電機(第80頁)產生電動勢；如果沒有極化(第72頁)，任何一種電池的電動勢都是一個常數。電動勢的縮寫爲 e.m.f.；符號爲 E。電動勢的計量單位爲伏特(↓)。

negative 負 　　　positive 正

direction of current 電流的方向

conductor 導體

insulated wire 絕緣電線

metal wire (conductor) 金屬線(導體)

plastic cover (insulator) 塑料包皮(絕緣體)

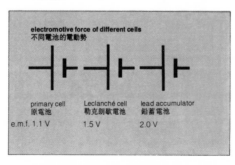

electromotive force of different cells 不同電池的電動勢

primary cell 原電池
e.m.f. 1.1 V

Leclanché cell 勒克朗歇電池
1.5 V

lead accumulator 鉛蓄電池
2.0 V

1 volt
1 伏特

current 電流

1 ampere
1 安培

1 joule of work
done in 1 second
1 秒內所做的功為 1 焦耳

power = 1 watt
功率 = 1 瓦特

**relation between volt,
ampere, joule and watt
伏特、安培、焦
耳和瓦特間的
關係**

potential difference the difference in electric force between any two points in an electrical circuit (p.76). Abbreviation: p.d.; symbol: *V*. An electric current always flows from a higher to a lower p.d.; it flows from a positive to a negative p.d. (This is a convention).

resistance[2] (*n*) the force which opposes the flow of an electric current through a conductor (↑). The resistance of an object, e.g. a wire, is measured in ohms (↓). Symbol: *R*. (See resistor p.76).

ampere (*n*) the S.I. unit of electric current. If a current of 1 ampere flows through each of two parallel conductors (↑), put 1 metre apart in a vacuum, then there will be a force of 2×10^{-7} newtons per metre of length of the conductors. Symbol: A.

coulomb (*n*) the S.I. unit of electric charge; the quantity of charge transferred by 1 ampere in 1 second between any two points of a circuit. Symbol: C. A current of 6A flowing for 3 seconds transfers 18C.

volt (*n*) the S.I. unit of electric potential; 1 volt is the difference of potential between two points if 1 joule of work is done, transferring 1 coulomb of charge between the points. **voltage** (*n*) a potential difference measured in volts.

ohm (*n*) the S.I. unit of resistance; a resistance of 1 ohm exists between two points in a circuit (p.76) when a potential difference of 1 volt between the points produces a current of 1 ampere. Symbol: Ω.

電勢差 電路(第 76 頁)中任意兩點間的電力差。縮寫爲 p.d.；符號爲 *V*。電流總是從高電位流向低電位；從正電位流向負電位(這是習慣規定)。

電阻(名) 阻止電流流過導體(↑)的力。物體的電阻，例如導線的電阻是以歐姆(↓)爲計量單位。電阻的符號爲 *R* (見第 76 頁" 電阻器")。

安培(名) 電流的國際制單位。如果 1 安培的電流通過兩個平行導體(↑)中的每一個導體，在真空中使它們相距 1 米，那麼每米長度導體上受的力爲 2×10^{-7} 牛頓。安培的符號爲 A。

庫倫(名) 電荷的國際制單位。電路中任意兩點間在 1 秒鐘內由 1 安培電流所傳遞的電荷量。庫倫的符號爲 C。6A 電流在 3 秒鐘內傳導 18C 的電量。

伏特(名) 電動勢的國際制單位。若兩點間傳導 1 庫倫的電荷時做 1 焦耳的功，則 1 伏特是兩點間的電位差。(名詞爲 voltage 指以伏特計算的電位差)

歐姆(名) 電阻的國際制單位。當兩點間 1 伏特電位差產生 1 安培電流時，電路中兩點間的電阻爲 1 歐姆。歐姆的符號爲 Ω。

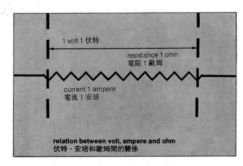

1 volt 1 伏特

resistance 1 ohm
電阻 1 歐姆

current 1 ampere
電流 1 安培

**relation between volt, ampere and ohm
伏特、安培和歐姆間的關係**

circuit (*n*) a continuous path of conductors and other electrical devices along which an electric current can flow; this is a *closed circuit*. If the circuit is broken at a point, so the current cannot flow, it is an *open circuit*.

terminal (*n*) a metal nut on a screw thread, to which a wire can be connected by tightening the nut. It joins wires to resistors (↓), cells, and other electric devices.

flex (*n*) a wire that bends to any shape to make electrical connections, i.e. to join parts of a circuit. It is always covered with insulating material.

switch (*n*) a device which is used to join (switch on) parts of a circuit and to break (switch off) parts of a circuit. The most common switch is for connecting and disconnecting an electric light. **switch on** (*v*), **switch off** (*v*).

open circuit see circuit (↑).

resistor (*n*) a device which offers resistance to an electric current. If the value of the resistor is known in ohms (p.75) and is constant, it is a *fixed resistor*.

variable resistor a resistor whose resistance can be changed. A contact, *see diagram*, slides round a length of wire, and thus varies the resistance.

電路(名) 由一些導體和電流可以沿着流過的其他電氣設備組成的連續通路；這是**閉路**。如果電路在某一點上被切斷，電流不能流過，就成爲**斷路**。

接線端(名) 一個安在螺紋上的金屬螺母，螺紋上可連接一條電線，再擰緊螺母。接線端可將導線接到電阻器(↓)、電池，及其他電器設備上。

花線；皮線(名) 可變成任何形狀以連接電路，即可將電路上各個部分連接起來的導線。花線通常都包覆有絕緣材料。

開關；掣(名) 用於接通(開)和切斷(關)電路各部分的裝置。最常用的開關是用來開、關電燈的開關。(動詞爲 switch on，switch off)。

斷路 見"電路"(↑)。

電阻器(名) 對電流起電阻作用的裝置。如果電阻器的歐姆(第75頁)值己知，且爲一個常數，則爲"固定的電阻器"。

可變電阻器(名) 可改變電阻的電阻器。其接觸器(見圖)可沿一段繞線管滑動，以改變電阻。

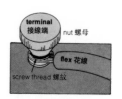

terminal
接線端　nut 螺母
flex 花線
screw thread 螺紋

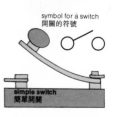

symbol for a switch
開關的符號

simple switch
簡單開關

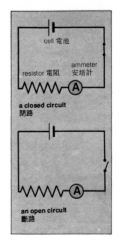

cell 電池

resistor 電阻　　ammeter 安培計
(A)

a closed circuit
閉路

(A)

an open circuit
斷路

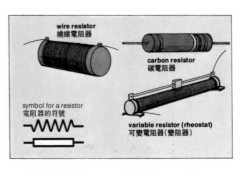

wire resistor
繞線電阻器

carbon resistor
碳電阻器

symbol for a resistor
電阻器的符號

variable resistor (rheostat)
可變電阻器(變阻器)

rheostat (*n*) another name for a variable resistor.

heating element a part of a heating appliance; it is a length of wire, of low resistance, that is heated when an electric current flows through it. It is used in electric fires, kettles, stoves, irons.

變阻器(名) 可變電阻的別稱。

發熱元件 加熱器具的一種零件；它是一段低電阻的導線，當電流通過時，導線被加熱。可用於電爐、電壺、電灶、電熨斗。

fuse
保險絲

fuse wire 保險絲

symbols for a fuse
保險絲的符號

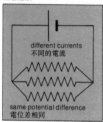

connected in parallel
並聯連接

different currents
不同的電流

same potential difference
電位差相同

centre-zero galvanometer
中心零位電流計

current can flow in
either direction
電流可在兩個
方向流動

symbols
符號

(I) galvanometer 電流計

(A) ammeter 安培計

(V) voltmeter 伏特計

filament[1] (*n*) a piece of very thin wire of high resistance which, when an electric current passes through it, becomes very hot and gives out light; used in an electric light bulb. A filament is also used in a thermionic valve (p.85) to give out electrons (p.106).

fuse (*n*) a device containing a piece of wire which melts if too great an electric current is passed through it. This breaks the circuit in which the fuse is placed; the fuse acts as a safety device.

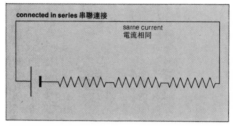

connected in series 串聯連接

same current
電流相同

series circuit a circuit in which all the parts are connected one after the other so that the same electric current flows through each part.

parallel circuit a circuit in which all the parts are connected to the same electromotive force (p.74) so that the same potential difference (p.75) is applied to each part.

Ohm's law the electric current flowing through a metallic conductor is proportional to the potential difference between the ends of the conductor, if the temperature is kept constant. The potential difference divided by the current gives the resistance of the conductor. In symbols: $I \propto V$; $V = IR$, where V is in volts, I in amperes, and R in ohms.

galvanometer (*n*) an instrument for measuring electric currents, particularly small currents. It has a scale so that currents can be compared, but it does not measure current in amperes.

ammeter (*n*) an instrument for measuring electric current in amperes.

voltmeter (*n*) an instrument for measuring electromotive force or potential difference (p.75) in volts.

燈絲（名） 極細的高電阻金屬絲，通電流時，它變成熾熱並發光；它用於電燈泡中。燈絲也用於熱離子管（第 85 頁）中以放出電子（第 106 頁）。

保險絲（名） 裝有一段金屬絲的設備，通過的電流太強時金屬絲就熔化。從而切斷裝有保險絲的電路；保險絲起保險裝置的作用。

串聯電路 電路中各段都首尾相聯，同一電流可以流過每一段電路。

並聯電路 電路中的各段電路都與同一電動勢（第 74 頁）相聯，施加於每段電路的電位差（第 75 頁）都相同。

歐姆定律 溫度保持不變時，通過金屬導體的電流與導體兩端間的電位差成正比。電位差與電流之比等於導體的電阻。用符號表示爲 $I \propto V$, $V = IR$, 式中 V 以伏特爲單位，I 以安培爲單位，R 以歐姆爲單位。

電流計（名） 測量電流，特別是測量小電流的儀器。它有一個刻度盤，可以比較電流的大小，但不能以安培來計量電流。

安培計（名） 以安培爲單位的測量電流的儀器。

伏特計（名） 以伏特爲單位測量電動勢或電位差（第 75 頁）的儀器。

Seebeck effect if two wires of different metals are joined at their ends to form a closed circuit (p.76) with two **junctions**, and if the two junctions are kept at different temperatures, then an electric current flows in the circuit.

thermoelectric (*adj*) describes the production of electric current directly from heat, e.g. *thermoelectric couple*: two wires of different metals used in the Seebeck effect (↑); *thermoelectric junction*: the join between two wires in a thermoelectric couple.

thermocouple (*n*) another name for thermo-electric couple (↑).

thermopile (*n*) an instrument with thermocouples (↑) connected in series behind a cone. It detects heat radiation and produces a thermoelectric current, measured by a galvanometer.

Peltier effect if two wires of different metals are joined and an electric current is passed through the junction, the junction is either warmed or cooled depending on the direction of the flow of current.

electromagnetism (*n*) (1) the study of magnetic effects produced by electric currents and of electric currents produced by magnetic fields. (2) magnetism produced by an electric current. **electromagnetic** (*adj*).

coil (*n*) a wire wound round a solid object (a former) in rings. The length of a coil is usually much shorter than its diameter. Coils are used in circuits for their electromagnetic effects.

winding (*n*) wire wound round part of an electrical device so that it acts as a coil, e.g. the winding on an armature (p.80).

turns (*n.pl.*) the rings of wire in a coil, e.g. a coil of 20 turns has 20 complete circles of wire wound round the former, *see diagram*.

core (*n*) (1) the middle part of a solid object. (2) the middle part of a coil or solenoid (↓), e.g. an iron rod in the middle of a solenoid; if a coil has no metal core it has an air core.

solenoid (*n*) a kind of coil in which the length is very much greater than the diameter. When an electric current passes through it, the solenoid acts as a magnet.

塞貝克溫差電動勢效應 將兩根不同的金屬導線連接形成一個有兩個接頭的閉路（第76頁），當兩個接頭保持不同的溫度時，電路中有電流流動。

熱電的(形) 描述由熱直接產生電流。例如"熱電偶"：塞貝克效應(↑)中用的兩根不同的金屬導綫；"熱電接頭"：熱電偶中兩根導線間的連接處。

溫差電偶(名) 熱電偶(↑)的別稱。

熱溫差電堆(名) 由多個溫差電偶(↑)串聯在一個錐體後面的一種裝置。用於檢測熱輻射及產生熱電流，熱電流可用電流計測出。

帕耳帖效應 將兩種不同金屬的線連接，使電流通過連接點，則連接點將發熱或變冷，視電流方向而定。

電磁學；電磁(名) (1)研究電流產生磁效應和磁場產生電流效應的學科；(2)電流產生的磁力。(形容詞爲 electromagnetic)

線圈(名) 一圈一圈繞在物件(線圈架)上的導線。線圈的長度通常比它的直徑短得多。電路中使用線圈以產生電磁效應。

繞組(名) 繞在電氣裝置部件上起線圈作用的導線。例如電樞(第80頁)上的繞組。

線圈的匝數(名、複) 線圈中導線的圈數。例如一個20匝的線圈有20整圈的線繞在線圈架上(見圖)。

芯(名) (1)固體物件的中心部分；(2)線圈或螺線管(↓)的中心部分。例如螺線管中心的一根鐵棒；如果線圈沒有金屬芯，則必有空氣芯。

螺線管(名) 一種長度比直徑大得多的線圈。電流通過螺線管時可起磁體作用。

Seebeck effect
塞貝克光效應

copper 銅
iron 鐵
熱接頭 冷接頭
hot junction   cold junction
current flows in circuit
電流在電路中流動

Peltier effect
帕耳帖效應
iron 鐵
copper 銅
heat given out or taken in
放熱或吸熱

former
線圈架
20 turns
20 匝

coil of wire
金屬線線圈

coil with iron core
有鐵芯的線圈

solenoid
螺線管

symbol for a coil
線圈的符號

coil with an iron core
帶有鐵芯的線圈

electromagnet
電磁體

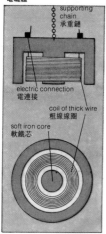

supporting chain
承重鏈

electric connection
電連接

coil of thick wire
粗線線圈

soft iron core
軟鐵芯

electromagnet (*n*) a coil with a soft iron core; when an electric current flows through the coil it acts as a magnet; when the current stops flowing, it loses its magnetism.

microphone (*n*) a device for transforming sound waves into an electric current. A diaphragm, *see diagram*, vibrates to sound waves, and pushes against carbon granules; the resistance of the granules varies with pressure, and causes a variation in current passing from the diaphragm to the iron plate. This varying current is passed to a receiver.

receiver (*n*) a device for transforming electric current into sound waves. A diaphragm is held above a permanent magnet; coils of wire are wound round the magnet, *see diagram*. Electric current (passed from the microphone) flows in the coils and the varying current causes the diaphragm to vibrate and give out sound waves.

電磁鐵(名) 有軟鐵芯的線圈。電流通過線圈時，線圈起磁鐵的作用；電流一中斷就失去磁性。

擴音器；送話器(名) 將聲波轉變爲電流的裝置。一個振動膜片(見圖)受聲波振動推撞碳粒；碳粒的電阻隨壓力而變化，引致從振動膜片流到鐵盤的電流發生變化。將變化的電流傳到受話器。

受話器；聽筒(名) 將電流轉變爲聲波的裝置。一個振動膜片裝在永磁鐵的上方；線圈繞在磁鐵上(見圖)。電流(來自擴音器)在線圈中流動，不斷變化的電流使膜片振動，發出聲波。

microphone
擴音器

plastic cover 塑料蓋

carbon granules 碳粒

carbon blocks 碳塊

electrical connections
電聯接線

receiver 受話器

metal diaphragm 金屬膜片
permanent magnet 永磁鐵
coils round pole pieces of magnet
繞磁鐵極的線圈

electrical connections
電聯接線

metal diaphragm
金屬膜片

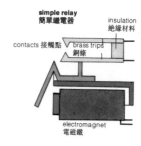

simple relay
簡單繼電器

insulation
絕緣材料

contacts 接觸點

brass trips
銅條

electromagnet
電磁鐵

relay (*n*) an electrical device which uses a small electric current to control a greater current in another circuit by switching (p.76) it on or off. Electrical relays use electromagnets to work a switch, *see diagram*.

繼電器(名) 一種電氣裝置，它用微小的電流藉開(第76頁)或關來控制另一電路中較大的電流。繼電器用電磁鐵來啟動開關(見圖)。

telephone (*n*) a circuit connecting a microphone, a receiver and a source (p.52) of electric energy, by wires. **telephonic** (*adj*), **telephony** (*n*).

telegraph (*n*) a device for sending messages over a distance by an intermittent (p.81) current conducted by wires. A key, *see diagram*, makes the intermittent current which is heard in a receiver. A code represents letters and numbers; it is called the Morse Code.

moving-coil (*adj*) describes an instrument which uses the motion of a coil in a magnetic field. When a current is passed through the coil, it acts as a magnet and turns to set in the direction of the magnetic field.

motor[1] (*n*) a device, other than an engine, which produces motion. An electric motor consists of a coil of wire wound round an armature (↓) which turns between the poles of a magnet.

armature (*n*) a piece of soft iron, turning on an axle, with a coil wound round it. When an electric current passes through the coil, the armature turns, producing mechanical motion. See generator (↓).

generator (*n*) a machine for transforming mechanical energy into electrical energy. The simplest generator has an armature, with a coil wound on it, turning inside the poles of a permanent magnet. When the armature is turned, a current flows in the coil. Larger generators use electromagnets with stators and rotors (↓); they provide electricity for houses.

dynamo (*n*) another name for generator, especially a small generator supplying direct current (↓).

rotor (*n*) the part of a generator, or turbine (p.44), that turns.

stator (*n*) the part of a generator, or turbine, that remains stationary.

direct current an electric current that flows in one direction only. Abbreviation: d.c.

alternating current an electric current which increases to a definite value, then decreases, finally changing direction and reaching the same value in the opposite direction, then increases again and repeats the changes. The frequency

電話機（名） 用電線將一個送話器、一個受話器和一個電源（第 52 頁）連接成的線路。（形容詞爲 telephonic，名詞爲 telephony）

電報機（名） 藉電線傳導的間歇（第 81 頁）電流將信息送到遠處的裝置。通過受話器可以聽見電鍵（見圖）產生的間歇電流。用電碼代表字母和數字；並轉向磁場的方向。

可動線圈（形） 描述一種利用線圈在磁場中運動的裝置。當電流通過線圈時，線圈起磁體的作用，並轉向磁場的方向。

電動機；馬達（名） 可以產生運動而又不是引擎（發動機）的一種裝置。電動機由繞在電樞（↓）上的線圈組成，電樞在磁體兩極間轉動。

電樞（名） 一塊繞軸轉動的軟鐵，上面繞有線圈。當電流通過線圈時，電樞轉動，產生機械運動。參見＂發電機＂（↓）。

發電機（名） 將機械能轉變爲電能的機器。最簡單的發電機有一個繞有線圈的電樞，電樞在永磁體的兩極間旋轉。電樞轉動時，有電流在線圈中流動。較大的發電機使用帶定子和轉子（↓）的電磁體；這種發電機給住宅供電。

電機（名） 發電機的別稱，尤指提供直流電（↓）的小發電機。

轉子（名） 發電機或渦輪機（第 44 頁）的轉動部件。

定子（名） 發電機或渦輪機中保持靜止的部件。

直流電 只單方向流動的電流。縮寫爲 d.c.。

交流電 上升至一定值後即下降並改變方向，在相反方向達到相同值，然後又上升，重複這一變化過程的電流。

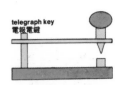

telegraph key
電報電鍵

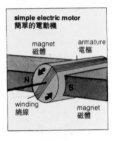

simple electric motor
簡單的電動機

magnet 磁體　　armature 電樞

N　　S

winding 繞線　　magnet 磁體

generator
發電機

rotor 轉子

stator 定子

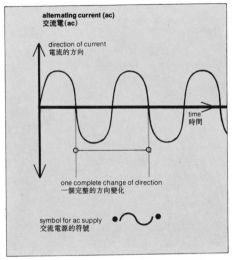

alternating current (ac)
交流電(ac)

direction of current
電流的方向

time
時間

one complete change of direction
一個完整的方向變化

symbol for ac supply
交流電源的符號

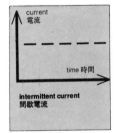

current
電流

time 時間

intermittent current
間歇電流

of the current is the number of complete
changes made in one second. Abbreviation: a.c.
Alternating current is produced by a generator.
rectifier (*n*) an electrical device which converts
alternating current to direct current. Thermionic
valves (p.85) are often used for this purpose.
rectified (*adj*), **rectify** (*v*).
intermittent (*adj*) describes an action or an effect
that starts and stops time after time, e.g. the
ringing sound of a telephone is intermittent.
fluctuating (*adj*) describes the values of a physical
quantity (p.13) which vary above and below
an expected value, e.g. the sound of a radio can
fluctuate above and below the expected
loudness.

電流頻率是一秒鐘內產生完整變化的次數。
交流電縮寫為 a.c.。交流電是發電機產生的
電流。

整流器(名) 將交流電轉變為直流電的電氣裝
置。熱離子管(第 85 頁)常作此用途(形容詞
為 rectified，動詞為 rectify)

間歇的(形) 描述一種時動時止的作用或效應。
例如，電話的鈴聲是間歇的。

波動的(形) 描述一個物理量(第 13 頁)的值在
預期值上下變化。例如，無線電的聲音能夠
在預期響度上下波動。

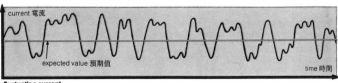

current 電流

expected value 預期值

time 時間

fluctuating current
波動電流

induced current if a conductor, e.g. a coil, is moved in a magnetic field, or there is a change in a magnetic field round a coil, then an induced current flows in the coil.

Faraday's law whenever there is a change in the magnetic field passing through a conductor an electromotive force is induced; the strength of the e.m.f. is proportional to the rate of change of the strength of the magnetic field.

Lenz's law the direction of the induced current is always such that it opposes the change producing it, e.g. if the north pole of a magnet enters a coil, the induced current in the coil makes the face of the coil a north pole, i.e. opposing entry of the magnet.

transformer (*n*) an apparatus which changes the voltage of alternating current (p.80). It consists of two coils of wire wound round the same iron core, *see diagram*. One coil contains a few turns of thick wire and the other coil many turns of thin wire. An alternating current is passed into one coil (the primary) and an induced alternating current, of the same frequency, is obtained from the other coil (the secondary).

turns ratio the ratio of the number of turns in the secondary winding to the number of turns in the primary winding of a transformer, e.g. 5000 secondary turns, 100 primary turns; turns ratio = 5000/100 = 50.

感應電流 導體(例如線圈)在磁場中移動,或者線圈周圍的磁場發生變化,線圈中均有感應電流流動。

法拉第定律 只要穿過導體的磁場發生變化,就會感應產生電動勢,其強度與磁場強度的變化率成正比。

楞次定律 感應電流的方向總是對抗產生此電流的磁場變化。例如,磁體北極進入線圈,則線圈中的感應電流使線圈的前端成爲北極,即對抗磁體的進入。

變壓器(名) 一種改變交流電(第 80 頁)電壓的裝置。由兩個繞在同一個鐵芯上的線圈組成(見圖)。其中一個線圈上有數匝粗金屬絲,另一線圈上有很多匝細金屬絲。當交流電輸入一線圈(原線圈),則可從另一線圈(副線圈)上獲得相同頻率的感應交流電。

匝數比 變壓器中副繞組匝數與原繞組匝數的比值。例如,副線圈匝數爲 5000,原線圈匝數爲 100;匝數比 = 5000/100 = 50。

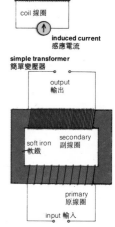

coil 線圈

induced current
感應電流

simple transformer
簡單變壓器

output
輸出

soft iron
軟鐵

secondary
副線圈

primary
原線圈

input 輸入

transformer
變壓器

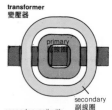

primary
原線圈

secondary
副線圈

secondary coil with
primary coil underneath
副線圈之下有原線圈

soft iron
軟鐵

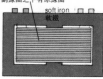

step-up transformer
升壓變壓器

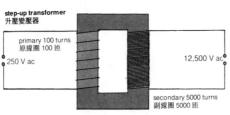

primary 100 turns
原線圈 100 匝

250 V ac

12,500 V ac

secondary 5000 turns
副線圈 5000 匝

step-up (*adj*) giving increased voltage, e.g. a step-up transformer has a turns ratio (↑) greater than 1.

step-down (*adj*) giving decreased voltage, e.g. a step-down transformer has a turns ratio (↑) less than 1.

升壓的(形) 使電壓升高的。例如,升壓變壓器的匝數比(↑)大於 1。

降壓的(形) 使電壓降低的。例如,降壓變壓器的匝數比(↑)小於 1。

symbol for a transformer
變壓器的符號

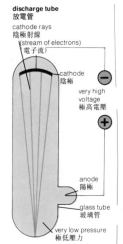

discharge tube
放電管

cathode rays
陰極射線
(stream of electrons)
電子流)

cathode
陰極

very high
voltage
極高電壓

anode
陽極

glass tube
玻璃管

very low pressure
極低壓力

discharge tube a glass vessel with electric current passing from anode to cathode in a high vacuum.

radiation[2] (*n*) the spreading of energy by electromagnetic waves, i.e. light, radiant heat, X-rays (p.84), radio and gamma rays (p.108). **radiate** (*v*).

infra-red rays electromagnetic waves with wavelengths longer than those of red light in the visible spectrum. These are the rays of radiant heat; they pass through mist and some solid substances which are opaque (p.53) to light rays.

ultra-violet rays electromagnetic waves with wavelengths shorter than those of violet light in the spectrum. These rays cannot be seen, but they act on photographic film, and cause burning of the skin when in sunshine. Abbreviation: u.v. rays; u.v. light.

cathode rays a stream of electrons (p.106) from the cathode (p.100) of a cathode-ray tube (p.86).

bombard (*v*) to send a stream of particles, either charged or not, with force against a surface, e.g. a stream of electrons bombards the screen of a television set (p.86). **bombardment** (*n*).

target (*n*) any surface at which a stream of particles is aimed or directed.

放電管　一種有電流從陽極流向陰極的高真空玻璃管。

輻射（名）　以電磁波形式，即以光、輻射熱、X射線（第 84 頁）、無線電和 γ 射線（第 108頁）形式傳播能量。（動詞為 radiate）

紅外線　波長長於可見光譜中紅光波長的電磁波。紅外線是輻射熱的射線；它能穿透雲霧和一些不透光的（第 53 頁）固體物質。

紫外線　波長短於光譜中紫光的電磁波。紫外線是看不見的，但可對照相底片起作用，日曬中的紫外線可引起皮膚灼傷。紫外線縮寫為u.v. 射線；u.v. 光。

陰極射線　從陰極射線管（第 86 頁）的陰極（第100 頁）發出的電子（第 106 頁）流。

轟擊（動）　放射出帶電或不帶電的粒子流，向一平面衝擊。例如：電子流轟擊電視機（第 86頁）的屏幕。（名詞為 bombardment）

靶（名）　粒子流對準或指向的任何平面。

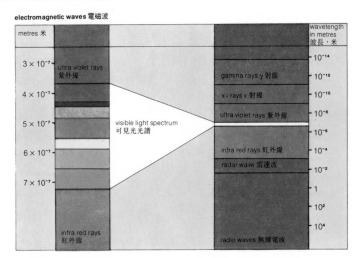

electromagnetic waves 電磁波

metres 米			wavelength in metres 波長，米
3×10^{-7} ultra violet rays 紫外線			10^{-14}
		gamma rays γ 射線	10^{-12}
4×10^{-7}		x - rays x 射線	10^{-10}
	visible light spectrum 可見光譜	ultra violet rays 紫外線	10^{-8}
5×10^{-7}			10^{-6}
		infra red rays 紅外線	10^{-4}
6×10^{-7}		radar wave 雷達波	10^{-2}
7×10^{-7}			1
			10^{2}
			10^{4}
infra red rays 紅外線		radio waves 無線電波	

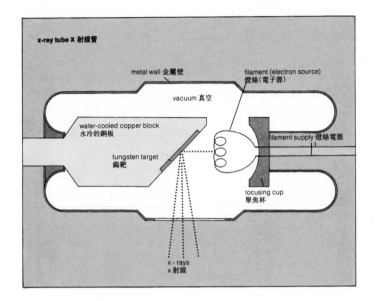

x-ray tube X 射線管

metal wall 金屬壁

filament (electron source)
燈絲（電子源）

vacuum 真空

water-cooled copper block
水冷的銅板

filament supply 燈絲電源

tungsten target
鎢靶

focusing cup
聚焦杯

x - rays
x 射線

X-ray tube an apparatus for producing X-rays, *see diagram*. A heated filament gives out a stream of electrons; these are focused, by a cup, onto a target. The filament is the cathode (negative electrode) and the target is in a copper block which is the anode (positive electrode). The electrons have a negative charge and are accelerated by repulsion from the cathode. When the electrons hit the target, X-rays are produced, together with a large quantity of heat.

X-rays electromagnetic waves with a very short wavelength, between approximately 10^{-9} and 10^{-11} of a metre. The greater the potential difference between anode and cathode of an X-ray tube, the shorter the wavelength of the X-rays. X-rays pass through many materials opaque to light, e.g. flesh and bones, with bones absorbing more than flesh. The shorter the wavelength of an X-ray, the more easily it passes through a material.

X 射線管 產生 X 射線的裝置（見圖）。熾熱燈絲發出的電子流被一個杯狀物聚到靶上。燈絲爲陰極（負極），靶嵌在銅板裏，銅板爲陽極（正極）。電子帶負電荷因此受陽極排斥而加速。當電子撞擊靶時，產生 X 射線及大量的熱。

X 射線 波長極短，大約在 10^{-9} 至 10^{-11} 米之間的電磁波。X 射線管的陽極和陰極之間的電位差越大，X 射線的波長就越短。X 射線能穿透許多不透光的物質。例如肌肉和骨骼，而骨骼吸收的 X 射線比肌肉多。X 射線的波長越短，越容易穿透物質。

diode valve
二極管

plate (anode)
板極(陽極)

filament (cathode)
燈絲(陰極)

vacuum
真空

valve² (n) an electrical device which allows electric current to pass through in one direction only. Valves are called diode, triode, pentode, etc. according to the number of electrodes (↓) they have.

thermionic valve another name for valve² (↑).

electron tube another name for valve² (↑).

tube (n) another name for valve², used in America.

electrode¹ (n) a wire, rod, or plate conducting electric current into, or out of, any device.

diode (n) a valve² (↑) with two electrodes. The cathode (negative electrode) is a heated filament, which gives off electrons (p.106). The anode (positive electrode) is a plate. The anode attracts electrons, and an electron current flows from cathode to anode. This is a flow of negative charge, and is opposite to the usual way of saying current flows from positive to negative. If an alternating voltage is applied to the anode and cathode, then current flows in one direction only and the alternating current is rectified (p.81). This is the main use of diodes.

電子管(名) 只允許電流向一個方向流動的電氣裝置。根據電子管所具的電極(↓)數目不同,分爲二極管、三極管、五極管。

熱離子管 電子管(↑)的別稱。

電子管(↑)的另一英文名稱爲 **electron tube**。

電子管在美國也稱爲 **tube**。

電極(名) 指將電流導入或導出任何裝置的金屬線、棒或板。

二極管(名) 有兩個電極的電子管(↑)。陰極(負電極)是一個放射出電子(第106頁)的熾熱燈絲。陽極(正電極)是一個板。陽極吸引電子使電子流從陰極流向陽極。這是一個負電荷流,與一般所說的電流從正極流向負極相反。交流電壓加於陽極和陰極時電流只向一個方向流動,而交流電被整流(第81頁)爲直流。這就是二極管的主要用途。

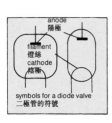

anode
陽極

filament
燈絲
cathode
陰極

symbols for a diode valve
二極管的符號

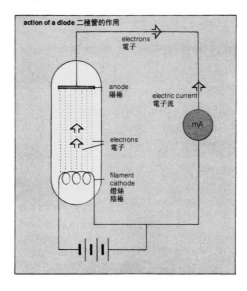

action of a diode 二極管的作用

electrons
電子

anode
陽極

electric current
電子流

mA

electrons
電子

filament
cathode
燈絲
陰極

triode (*n*) a valve² (p.85) with three electrodes; a grid (a wire net) is put between the anode and cathode. The grid is given a negative voltage, and a small change in this voltage has a big effect on the electron current, and gives a big voltage at the anode. The triode thus steps-up (p.82) voltage.

cathode-ray tube (*see diagram*). A heated filament gives out electrons which are focused into a narrow stream, accelerated by the anode, and then pass between two sets of parallel plates. The electrons produce light when they hit the screen. The X and Y parallel plates can make the electrons hit any part of the screen because they are given an electric charge which repels or attracts the electrons.

三極管(名)　具有三個電極的一種電子管(第 85 頁)；柵極(金屬絲網)置於陽極和陰極之間。給柵極施以負電壓，微小的電壓變化都可對電子流產生大的效應，並在陽極產生高的電壓。因此三極管能升高(第 82 頁)電壓。

陰極射線管(見圖)　熾熱燈絲放射出電子，這些電子聚集成爲一條狹窄的電子流，經陽極加速後，在兩組平行板之間通過。電子碰撞屏幕時產生光。由於 X 和 Y 平行板帶有排斥或吸引電子的電荷，所以能使電子撞擊屏幕的任何部分。

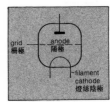

symbol for a triode valve
三極管的符號

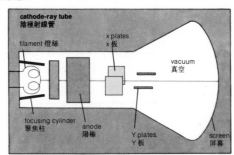

oscilloscope (*n*) an instrument using a cathode-ray tube which produces an image on the screen of varying electrical voltages. It is also used in radar (↓) to show the position of objects.

radio (*n*) the use of electromagnetic waves to send messages from a microphone in a radio station to a receiver in a radio set.

television (*n*) the use of radio waves to send pictures to a television set. The set has a cathode-ray tube which builds the picture from 625 lines (or 405 lines) each of 400 small spots of light.

radar (*n*) an abbreviation of Radio Detection And Ranging. Electromagnetic waves, of wavelengths of a few centimetres, are reflected from distant objects and the reflections are recorded on the screen of a cathode-ray tube.

示波器(名)　一種使用陰極射線管在屏幕上產生不同電壓的圖像的儀器。示波器也用於雷達(↓)以顯示物體的位置。

無線電廣播(名)　從無線電台的發送器將信息以電磁波傳送到收音機的接收器。

電視(名)　將圖像以無線電波送向電視接收機。電視機有一個陰極射線管，它以 625(或 405)條線，每條線 400 個小光點組成圖像。

雷達(名)　爲 Radio Detection And Ranging(無線電探測和定位)的縮寫。波長爲幾厘米的電磁波從遠處物體上反射回來，反射的映像顯示在一個陰極射線管的屏幕上。

Chemistry 化 學

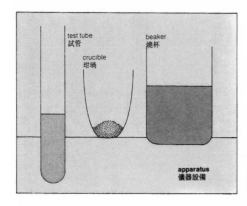

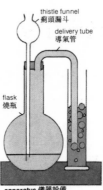

apparatus 儀器設備

apparatus (n) any instrument, or collection of instruments, or objects such as test-tubes, beakers, flasks, pipettes, put together for work in science.

test-tube (n) a tube, closed at one end used to make chemical tests.

beaker (n) a vessel used to hold liquids or solutions, and to heat them.

crucible (n) a small cup used for heating solids strongly.

flask (n) a vessel used for boiling liquids, preparing gases.

Woulfe-bottle a vessel used for preparing gases.

funnel (n) a device used for filtering solutions.

thistle funnel a funnel used for adding liquids to flasks.

delivery tube a tube used for conducting gases.

gas-jar (n) a vessel used for collecting gases.

pneumatic trough a vessel used with a gas-jar for collecting gases.

eudiometer tube a glass tube used for experiments on gases.

Kipp's apparatus an apparatus used for supplying a gas.

burette (n) a glass tube used for measuring liquids to 0.1 cm³.

pipette (n) a piece of apparatus used for measuring a fixed volume of liquid, usually 10 cm³, 25 cm³ or 50 cm³.

儀器設備(名)　任何儀器或一套儀器，或組合一起作科學實驗用的試管、燒杯、燒瓶、移液管等物件。

試管(名)　一端封閉，用於做化學試驗的管子。

燒杯(名)　用於盛液體或溶液並可將之加熱的容器。

坩堝(名)　用於強烈加熱固體物質的一種小杯。

燒瓶(名)　用於煮沸液體製備氣體的瓶子。

渥耳夫瓶　製備氣體用的瓶子。

漏斗(名)　過濾溶液用的器具。

薊頭漏斗　把液體加入燒瓶中用的漏斗。

導氣管　氣體用的管子。

集氣瓶(名)　用於收集氣體的瓶子。

集氣槽　和集氣瓶一起用於收集氣體的器皿。

量氣管　氣體實驗用的管子。

基普氏氣體發生器　供給氣體用的裝置。

滴定管(名)　測量液體至 0.1 cm³ 用的玻璃管。

吸移管(名)　測量固定量(通常為 10cm³、25cm³ 或 50cm³)液體用的儀器。

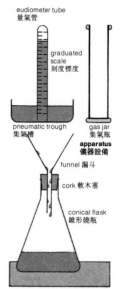

apparatus 儀器設備

Kipp's apparatus
基普氏氣體發生器

dissolve (v) to make a solid or a gas disappear, or for it to disappear, into a liquid, e.g. to dissolve sugar in coffee, to dissolve common salt in water. Compare **melt**: when sugar is heated, it melts, when added to water, it dissolves without heating.

solute (n) any solid or gas which, when added to water, or other liquid, will dissolve, e.g. (a) when common salt is added to water, the salt is the solute; (b) when carbon dioxide gas is added to water to make soda water, the gas is the solute.

solvent (n) the liquid in which a solute is dissolved, e.g. (a) water is the solvent and common salt is the solute when salt dissolves in water; (b) paint is the solute and turpentine is the solvent when paint is dissolved in turpentine.

溶解(動)　使固體或氣體物質消散或它們自行消散在一種液體中。例如，使糖溶解於咖啡內，使鹽溶解於水裏。與熔化作比較：糖受熱時熔化，而放到水裏，不需加熱就溶解了。

溶質(名)　放入水裏或其他液體裏都會溶解的任何固體或液體物質。例如：(a)當食鹽放入水裏時，食鹽是溶質；(b)二氧化碳氣放入水中製成蘇打水，二氧化碳氣體為溶質。

溶劑(名)　溶質溶解於其中的液體。例如：(a)食鹽溶解於水時，水為溶劑，食鹽為溶質；(b)油漆溶解於松節油時，油漆為溶質，松節油則為溶劑。

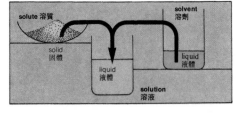

pipette 吸移管　　**burette** 滴定管

graduation mark
刻度標記

graduated/scale
刻度標度

solution (n) the result of dissolving a solute in a solvent, e.g. salt solution (usually called brine) is formed when common salt is dissolved in water. Unless described in some other way a solution is always in water.

saturated (adj) of a solid, cannot absorb (p.163) more liquid; of a solution, cannot dissolve more solute; of a gas, cannot contain more vapour, e.g. (a) when the air is saturated it cannot contain more water vapour, so water cannot evaporate into the air; (b) a saturated solution of common salt cannot dissolve more salt.

unsaturated (adj) of a solution, it can dissolve more solute; of a gas, it can contain more vapour, e.g. (a) unsaturated air can contain more water vapour; (b) an unsaturated solution can dissolve more solute.

溶液(名)　溶質溶解於溶劑中的產物。例如：食鹽溶液(通常稱為鹽水)為食鹽溶於水中形成的產物。如無另外說明，溶液總是指水溶液。

飽和的(形)　指固體不能再吸收(第163頁)更多的液體；溶液不能溶解更多的溶質；氣體不能含有更多水汽。例如：(a)空氣飽和時，不能含有更多的水蒸汽，以致水不能蒸發入空氣中；(b)食鹽的飽和溶液不能再溶解更多的食鹽。

不飽和的(形)　指溶液還能溶解更多的溶質；氣體還能含有更多的水汽。例如：(a)不飽和空氣能含有更多的水蒸汽；(b)不飽和溶液能溶解更多的溶質。

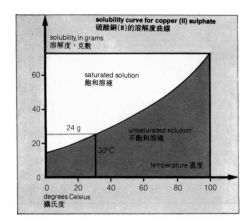

solubility (*n*) the mass of a solute in grammes that can be dissolved in 100 g of a solvent to form a saturated solution at a given temperature. Water is considered the usual solvent, any other solvent must be named, e.g. the solubility of copper (II) sulphate is 24 g at 30°C, so 24 g of the solute dissolved in 100 g of water at 30°C forms a saturated solution. **soluble** (*adj*).

insoluble (*adj*) describes a solid or a gas which does not dissolve in a named solvent, e.g. hydrogen (a gas) is insoluble in water.

concentration (*n*) (1) a measure of the amount of solute, dissolved in a particular volume of its solution. The measurement can be (a) grammes per dm³; (b) moles (p.105) per dm³; (c) cm³ of gas per cm³ of solution, e.g. a concentration of 125 g of copper (II) sulphate in 1 dm³ of solution. (2) the action of increasing the concentration of a solution by evaporating (p.42) the solvent with heat; the result is a more concentrated solution. **concentrate** (*v*), **concentrated** (*adj*).

dilute (*v*) to make a solution less concentrated by adding more solvent. **dilution** (*n*).

dilute (*adj*) describes a solution which has been diluted, or one which has a low concentration.

溶解度(名) 在一定溫度下，能溶解於 100 克溶劑中形成飽和溶液的溶質的克數。水是常用的溶劑，如用任何其他溶劑則必須指明。例如硫酸銅(II)在 30°C 時的溶解度為 24 克，所以在 30°C 時 24 克溶質溶於 100 克水中形成飽和溶液。(形容詞為 soluble)

不溶解的(形) 描述不溶於指定溶劑中的固體或氣體物質。例如，氫(一種氣體)不溶於水中。

濃度；濃縮(名) (1)溶解於一定體積的溶液中的溶質量度。此量度可以是：(a)每立方分米的克數；(b)每立方分米的摩爾數(第 105 頁)；(c)每立方厘米溶液中氣體的立方厘米數。例如：1 立方分米溶液中含 125 克硫酸銅(II)的濃度；(2)藉加熱蒸發(第 42 頁)去掉溶劑而增加溶液濃度；結果得到更加濃縮的溶液。(動詞為 concentrate，形容詞為 concentrated)

稀釋(動) 加入更多的溶劑，降低溶液的濃度。(名詞為 dilution)

稀釋的(形) 描述一種已稀釋的溶液或低濃度的溶液。

immiscible liquids
不溶混液體

oil 油

water 水

precipitation 沉澱

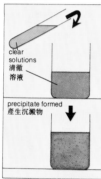

clear
solutions
清澈
溶液

precipitate formed
產生沉澱物

miscible (adj) describes a liquid which will mix with another named liquid, e.g. alcohol is miscible in water.

immiscible (adj) describes a liquid which will not mix with another named liquid, e.g. olive oil is immiscible in water; it forms two layers of liquid with the oil on top of the water.

separate (v) (1) of immiscible liquids, to form two layers. (2) to obtain each substance from a mixture (p.95). **separation** (n), **separate** (adj), **separable** (adj).

precipitate (n) a solid thrown out of solution (p.89), when one solution is added to another solution or a gas is passed into a solution, because the substance formed is insoluble, e.g. (a) adding silver nitrate solution to sodium chloride solution forms a precipitate of insoluble silver chloride; (b) passing carbon dioxide gas into lime water forms a precipitate of insoluble calcium carbonate. **precipitate** (v).

precipitation[2] (n) the forming of a precipitate (↑).

filter (v) to remove solid material from a mixture of liquid and solid, e.g. to filter off a precipitate from a mixture of liquid and precipitate. **filter** (n).

可溶混的(形) 描述一種液體能與另一種指定液體相混合，例如酒精可以溶混於水。

不互混的(形) 描述一種液體不與另一種指定液體相混合。例如橄欖油能與水不互混；它形成兩層液體，油浮在水面上。

分離(動) (1)不互混的液體形成兩層液體；(2)從一種混合物(第 95 頁)中取得每種物質。(名詞爲 separation，形容詞爲 separate，separable)

沉澱物(名) 一種液體加入另一種液體中，或一種氣體注入液體時，由於生成不溶解的物質，而從液體中分離出的固體。例如：(a)將硝酸銀溶液加入氯化鈉溶液中，可生成不溶解的氯化銀沉澱物；(b)將二氧化碳氣注入石灰水中可生成不溶解的碳酸鈣沉澱物。(動詞爲 precipitate)

沉澱作用(名) 形成沉澱物(↑)。

過濾(動) 從液體和固體的混合物中去除固體物質。例如從液體和沉澱物混合物中濾去沉澱物。(名詞爲 filter)

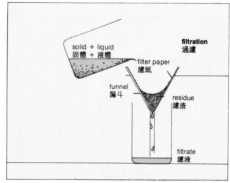

solid + liquid
固體 + 液體

filtration
過濾

filter paper
濾紙

funnel
漏斗

residue
濾渣

filtrate
濾液

filtrate (n) the liquid part which has passed through a filter.

residue (n) the solid material separated when a mixture of liquid and solid is filtered.

濾出液(名) 流過濾器的液體部分。

濾渣(名) 液體和固體的混合物過濾後分離出的固體物質。

evolve (v) (1) to form a gas as a result of a chemical reaction (p.96) with the gas escaping into the atmosphere, e.g. when zinc is added to sulphuric acid, hydrogen is evolved, i.e. formed, and escapes from the solution into the air. (2) to set free heat, or sparks. **evolution** (n).

collect (v) (1) to gather, or to come, together in one place, e.g. to collect different kinds of rocks. (2) to lead a gas into a vessel, e.g. to collected hydrogen in a gas-jar. **collection** (n).

放出（動）（1）化學反應（第 96 頁）產生的氣體逸入大氣中。例如，將鋅加入硫酸時，放出氫，即生成氫氣，並從溶液逸入空氣中；（2）釋放出熱或火花。（名詞爲 evolution）

收集（動）（1）聚集於一處。例如：收集不同類的岩石；（2）將氣體引入一個容器內。例如：收集氫氣於集氣瓶內。（名詞爲 collection）

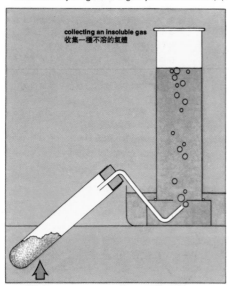

collecting an insoluble gas
收集一種不溶的氣體

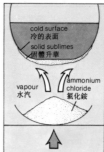

sublimation 昇華

cold surface
冷的表面
solid sublimes
固體升華

vapour
水汽

ammonium chloride
氯化銨

bubble (n) a ball of gas either in a liquid, or with a thin skin of liquid round it, e.g. the bubbles in soda water.

bubble (v) to pass bubbles of a gas through a liquid, e.g. to bubble carbon dioxide gas through lime water.

sublime (v) to change a solid substance to a vapour by heat, and then change the vapour back to a solid by cooling it, e.g. ammonium chloride vaporizes on heating, and sublimes back to a solid on cooling.

氣泡（名） 在液體中的氣體小球或者包着一層液體薄膜氣體小球。例如蘇打水裏的氣泡。

作泡狀通過；起泡（動） 使氣泡通過液體。例如，使二氧化碳氣泡通過石灰水。

昇華（動） 加熱使固體物質變成蒸汽，之後再使蒸氣冷卻變回固體。例如，氯化銨加熱時汽化昇華，冷卻時還原爲固體。

experiment (*n*) an exercise, using apparatus (p.88) or instruments (p.10), to observe the behaviour of objects, materials, or organisms (p.147) under controlled conditions, e.g. (a) an experiment to find the refractive index of glass; (b) an experiment to see how chalk behaves when heated. **experiment with** (*v*).

preparation (*n*) the making and collecting of a quantity of a substance using a chemical reaction (p.96), e.g. the preparation of oxygen by heating potassium chlorate, and collecting the gas in a gas-jar.

investigate (*v*) to carry out, for a particular purpose, an experiment (↑) on an object or a material and to record carefully all observations (↓), e.g. to investigate the chemical and physical properties of carbon.

chemical property a property (p.27) which describes the way in which a substance acts when heated, electrolysed (p.100), or added to other substances so that a chemical change (p.94) takes place. All the chemical properties of a substance make up its chemical nature.

test (*n*) a simple exercise carried out: (1) to see if apparatus or instruments are working correctly; (2) to find out whether a particular substance is present or absent. **test** (*v*).

observation (*n*) the intentional use of seeing, hearing, smelling, tasting, or touching, using knowledge gathered in the past to know what to look for, e.g. to observe that hydrogen is given off by testing with a lighted splint (past knowledge of the effect together with hearing the result).

identification (*n*) (1) the action, carried out by experiment, of determining the properties (p.27), hence the name, of a substance. (2) the action of naming a process (p.129) or a form of energy by comparing its characteristics (p.147) with those of a known process or form of energy, e.g. (a) the identification of a metal as magnesium by investigating (↑) its properties; (b) the identification of a radiation as X-rays by investigating its characteristics. **identify** (*v*), **identifiable** (*adj*).

實驗(名) 使用各種裝置(第88頁)或儀器(第10頁)在可控制的條件下觀察物體、物質或有機體(第147頁)性狀的一種做法。例如：(a)做實驗求出透鏡折射率；(b)做實驗觀察白堊在加熱時的變化。(動詞爲 experiment with)

製備(名) 利用化學反應(第96頁)製作並收集一定量的物質。例如，通過加熱氯酸鉀製備氧氣，並將之收集於集氣瓶。

調查研究(動) 爲一個特定目的而對物體或物質進行實驗(↑)，並仔細記錄所有觀察結果(↓)。例如，研究碳的化學和物理性質。

化學性質 描述物質在受熱、電解(第100頁)或加入其他物質所發生化學變化(第94頁)的一種性質(第27頁)。物質的全部化學性質構成該物質的化學本質。

試驗(名) 一種簡單的做法，藉以：(1)觀察裝置或儀器是否運轉正常；(2)查明一種特定物質是否存在。(動詞爲 test)

觀察(名) 利用已有的知識，有目的地使用視覺、聽覺、嗅覺、味覺或觸覺來了解需要探索的事物。例如，用點燃的細木片作試驗以觀察放出的氫氣(通過已知的效果和聽到的聲響。)

鑑別(名) (1)進行實驗以確定物質性質(第27頁)並據此性質確定其名稱的活動；(2)將能的一種過程(第129頁)或形式的特徵(第147頁)和另一種已知能的過程或形式的特徵相比較而給能的過程或形式命名的活動。例如(a)藉研究(↑)金屬的性質，鑑別它爲鎂；(b)藉研究放射特性，鑑別它爲X射線。(動詞爲 identify，形容詞爲 identifiable)

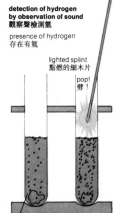

detection of hydrogen by observation of sound
觀察聲檢測氫

presence of hydrogen
存在有氫

lighted splint
點燃的細木片

pop!
劈！

zinc + hydrochloric acid
鋅 + 鹽酸

identical (*adj*) describes objects, substances, processes (p.129) and radiations (p.83) which have exactly the same number of properties (p.27) or characteristics (p.147) which are exactly the same, e.g. two crystals are identical if they both have the same colour, shape, and size, and are the same substance. **identity** (*n*).

similar (*adj*) describes objects, substances, processes (p.129) and radiations if they have many common properties or characteristics, but have a few different ones. **similarity** (*n*).

physical change a change in the state or in a physical property of a material while remaining the same material, e.g. (a) ice melting to water; (b) brass expanding on heating; (c) iron being magnetized.

chemical change a change in which new substances are formed with different properties, e.g. chalk, when heated, is chemically changed to lime and carbon dioxide.

相同的(形)　描述物體、物質、過程(第 129 頁)和輻射(第 83 頁)所具有的性質(第 27 頁)數目完全相同或特徵(第 147 頁)完全相同。例如，兩個晶體都具有相同的顏色、形狀和大小而且是同一種物質，則它們是相同的。(名詞爲 identity)

類似的(形)　描述物體、物質、過程(第 129 頁)和放射，雖然它們具有許多共同的性質或特徵，但仍有一些不同的方面。(名詞爲 similarity)

物理變化　雖然物質的狀態或物理性質發生變化，但仍爲同一物質的變化。例如：(a)冰熔化爲水；(b)黃銅加熱膨脹；(c)鐵被磁化。

化學變化　生成具有不同性質的新物質的變化。例如，白堊加熱時，發生化學變化，生成石灰和二氧化碳。

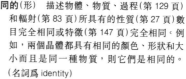

identical crystals 同樣品的晶體

same colour 相同的顏色
same shape 相同的形狀
same size 相同的大小
same compound 同一種化合物

similar crystals 類似的晶體

same colour 相同的顏色
same shape 相同的形狀
same compound 同一種化合物
different size 不同的大小

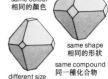

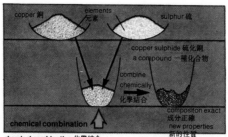

copper 銅　elements 元素　sulphur 硫

copper sulphide 硫化銅　a compound 一種化合物

combine chemically 化學結合

compositon exact 成分正確　new properties 新的性質

chemical combination 化學結合

chemical combination 化學結合

combination (*n*) the chemical union of two substances to form a new substance with different properties, e.g. the combination of iron and oxygen to form iron oxide (rust). **combine** (*v*).

formation (*n*) (1) the action of bringing a material into being by a chemical change, e.g. the formation of ammonia gas on heating an ammonium salt with an alkali. (2) bringing any object into being by a physical change or effect, e.g. (a) the formation of dew by condensation; (b) the formation of an image by a lens. **form** (*v*).

化合作用(名)　兩種物質化學結合成一種具有不同性質的新物質。例如，鐵和氧化合生成氧化鐵(鐵銹)。(動詞爲 combine)

形成；生成(名)　(1)通過化學變化產生一種物質的作用。例如：將銨鹽和鹼一起加熱生成氨氣；(2)通過物理變化或物理效應產生任何物體。例如：(a)通過冷凝作用形成露水；(b)透鏡成像。(動詞爲 form)

composition (n) the elements (p.103) with their proportions (p.23) in a substance (↓) form its chemical composition. **be composed of** (v).

compound (n) a substance (↓) with a known composition of elements which cannot be separated by physical means, e.g. lime is a compound of calcium and oxygen (2 elements) in the proportion 40 parts calcium : 16 parts oxygen by mass. An exact formula (p.105) can be given for a compound.

substance (n) a material whose composition does not vary, but it may not be possible to know its exact formula (p.105), e.g. starch is a substance because its composition is known though its exact formula is not. Its formula is $(C_6H_{10}O_5)n$ where n is a large unknown number.

decompose (v) to break a substance or compound into simpler substances or compounds by chemical action, heat, or electric current, e.g. heat decomposes chalk into lime and carbon dioxide. **decomposition** (n).

mixture (n) a material made by mixing substances together; the substances can be in any proportion, and can be readily separated from each other by physical methods. Each substance keeps its own properties.

組成（名） 物質（↓）中各種元素（第 103 頁）按一定比例（第 23 頁）構成該物質的化學組成。（動詞 be composed of 意為由⋯組成）

化合物（名） 已知各元素的組成且不能用物理方法將之分開的物質（↓）。例如，石灰是鈣和氧（兩種元素）的化合物，這兩種元素的質量比為 40 份鈣：16 份氧。化合物可用準確的化學式（第 105 頁）表示。

物質（名） 一種組成不變，且也許不可能知其準確化學式（第 105 頁）的物料。例如澱粉是一種物質，其組成已知卻不知其準確化學式。澱粉的化學式為 $(C_6H_{10}O_5)n$，式中 n 為一個大的未知數。

分解（動） 通過化學作用、加熱或通電流將一物質或一化合物分成較簡單的數種物質或化合物。例如，加熱將白堊分解成石灰和二氧化碳。（名詞為 decomposition）

混合物（名） 將多種物質混合在一起所製成的物料；這些物質可以任何比例混合，並可用物理方法容易地分離開來。其中每一種物質都保持其本身的屬性。

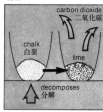

decomposition
分解

carbon dioxide
二氧化碳

chalk
白堊

lime

decomposes
分解

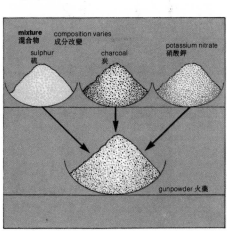

mixture
混合物　composition varies
成分改變

sulphur
硫

charcoal
炭

potassium nitrate
硝酸鉀

gunpowder 火藥

constituent (*n*) a single substance or an element in a compound or mixture, e.g. (a) sulphur is a constituent of copper sulphide; (b) sulphur is a constituent of gunpowder. **constituent** (*adj*).

contain (*v*) to have as a constituent when some, but not all, the constituents are named, e.g. gunpowder contains sulphur and carbon.

consist of (*v*) to have as a constituent when all the constituents present in a mixture or compound are named, e.g. gunpowder consists of sulphur, charcoal and potassium nitrate.

chemical reaction a chemical change (p.94) that takes place when two or more substances are put together, e.g. the chemical reaction between zinc and sulphuric acid. **chemical reactivity, chemically reactive.**

組分；成分（名）　化合物或混合物中的單一物質或元素。例如：（a）硫是硫化銅的組分；（b）硫是火藥的組分。（形容詞為 constituent）

含有（動）　這個詞用於只須提及組成中的一些而非全部組分的名字。例如火藥含有硫和碳。

由…組成（動）　這個片語用於必須提及混合物或化合物中所存在的全部組分的名字。例如，火藥是由硫、炭和硝酸鉀組成的。

化學反應　兩個或多個物質放在一起發生的化學變化（第 94 頁）。例如鋅和硫酸的化學反應。chemical reactivity 義爲化學反應性，chemically reactive 義爲化學活潑的。

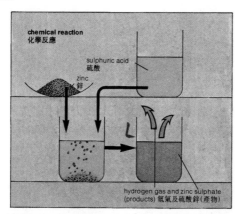

chemical reaction
化學反應

sulphuric acid
硫酸

zinc
鋅

hydrogen gas and zinc sulphate
(products) 氫氣及硫酸鋅（產物）

product (*n*) a substance formed by a chemical reaction, e.g. hydrogen is a product of the reaction between zinc and sulphuric acid.

reversible reaction a chemical reaction in which the products can react chemically with each other to form the original substances, so the reaction can go in either direction, e.g. steam and iron react to form hydrogen and an oxide of iron; the oxide of iron and hydrogen react to form steam and iron.

產物（名）　化學反應所生成的物質。例如，氫是鋅和硫酸反應的產物。

可逆反應　指一種化學反應的產物能相互起化學反應生成原來的物質，因而這種反應可以同時往兩個方向進行。例如：蒸汽和鐵反應生成氫和鐵的氧化物；鐵的氧化物和氫反應又形成蒸汽和鐵。

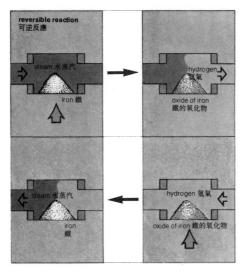

reversible reaction
可逆反應

steam 水蒸汽

iron 鐵

hydrogen
氫氣

oxide of iron
鐵的氧化物

steam 水蒸汽

iron
鐵

hydrogen 氫氣

oxide of iron 鐵的氧化物

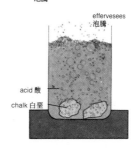

effervescence
泡騰

effervesees
泡騰

acid 酸

chalk 白堊

chain reaction a chemical reaction which produces a product and the product causes a second reaction producing a second product which causes a third reaction and so on, e.g. in nuclear fission (p.109) an atom produces a product and two neutrons (p.106); each neutron reacts with another atom to produce a product and two neutrons, so the reaction continues faster and faster.

effervesce (*v*) to produce a large quantity of bubbles (p.92) of gas by a chemical reaction, e.g. when an acid acts on chalk the liquid effervesces. **effervescence** (*n*), **effervescent** (*adj*).

replaceable (*adj*) describes, in an acid, hydrogen atoms (p.103) which can be displaced and metal atoms put in their place, e.g. the hydrogen in sulphuric acid is replaceable by zinc.

dehydrate (*v*) to remove water, whether present as moisture or chemically combined, by heat or by a chemical reaction, e.g. to dehydrate crystals of copper (II) sulphate by heat or by the action of concentrated sulphuric acid.

連鎖反應　一次化學反應產生一種產物，此產物又引起第二次反應產生第二種產物，第二種產物又引起第三次反應，如此繼續下去。例如，在原子核裂變（第 109 頁）中，一個原子產生一個產物和兩個中子（第 106 頁）；每個中子和另外一個原子反應產生一個產物和兩個中子。結果反應越來越快地繼續下去。

泡騰（動）　由於化學反應而產生大量氣泡（第 92 頁）。例如，當酸作用於白堊時，液體泡騰。（名詞爲 effervescence，形容詞爲 effervescent）

可置換的（形）　描述酸中氫原子（第 103 頁）可被金屬原子取代其位置。例如，硫酸中的氫可爲鋅所置換。

脫水（動）　通過加熱或化學反應去除濕氣和化學結合的水分。例如，通過加熱或濃硫酸作用使硫酸銅(II)晶體脫水。

volatile (*adj*) of a liquid, evaporates readily, e.g. petrol is a volatile liquid. **volatility** (*n*).

odour (*v*) the property of a substance recognized by smell, e.g. the odour of petrol. **odorous** (*adj*), **odoriferous** (*adj*).

odourless (*adj*) describes a substance without odour.

deliquescent (*adj*) describes a solid which tends to absorb so much water vapour from the atmosphere that it forms a solution. **deliquescence** (*n*).

hygroscopic (*adj*) describes a solid which tends to absorb water vapour from the atmosphere and becomes wet, e.g. common salt left in humid air.

efflorescent (*adj*) describes crystals which tend to lose chemically combined water to the atmosphere, e.g. crystals of sodium carbonate are efflorescent and decompose to a white powder. **efflorescence** (*n*).

reagent (*n*) a substance which produces a chemical reaction with a certain chemical, and can be used in testing to discover whether that chemical is present, e.g. silver nitrate is a reagent for testing for the presence of a chloride.

agent (*n*) a substance or solution used for a particular chemical process, e.g. an oxidizing (↓) agent to oxidize iron (II) sulphate.

oxidation (*n*) (1) the addition of oxygen to a substance. (2) the removal of hydrogen from a substance. (3) the increasing of positive electrovalency (p.109), e.g. (a) the oxidation of copper to copper (II) oxide; (b) the oxidation of hydrogen chloride to chlorine; (c) the oxidation of iron (II) sulphate to iron (III) sulphate. **oxidize** (*v*).

reduction (*n*) (1) the removal of oxygen from a substance. (2) the addition of hydrogen to a substance. (3) the decreasing of positive electrovalency (p.109), e.g. (a) the reduction of copper (II) oxide to copper; (b) the reduction of chlorine to hydrogen chloride; (c) the reduction of iron (III) sulphate to iron (II) sulphate. **reduce** (*v*).

揮發性的(形) 指液體容易汽化。例如汽油是一種揮發性液體。(名詞為 volatility)

氣味 可由嗅覺辨認出的物質性質。例如汽油的氣味。(形容詞為 odorous，odoriferous)

無氣味的(形) 描述一種物質沒有氣味。

潮解的(形) 描述一種固體易吸收大氣中的大量水蒸汽形成一種溶液。(名詞為 deliquescence)

吸濕的(形) 描述一種固體易吸收大氣中的水蒸汽變潮濕。例如置於濕空氣中的食鹽是吸濕的。

風化的(形) 描述晶體的化合水易散失於大氣中。例如，碳酸鈉晶體風化後分解成白色粉末。(名詞為 efflorescence)

試劑(名) 可以和化學劑發生化學反應的物質，可用於檢測該化學劑是否存在。例如，硝酸銀是一種檢驗氯化物是否存在的試劑。

劑；試劑(名) 一種用於特定化學過程的物質或溶液。例如，可使硫酸亞鐵 (II) 氧化的氧化 (↓)劑。

氧化作用(名) (1)給物質加氧；(2)除去物質的氫；(3)增加正電價(第 109 頁)。例如：(a) 銅氧化成一氧化銅(II)；(b) 氯化氫氧化成氯；(c) 硫酸亞鐵(II)氧化成為硫酸鐵(III)。(動詞為 oxidize)

還原作用(名) (1)除去物質中的氧；(2)給物質加氫；(3)減少正電價(第 109 頁)。例如：(a) 氧化銅(II)還原為銅；(b) 氯還原為氯化氫；(c) 硫酸鐵(III)還原為硫酸亞鐵(II)。(動詞為 reduce)

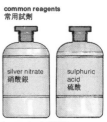

common reagents
常用試劑

silver nitrate 硝酸銀

sulphuric acid 硫酸

reagent bottle 試劑瓶

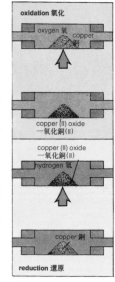

oxidation 氧化

oxygen 氧　copper 銅

copper (II) oxide 一氧化銅(II)

copper (II) oxide 一氧化銅(II)

hydrogen 氫

copper 銅

reduction 還原

rods in air
在空氣中的棒

copper 銅　silver 銀　iron 銅

tarnished
使失光澤

corrodes
腐蝕

rusts
生銹

bleaching 漂白

coloured 有色的

white 白色

chlorine 氯氣

thermal dissociation
熱離解作用
ammonium chloride
氯化氨

ammonium chloride
氯化銨

ammonia gas 氨氣

cold 冷卻

hydrogen chloride gas
氯化氫氣體

hot
加熱

corrosion (n) the slow destruction of a metal by chemical action such as an acid or atmospheric oxygen, e.g. copper corrodes in the air and forms a green coat. **corrode** (v).

corrosive (adj) describes an agent of corrosion of a metal, and also a substance which attacks animal tissues (p.140).

rust (n) the coat of oxide in iron formed by corrosion. **rust** (v).

tarnish (v) of bright, shining metals to corrode, e.g. silver tarnishes and becomes dull.

bleach (v) to make an object white by destroying its colour, e.g. chlorine bleaches cotton to make it white; chlorine is a bleaching agent.

thermal dissociation the temporary breaking down by heat of a substance into simpler substances; on cooling, these simpler substances combine to form the original substance, e.g. ammonium chloride under thermal dissociation forms ammonia and hydrogen chloride gases; on cooling, these two gases combine to form ammonium chloride.

pyrolysis (n) the decomposition by heat of a substance into simpler substances which do not combine on cooling. **pyrolitic** (adj).

catalysis (n) the increasing of the rate of a chemical reaction by a **catalyst**, a substance which is not itself changed, e.g. platinum increases the rate of sulphur dioxide combining with oxygen; catalysis has increased the rate of reaction. **catalyst** (n), **catalyze** (v), **catalytic** (adj).

violent (adj) describes a chemical reaction which is almost explosive in its rate. **violence** (n).

impurity (n) a small amount of a foreign substance in a large amount of another substance, e.g. lead is often an impurity in silver, i.e. silver contains a small amount of lead.

purify (v) to remove impurities.

腐蝕（名）　由於化學作用如酸或大氣氧使金屬緩慢損壞。例如，銅在空氣中受腐蝕生成銅綠。（動詞爲 corrode）

腐蝕性的（形）　描述一種能腐蝕金屬的藥劑，也描述一種能侵蝕動物肌體組織（第 140 頁）的物質。

鐵銹（名）　鐵受腐蝕生成的氧化物外層。（動詞爲 rust）

失去光澤（動）　指光亮或閃亮的金屬受腐蝕。例如，銀失去光澤顏色變暗淡。

漂白（動）　消除物體的顏色使之潔白。例如，用氯氣漂白棉花使之變白；氯是一種漂白劑。

熱離解作用　物質受熱時分解成較簡單的物質；冷卻後，這較簡單的物質又化合成原來的物質。例如，氯化銨熱離解生成氨和氯化氫氣體；冷卻後，這兩種氣體又化合成氯化銨。

熱解作用（名）　藉加熱使物質分解爲較簡單物質而冷卻後不能再化合成原物質的作用。（形容詞爲 pyrolitic）

催化作用（名）　利用催化劑（一種本身不發生變化的物質）加速化學反應的作用。例如，用鉑提高二氧化硫和氧化合的速度；催化作用提高反應的速度。（名詞爲 catalyst，動詞爲 catalyze，形容詞爲 catalytic）

劇烈的（形）　描述一種化學反應的速度幾乎是爆炸性。（名詞爲 violence）

雜質（名）　在一種大量的物質中存在的少量異物。例如，鉛通常是銀中存在的雜質，即銀含有少量鉛。

純化（動）　除去雜質。

electrolysis (*n*) the chemical decomposition (p.95) of a substance by an electric current; the substance is either in solution in water or is molten (p.40). **electrolyze** (*v*), **electrolytic** (*adj*).

electrolyte (*n*) (1) any substance which, either in solution or when molten, conducts (p.74) an electric current and is decomposed by the current. (2) a liquid in an electric cell or a voltameter (↓). Soluble inorganic salts (p.115) are electrolytes, e.g. sodium chloride.

non-electrolyte (*n*) any solid substance in solution, or molten, or any liquid substance which does not conduct an electric current, e.g. sugar, petrol, alcohol are non-electrolytes.

電解作用（名） 物質被電流化學分解（第 95 頁）；該物質或者是在水溶液中，或者是處於熔融（第 40 頁）狀態。（動詞爲 electrolyze，形容詞爲 electrolvtic）

電解質；電解液（名） (1)任何在溶液中或熔化時能導（第 74 頁）電並被電流分解的物質；(2)電解池或電量計（↓）中的液體。可溶性無機鹽（第 115 頁），例如氯化鈉是電解質。

非電解質（名） 任何在溶液中或熔化時不能導電的固體物質或任何不導電的液態物質。例如：糖、汽油、酒精都是非電解質。

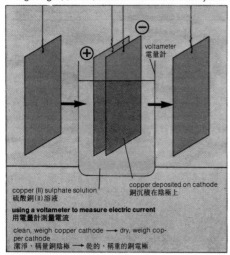

copper (II) sulphate solution
硫酸銅(II)溶液

voltameter
電量計

copper deposited on cathode
銅沉積在陰極上

using a voltameter to measure electric current
用電量計測量電流

clean, weigh copper cathode ⟶ dry, weigh copper cathode
潔淨、稱量銅陰極 ⟶ 乾的、稱重的銅電極

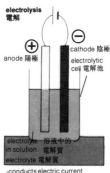

electrolysis
電解

anode 陽極

cathode 陰極

electrolytic cell 電解池

electrolyte in solution 溶液中的 電解質
electrolyte 電解質

-conducts electric current
-傳導電流
-is decomposed by current
-爲電流所分解

voltameter (*n*) a vessel in which electrolysis takes place for the purpose of measuring electric current by weighing the mass of a metal deposited (p.102) on a cathode (↓).

electrolytic cell any vessel used for electrolysis.

anode (*n*) the positive electrode (p.85) of an electrolytic cell. **anodic** (*adj*).

cathode (*n*) the negative electrode of an electrolytic cell. **cathodic** (*adj*).

電量計（名） 一種進行電解的容器，它通過稱量陰極（↓）上沉積（第 102 頁）出的金屬量來測量電流的值。

電解池 指任何用於電解的容器。

陽極（名） 即電解池的正電極（第 85 頁）。（形容詞爲 anodic）

陰極（名） 即電解池的負電極。（形容詞爲 cathodic）

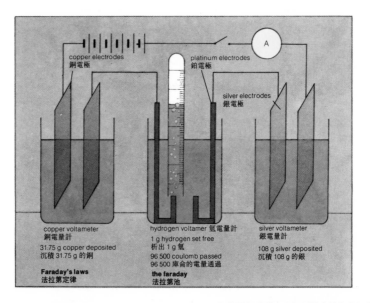

copper electrodes
銅電極

platinum electrodes
鉑電極

silver electrodes
銀電極

copper voltameter
銅電量計
31.75 g copper deposited
沉積 31.75 g 的銅
Faraday's laws
法拉第定律

hydrogen voltamer 氫電量計
1 g hydrogen set free
析出 1 g 氫
96 500 coulomb passed
96 500 庫侖的電量通過
the faraday
法拉第池

silver voltameter
銀電量計
108 g silver deposited
沉積 108 g 的銀

Faraday's laws of electrolysis (1) the mass of
a substance set free or deposited (p.102) is
proportional to the strength of the electric
current and to the time the current flows, i.e.
to the quantity of electric charge passed. (2)
the masses of different substances set free by
the same quantity of electric charge (measured
in coulomb) are proportional to their chemical
equivalents (p.104), e.g. 96 500 coulomb of
current will set free 1 g of hydrogen, or 8 g
oxygen, or 108 g silver, or 12 g magnesium.

ion (*n*) an atom (p.103), or a group of chemically
combined atoms, carrying an electric charge,
either positive or negative. A positive ion is
formed by an atom, or group of atoms, losing
one or more electrons (p.106). A negative ion is
formed by an atom, or group of atoms, gaining
one or more electrons. Ions carry the electric
current between the electrodes in electrolysis.
Symbols: Cu^{2+}, Na^+, Cl^-, SO_4^{2-}. **ionic** (*adj*).

ionize (*v*) to form ions. **ionization** (*n*).

法拉第電解定律 （1）電解時析出或沉積出（第
102 頁）的物質質量與電流強度及電流流過的
時間（即通過的電荷量）成正比；（2）相同的
電荷量（以庫倫計量）所析出的不同物質的質
量與其化學當量（第 104 頁）成正比。例如，
96 500 庫倫的電流，可析出 1 克氫，或 8 克
氧，或 108 克銀，或 12 克鎂。

離子（名） 帶正電荷或負電荷的一個原子（第 103
頁）或化學結合的原子團。正離子是失去一
個或多個電子（第 106 頁）的原子或原子團所
形成。負離子是獲得一個或多個電子的原子
或原子團所形成。電解時離子在兩極之間帶
電。離子的符號如：Cu^{2+}、Na^+、Cl^-、
SO_4^{2-}。（形容詞為 ionic）

離子化（動） 形成離子。（名詞為 ionization）

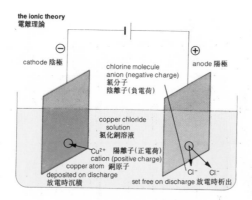

the ionic theory
電離理論

cathode 陰極

chlorine molecule
anion (negative charge)
氯分子
陰離子(負電荷)

anode 陽極

copper chloride
solution
氯化銅溶液

Cu²⁺ 陽離子(正電荷)
cation (positive charge)
copper atom 銅原子
deposited on discharge
放電時沉積

Cl⁻   Cl⁻
set free on discharge 放電時析出

anion (*n*) an ion with a negative charge; it is attracted to the anode (p.100) in electrolysis, and is discharged at the anode, e.g. chlorine ion, Cl⁻.

cation (*n*) an ion with a positive charge; it is attracted to the cathode (p.100) in electrolysis, and is discharged at the cathode, e.g. copper ion, Cu²⁺.

electrochemical equivalent the mass of a substance set free or deposited (↓) by 1 coulomb of electric charge, i.e. by a current of 1 ampere flowing for 1 second. Abbreviation: e.c.e.

deposit (*n*) (1) a layer of a metal put down on the cathode during electrolysis (p.100), e.g. a deposit of copper formed on the cathode in the electrolysis of copper (II) sulphate. (2) solid material left on a surface by a current of a fluid, e.g. (a) a deposit of mud at the mouth of a river; (b) a deposit of dust, from air, on surfaces. **deposit** (*v*), **deposition** (*n*).

faraday (*n*) a constant quantity of electric charge: approximately 96 500 coulombs. 1 faraday deposits 108 g silver or 31.7 g copper or sets free 1 g of hydrogen. It is equivalent to 1 mole (p.105) of electrons (p.106). Symbol: *F*.

electroplating (*n*) the deposition of a thin layer of a metal on another metal, for decoration or protection using an electrolytic process, e.g. the electroplating of steel with chromium.

陰離子(名) 帶負電荷的離子，電解時被吸向陽極(第 100 頁)並在陽極放電。例如氯離子Cl⁻。

陽離子(名) 帶正電荷的離子，電解時被吸向陰極(第 100 頁)並在陰極放電。例如銅離子Cu²⁺。

電化當量 由 1 庫倫電荷(即 1 安培電流流過 1 秒鐘的電量)所析出或沉積(↓)的物質質量。縮寫爲：e.c.e.。

沉積物(名) (1)電解(第 100 頁)過程中附在陰極上的金屬層。例如，在電解硫酸銅(II)過程中陰極上形成的銅沉積物；(2)流體流過時所餘留下的固體物質。例如：(a)河口處的淤泥沉積物；(b)由空氣帶來的留在物體表面上的塵埃沉積物。(動詞爲 deposit，名詞爲 deposition)

法拉第(名) 恆量的電荷：近似 96 500 庫倫。1 法拉第電量可沉積 108 克銀或 31.7 克銅或 1 克氫。它等於 1 摩爾(第 105 頁)的電子(第 106 頁)。符號爲 *F*。

電鍍(名) 用電解法進行裝飾或防護，在別一種金屬上鍍上一薄層金屬。例如將鉻電鍍在鋼上。

allotropes of carbon 碳的同素異形體

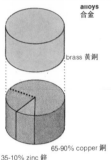

alloys
合金

brass 黃銅

65-90% copper 銅
35-10% zinc 鋅

bronze 青銅

95% copper 銅
4% tin 錫　1% zinc 鋅

70% mercury 汞
30% copper 銅

fillings
填縫料

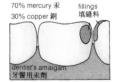

dentist's amalgam
牙醫用汞劑

element (n) a substance which cannot be decomposed (p.95) by chemical reaction into simpler substances. It consists of atoms (↓) of the same atomic number (p.107). (See periodic table on front endpapers.)

allotrope (n) one of the different physical forms of an element, but possessing the same chemical properties as other allotropes, e.g. the allotropes of carbon include diamond, graphite and charcoal, all with the same chemical properties. **allotropic** (adj).

metal (n) an element (↑) whose atoms (↓) form positive ions and which generally has the properties of lustre (↓), ductility (↓) and malleability (↓), and of being a good conductor of heat and electric current. **metallic** (adj).

non-metal (n) an element not possessing the general properties of a metal. Many non-metals are gases, some are solids and one is a liquid, at room temperature. Some form negative ions.

ductile (adj) describes a substance that can be pulled into a wire while cold and under pressure. **ductility** (n).

malleable (adj) describes a substance that can be beaten or rolled into thin sheets. **malleability** (n).

lustre (n) the property of being bright enough to be used as a mirror, e.g. silver has a lustre. **lustrous** (adj).

atom (n) the smallest particle (p.26) of an element which has the properties of that element, and takes part in chemical reactions.

molecule² (n) the smallest particle of an element or a compound which can exist free and has the properties of that element or compound. A molecule of an element consists of one or more atoms; a molecule of a compound consists of one or more atoms of each element composing the compound, e.g. a molecule of hydrogen consists of two atoms of hydrogen; a molecule of water consists of two atoms of hydrogen and one atom of oxygen. **molecular** (adj).

alloy (n) a material composed (p.95) of two or more metals, or composed of a metal and a non-metal, e.g. brass, steel, bronze are alloys. The composition of an alloy can vary slightly.

元素（名）　不能利用化學反應分解（第 95 頁）成更簡單物質的物質。元素由同一原子序數（第 107 頁）的原子（↓）組成。（見封面內頁的週期表）。

同素異形體（名）　元素的不同的物理形態之一，其化學性質和其他同素異形體相同。例如，碳的同素異形體包括金剛石、石墨和炭，它們都具有相同的化學性質。（形容詞爲 allotropic）

金屬（名）　一種元素（↑），其原子（↓）可形成正離子，金屬一般都具有光澤（↓）、延性（↓）和展性（↓），並且都是熱和電的良導體。（形容詞爲 metallic）

非金屬（名）　不具有金屬的一般性質的元素。許多非金屬是氣體，有些是固體，有一些在室溫下是液體。有些非金屬可形成負離子。

延性的（形）　描述在低溫和壓力下能被拉延成爲線的物質。（名詞爲 ductility）

有展性的（形）　描述能被錘打或軋成薄片的物質。（名詞爲 malleability）

光澤（名）　明亮如鏡的性質。例如銀具有光澤。（形容詞爲 lustrous）

原子（名）　具有元素的性質並參加化學反應的組成該元素的最小粒子（第 26 頁）。

分子（名）　能獨立存在並具有元素或化合物性質的組成該元素或化合物的最小粒子。元素的分子是由一個或多個原子組成的；化合物的分子是由構成該化合物的每一元素的一個或多個原子組成的。例如，氫分子由兩個氫原子組成；水分子由兩個氫原子和一個氧原子組成。（形容詞爲 molecular）

合金（名）　由兩種以上的金屬組成的材料，或由一種金屬和一種非金屬組成的材料。例如黃銅、鋼、青銅都是合金。合金的組成可略有變化。

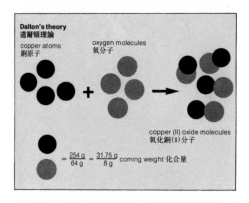

Dalton's theory 道爾頓理論

copper atoms 銅原子

oxygen molecules 氧分子

copper (II) oxide molecules 氧化銅(II)分子

$$= \frac{254\,g}{64\,g} = \frac{31.75\,g}{8\,g} \text{ coming weight 化合量}$$

Dalton's theory the theory explains the laws of chemical combination (p.94) by saying that all substances are composed of atoms. The atoms of any one element are identical (p.94), but are different from the atoms of all other elements, at least in mass. Compounds are formed by the chemical combination of atoms in simple proportions. Present theories have different descriptions of atoms (p.103), but Dalton's theory is suitable for explaining the laws of chemical combination.

relative atomic mass the ratio of the average mass of an atom of an element to 1/12 of the mass of a carbon atom. Average mass is used because there may be isotopes (p.106) of the atoms.

relative molecular mass the ratio of the mass of a molecule of a substance to 1/12 of the mass of a carbon atom. The relative molecular mass of a compound is equal to the sum of the relative atomic masses of all the atoms in the molecule

combining weight the mass in grammes of an element that will combine with or replace 1g of hydrogen or 8g of oxygen.

chemical equivalent (1) the combining weight of an element. (2) of an acid, the mass of acid (p.114) containing 1g of replaceable hydrogen; of a base (p.114), the mass that neutralizes (p.114) the chemical equivalent of an acid.

道爾頓理論 這一理論闡述化合作用(第 94 頁)定律，認爲一切物質都是由原子組成的。任何一種元素的原子都是相同的(第 94 頁)，但至少在質量方面和其他一切元素的原子不同。化合物是由原子按簡單比例化合而成。現代理論對於原子(第 103 頁)的描述不同，而道爾頓的理論適用於闡釋化合作用定律。

相對原子質量 元素的原子的平均質量與碳原子質量的 $\frac{1}{12}$ 之比。使用平均質量是因爲這些原子有可能是其同位素(第 106 頁)。

相對分子質量 物質的分子的質量與碳原子質量的 $\frac{1}{12}$ 之比。化合物的相對分子質量等於分子中所有原子的相對原子質量的總和。

化合量 指能化合或置換 1 克氫或 8 克氧的元素的質量克數。

化學當量 (1)指元素的化合量；(2)指酸：含有 1 克可置換氫的酸(第 114 頁)的量；指鹼(第 114 頁)：能中和(第 114 頁)酸的化學當量的鹼量。

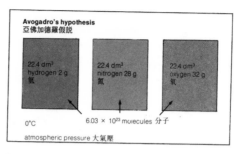

Avogadro's hypothesis
亞佛加德羅假說

22.4 dm³
hydrogen 2 g
氫

22.4 dm³
nitrogen 28 g
氮

22.4 dm³
oxygen 32 g
氧

0°C

6.03 × 10²³ molecules 分子

atmospheric pressure 大氣壓

Avogadro's hypothesis equal volumes of all gases, under the same conditions of temperature and pressure, contain the same number of molecules.

mole (*n*) the S.I. unit for amount of substance. 1 mole is the amount of substance which contains as many elementary units (i.e. particles) as there are atoms (p.103) of carbon in 0.012 kg of carbon-12. The elementary units must be named and can be molecules, atoms, ions (p.101), radicals (p.116), electrons (p.106) or other particles. Carbon-12 is an isotope (p.106) of carbon.

monatomic (*adj*) describes a molecule of an element which consists of one atom only, e.g. helium and neon are monatomic gases.

diatomic (*adj*) describes a molecule of an element which consists of two atoms, e.g. the molecules of hydrogen, oxygen, chlorine.

symbol (*n*) a letter or a small line diagram which represents a quantity (p.13), an instrument or device (p.10), a chemical element (p.103), or a process in mathematics, e.g. *p* stands for pressure;→•—stands for a switch; Fe stands for iron; ÷ stands for divide.

formula (*n*) chemical symbols written together to show the atoms in a molecule of a compound or in an ion, e.g. (a) the formula MgO stands for a molecule of magnesium oxide and shows it is composed of 1 atom of magnesium combined with 1 atom of oxygen; (b) the formula NO_3^- stands for a nitrate ion. The formula gives the composition of a substance.

亞佛加德羅假說 在相同的溫度和壓力條件下，等體積的一切氣體都含有同數的分子。

摩爾（名） 物質量的國際制單位。1摩爾是指物質總量中所含的基本單元（即粒子）和0.012千克碳-12原子（第103頁）相同。這些基本單元必須指明，可以是分子、原子、離子（第101頁）、原子團（第116頁）、電子（第106頁）或其他粒子。碳-12是碳的同位素（第106頁）。

單原子的（形） 描述只由一個原子組成元素的分子。例如，氦和氖都是單原子的氣體。

雙原子的（形） 描述由兩個原子組成元素的分子。例如，氫、氧、氯的分子。

符號（名） 表示量（第13頁）、儀器或裝置（第10頁）、化學元素（第103頁），或數學方法的字母或線條圖，例如*p*表示壓力；→•— 表示開關；Fe表示鐵；÷表示除。

化學式 為表示化合物分子中或離子中的原子而寫在一起的化學符號。例如：(a)分子式MgO代表氧化鎂分子並表示它是由一個鎂原子和一個氧原子化合而成；(b)NO_3^-代表一個硝酸根離子。化學式指示物質的組成。

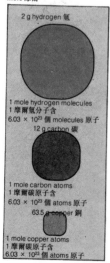

mole 摩爾

2 g hydrogen 氫

1 mole hydrogen molecules
1 摩爾氫分子含
6.03 × 10²³ 個 molecules 原子

12 g carbon 碳

1 mole carbon atoms
1 摩爾碳原子含
6.03 × 10²³ 個 atoms 原子

63.5 g copper 銅

1 mole copper atoms
1 摩爾碳原子含
6.03 × 10²³ 個 atoms 原子

electron (*n*) a very small particle, part of all atoms; it has the smallest possible negative electric charge (1.6 × 10⁻¹⁹ coulomb), and a mass approximately 1/1840 that of a hydrogen atom. An electron can be set free from an atom and travel by itself, e.g. as in cathode-ray tubes (p.86). **electronic** (*adj*).

proton (*n*) a very small particle, part of all atoms; it has the smallest possible positive charge, exactly equal to that of an electron; and a mass approximately 1839/1840 that of a hydrogen atom. A proton is a hydrogen ion.

neutron (*n*) a very small particle, part of all atoms except hydrogen; it has no electric charge; its mass is approximately equal to that of

電子(名)　一種極小的粒子，爲一切原子的組成部分；電子帶有負電荷，電荷量盡可能最小 (1.6 × 10⁻¹⁹ 庫倫)，質量近似於氫原子的 1/1840。電子可以從原子中放出並自身運行，就像在陰極射線管(第 86 頁)中那樣。(形容詞爲 electronic)

質子(名)　一種極小的粒子，爲一切原子的組成部分；帶有正電荷，電荷量盡可能最小；正好和電子的電量相等；質量近似於氫原子的 1839/1840 倍。質子是一個氫離子。

中子(名)　一種極小的粒子，除氫原子不含中子外，它是一切原子的組成部分；中子不帶電荷，質量大致與質子相等。

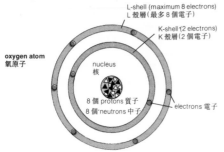

oxygen atom
氧原子

L-shell (maximum 8 electrons)
L 殼層(最多 8 個電子)

K-shell (2 electrons)
K 殼層(2 個電子)

nucleus
核

8 個 protons 質子
8 個 neutrons 中子

electrons 電子

two isotopes of carbon
碳的兩個同位素

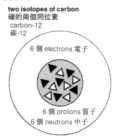

carbon-12
碳-12

6 個 electrons 電子

6 個 protons 質子
6 個 neutrons 中子

carbon-13
碳-13

6 protons 質子
6 neutrons 中子

6 個 electrons 電子

nucleus¹ (*n*) (*nuclei*) the central part of an atom, containing protons and neutrons, except for hydrogen which has a nucleus of one proton. The mass of an atom is almost all in its nucleus, as the electrons which surround the nucleus have a very much smaller mass. The nucleus has a positive charge (from its protons) which, in an atom, is exactly equal to the total charge of the electrons which surround it. **nuclear** (*adj*).

isotope (*n*) one, of two or more atoms of an element, which contains the same number of protons as the other atoms, i.e. has the same atomic number, and the same chemical properties, but has a different number of neutrons and hence a different atomic mass. **isotopic** (*adj*).

原子核(名)　原子的中心部分，它含有多個質子和中子；但氫的原子核內只有一個質子，沒有中子。原子的質量幾乎全在核內，繞核旋轉的電子質量就微小得多。核正電荷(來自其質子)，此正電荷和原子中繞核旋轉的電子電荷總數正好相等。(形容詞爲 nuclear)

同位素(名)　指一種元素有兩個或多個相同質子數目的原子，這些原子即爲同位素，亦即同素位的原子序數和化學性質相同，但中子數目不同，因而原子質量不同。(形容詞爲 isotopic)

electron shell electrons surround the nucleus (↑), and are grouped together in different shells. Each shell is at a different distance from the nucleus, and is filled by a definite number of electrons. The shells are considered to be spheres surrounding the nucleus and the outer shell gives an atom its volume. The diameter of an outer shell is about 100 000 times greater than the diameter of the nucleus.

K-shell the shell of electrons nearest the nucleus; it is filled by 2 electrons.

L-shell the next shell, after the K-shell; it is filled by 8 electrons.

M-shell the next shell after the L-shell; it is filled by 18 electrons.

atomic number the number of protons in a nucleus (↑). It determines the chemical nature of the atom, i.e. it determines to which element an atom belongs. The symbol for atomic number is Z.

periodic system the elements, when arranged along rows in a table of increasing atomic number, form groups in the columns of the table; the elements in a group have similar chemical properties. This regular repeating of properties is the periodic system. (See periodic table on front endpapers.)

transition elements these elements are in the middle of a row in the periodic system. They use an inner shell to provide valency electrons (p.106). The elements are metals with more than one electrovalency (p.109).

radioactivity (*n*) the property of atomic nuclei (↑) of spontaneous (p.112) disintegration (↓). As the nuclei disintegrate, they emit (p.45) alpha or beta particles (p.108) or gamma rays (p.108). Radioactivity takes place mainly in elements of high atomic number. It is not altered by changes in pressure or temperature. **radioactive** (*adj*).

disintegrate (*v*) (1) to break into small pieces because of physical force being applied, e.g. rocks disintegrate because of the effects of wind and water. (2) of radioactive nuclei to emit one or more atomic particles, e.g. protons. **disintegration** (*n*).

table 表

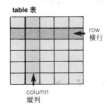

row 橫行

column 縱列

電子殼層 電子環繞着原子核(↑),並聚集在不同的殼層中。每一殼層與核的距離都不相同,層充滿一定數量的電子。電子層是環繞核的球形體,其外層確定原子的體積。外層的直徑比核的直徑約大 100 000 倍。

K 殼層 離核最近的一個電子殼層,填滿兩個電子。

L 殼層 K 殼層外的一個殼層,填滿 8 個電子。

M 殼層 L 殼層外的一個殼層,填有 18 個電子。

原子序數 原子核(↑)中的質子數,它決定原子的化學性質,即它決定原子屬於哪種元素。原子序數的符號爲 Z。

週期系 元素依原子序數遞增順序在表中排列成橫行時,形成表中縱列的各個族,同一族的各元素具有相似的化學性質。這種有規則的性質重複就是週期系。(見封面內頁的週期表)

過渡元素 這些元素位於週期系中行的中間。過渡元素以內殼層提供價電子(第 106 頁)。這些元素都是電價(第 109 頁)大於 1 的金屬。

放射性(名) 原子核(↑)自發(第 112 頁)裂變(↓)的性質。核裂變時,放射(第 45 頁)出 α 或 β 粒子(第 108 頁)或 γ 射線(第 108 頁)。放射性主要發生在原子序數大的元素中。它不隨壓力或溫度而變化。(形容詞爲 radioactive)

破裂;裂變(動) (1)因受到物理力而破成小碎片。例如,岩石受風和水的作用發生破裂;(2)放射性原子核放射出一個或多個原子微粒。例如質子。(名詞爲 disintegration)

range (n) (1) the distance to which an object or particle can travel, e.g. the range of a bullet from a gun. (2) the difference between the highest and the lowest of a set of values, e.g. the range of an ammeter. (3) the land area over which an organism is found.

alpha particle a postively charged particle consisting of two protons and two neutrons, identical with the nucleus of a helium atom. Its range in air is about 6 cm, and it is emitted (p.45) by some radioactive (p.107) nuclei. Alpha particles form alpha rays.

beta particle an electron with a high velocity emitted (p.45) from a radioactive nucleus; this happens when a neutron in the nucleus changes into a proton and an electron, and the electron is emitted as a beta particle. The range in air is about 750 cm. Beta particles form beta rays.

gamma rays radiation (p.83) with wavelengths shorter than those of X-rays. The rays are emitted (p.45) by radioactive nuclei together with either alpha or beta particles. There is no limit to their range in air. Their energy is decreased by half each time they pass through lead 1 cm thick.

Geiger counter an instrument which tests for radioactivity and measures the strength of alpha, beta or gamma rays, *see diagram*. Rays enter the window of the counter and discharge the potential between anode and cathode. The discharge is heard in a telephone receiver. The number of dis
strength of the

程；範圍（名）（1）物體或粒子運行所能及的距離，例如子彈的射程；（2）在一組值中最高值和最低值之間的差值，例如安培計的量程；（3）有生物存在的陸地區域。

α粒子　由兩個質子和兩個中子組成的正電荷的粒子，它和氦原子核相同。它在空氣中的能量範圍約爲 6 cm，它是由某些放射性（第107頁）核發射（第45頁）出來的。α粒子形成α射線。

β粒子　從放射性核發射（第45頁）的高速電子；這發生於核內的一個中子變爲一個質子和一個電子時，此電子就作爲β粒子發射出。它在空氣中的能量範圍約爲 750 cm。β粒子形成β射線。

γ射線　波長短於 X 射線的輻射線（第83頁）。這種射線是由放射性核隨α粒子或β粒子一起發射出（第45頁）。它們在空氣中的能量範圍是沒有限制。γ射線每穿透 1 cm 厚的鉛時，其能量被減弱一半。

蓋革計數器　檢驗放射性並測量 α、β 或 γ 射線強度的儀器（見圖）。射線射入計數器窗口並在陽極和陰極之間放電。在電話聽筒中可以聽出其聲音。放電次數與射線強度成正比。

effect of a magnet on radioactive radiation
磁體對放射性射線的影響

radioactive source 放射性源

lead screen 鉛屏蔽

magnet 磁體

N S

alpha rays α射線

beta rays β射線

gamma rays no deviation γ射線不偏轉

Alpha and beta rays are deflected by magnetic fields because they are electrically charged. Gamma rays are not deflected as they have no electric charge. The illustration shows the deviation of rays caused by a magnet. 磁場使 α 射線和 β 射線偏向，因爲這兩種射線都帶電荷。γ 射線不偏向，因爲它不帶電荷。圖示磁體引起的射線偏向。

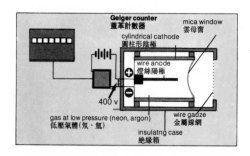

Geiger counter 蓋革計數器

mica window 雲母窗

cylindrical cathode 圓柱形陰極

wire anode 燈絲陽極

400 v

gas at low pressure (neon, argon) 低壓氣體（氖、氬）

wire gadze 金屬線網

insulatng case 絕緣箱

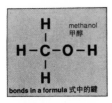

methanol
甲醇

H–C–O–H

bonds in a formula 式中的鍵

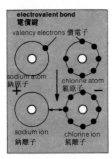

electrovalent bond
電價鍵

valency electrons 價電子

sodium atom
鈉原子

chlorine atom
氯原子

sodium ion
鈉離子

chlorine ion
氯離子

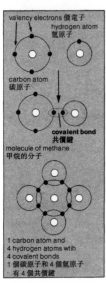

valency electrons 價電子
hydrogen atom
氫原子

carbon atom
碳原子

covalent bond
共價鍵

molecule of methane
甲烷的分子

1 carbon atom and
4 hydrogen atoms wtih
4 covalent bonds
1 個碳原子和 4 個氫原子
有 4 個共價鍵

half-life the time taken for half the atoms in a piece of radioactive material to disintegrate. Each radioactive element has its own half-life period and the values have a range (p.108) from one millionth of a second to more than one million years.

nuclear fission the breaking of the nucleus of an element with a high atomic number (e.g. uranium) into two approximately equal parts, with the emission (p.45) of neutrons (p.106) and the setting free of large quantities of energy. Fission can be spontaneous (p.112) or caused by the bombardment (p.83) of nuclei by neutrons.

nuclear fusion the forming of an atomic nucleus from two nuclei of low atomic number; large quantities of energy are set free. The reaction takes place only at very high temperatures.

bond (*n*) the force which holds atoms or ions together to make molecules or crystal structures. There are two main kinds of bonds, electrovalent (↓) and covalent (↓).

valency electron an electron which takes part in a bond. Except for the transitional elements (p.107), valency electrons are in the outermost electron shell.

electrovalency (*n*) (1) the joining together of two atoms by electrostatic (p.71) charges. One atom loses one or more electrons and becomes positively charged, the other atom gains one or more electrons and becomes negatively charged. The unlike charges attract and form an electrovalent bond. (2) the number of electrons that an atom, or group of atoms, loses or gains; the number of electrovalent bonds it can form. **electrovalent** (*adj*).

shared (*adj*) of an object, held by two or more other objects, e.g. an electron shared by two atoms.

covalency (*n*) (1) the joining together of two atoms by the sharing of a pair of electrons. Each atom gives one electron to form the pair. The electron pair forms a covalent bond. (2) the number of covalent bonds an atom can form.

半衰期 一塊放射性物質裂變其所含半數原子所需的時間。每種放射性元素都有自己的半衰期，其數值範圍（第 108 頁）從百分之一秒到一百萬年以上。

核裂變 原子序數大的元素（例如鈾），其原子核分裂成爲近似相等的兩個部分，同時發射出（第 45 頁）中子（第 106 頁），釋放大量能量。裂變可以是自發的（第 112 頁）或者是用中子轟擊（第 83 頁）原子核所引起。

核聚變 由兩個原子序數小的核合併成一個原子核；並釋放出大量能量。反應只在極高的溫度下才發生。

鍵(名) 使原子或離子結合形成分子或晶體結構的力。主要有兩種：電價(↓)鍵和共價(↓)鍵。

價電子 指參與形成鍵的電子。除過渡元素（第 107 頁）外，其他元素的價電子都位於最外面的電子殼層。

電價(名) (1)指兩個原子靠靜電(第 71 頁)荷結合在一起，其中一個原子失去一個或多個電子而帶正電，另一個原子得到一個或多個電子而帶負電。異電相吸而形成的電價鍵；(2)指一個原子或原子團失去或得到的電子數即它能夠形成的電價鍵數。（形容詞爲 electrovalent）

共用的(形) 指一個物體歸兩個或多的別的物體所共有。例如，爲兩個原子所共用的一個電子。

共價(名) (1)指兩個原子靠共用一對電子結合在一起。每個原子各出一個電子形成一對電子。這對電子形成一個共價鍵；(2)一個原子能形成的共價鍵數目。

crystal (*n*) a glass-like piece of a solid with a regular shape, e.g. crystals of sugar. A substance which forms crystals is called a crystalline solid; all its crystals have the same shape. Electrovalent (ionic) crystals are formed by electrolytes (p.101) from ions; covalent crystals are formed by non-electrolytes from molecules; metallic crystals are formed by metals from atoms. **crystallize** (*v*), **crystalline** (*adj*).

crystallization (*n*) the formation of crystals from a warm, saturated solution.

water of crystallization a definite molecular proportion of water, chemically combined with ions or molecules, in crystals formed from solutions in water, e.g. in copper (II) sulphate the proportion of compound to water of crystallization is 1:5; the formula of the compound is $CuSO_4, 5H_2O$.

hydrated (*adj*) describes crystals with water of crystallization (↑).

anhydrous (*adj*) describes crystals with no water of crystallization (↑) or amorphous (↓) solids.

amorphous (*adj*) without regular shape, not having a crystalline structure, e.g. an amorphous solid is like a powder, such as flour.

tetrahedral (*adj*) with a shape like a tetrahedron, which has four flat triangular faces, *see diagram*. A regular tetrahedron has three equal sides for each triangle.

晶體(名) 形狀規則的玻璃狀固體塊。例如糖的晶體。能形成晶體的物質稱爲結晶固體；此結晶固體的所有晶體都具有相同的形狀。電價(離子)晶體是由電解質(第 101 頁)的離子形成的；共價晶體是由非電解質的分子形成的；金屬晶體是由金屬原子形成的。(動詞爲 crystallize，形容詞爲 crystalline)

結晶作用(名) 由溫的飽和溶液形成晶體的作用。

結晶水 從水溶液所形成的晶體中，與離子或分子化學結合的，水有一定分子比例的水。例如，在硫酸銅(II)中，化合物與結晶水的比例爲 1:5；化合物的分子式爲 $CuSO_4 \cdot 5H_2O$。

水合的(形) 描述含有結晶水(↑)的晶體。

無水的(形) 描述不含結晶水(↑)的晶體或無定形的(↓)固體。

無定形的(形) 形狀不規則，沒有晶體結構的。例如，像麵粉此類粉末就是一種無定形的固體。

四面體的(形) 具有四面體形狀的，這種四面體有四個平面形三角，(見圖)。一個規則的四面體的每個三角形有三個等邊。

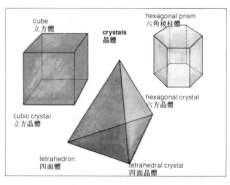

cube 立方體

crystals 晶體

hexagonal prism 六角稜柱體

cubic crystal 立方晶體

hexagonal crystal 六方晶體

tetrahedron 四面體

tetrahedral crystal 四面晶體

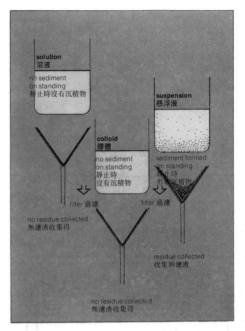

colloid (n) a substance that does not dissolve, nor is suspended (p.35) in a liquid, but is *dispersed* in a liquid. A colloid has particles larger than molecules or ions in solution, but smaller than those in suspensions (↓); the particles cannot be filtered (p.91) to form a residue, e.g. starch, glue, are colloids; they form **colloidal solutions**. colloidal (*adj*).

suspension (n) a liquid containing small solid particles suspended (p.35) in it. When filtered (p.91) the solid particles are collected as a residue; if left, the particles slowly **settle** to the bottom of the container and form a sediment, e.g. earth shaken up with water forms a suspension.

emulsion (n) a colloidal solution of one liquid in another liquid, e.g. oil shaken up with water forms an emulsion.

膠體(名) 既非溶解亦非懸浮(第 35 頁)而是分散在液體中的一種物質。膠體的粒子大於溶液中的分子或離子，而小於懸浮液(↓)中的分子或離子。膠體粒子不能濾過(第 91 頁)而形成濾渣。例如：澱粉、膠水都是膠體，可形成**膠態溶液**。(形容詞爲 colloidal)

懸浮液(名) 一種含有懸浮(第 35 頁)的細小固體粒子的液體。在過濾(第 91 頁)時，這些固體粒子作爲濾渣收集，如沒有濾掉，這些粒子就慢慢**沉降**到容器的底部成爲沉積物。例如，土壤放在水中搖動形成懸浮液。

乳狀液；乳膠(名) 一種液體在另一種液體中的膠態溶液。例如：油放在水中一起搖動形成乳狀液。

air (*n*) the mixture (p.95) of gases which surrounds the Earth and forms its atmosphere. Dry air consists of approximately 78% nitrogen, 21% oxygen, 0.03% carbon dioxide and the remainder the rare (also called noble or inert) gases such as helium, neon, argon.

combustion (*n*) a chemical reaction, between a substance and oxygen in the air, in which heat is given out, and a flame may be formed, i.e. burning. In *rapid combustion*, heat is given out at a high temperature, but in *slow combustion*, heat is given out at a lower temperature, and never with a flame. **combustible** (*adj*).

flame (*n*) a mass of gas so hot that it gives out light; the heat is produced by combustion.

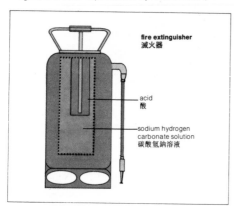

fire extinguisher
滅火器

acid
酸

sodium hydrogen carbonate solution
碳酸氫鈉溶液

fire-extinguisher (*n*) a device that supplies a fluid which stops combustion, e.g. an extinguisher using carbon dioxide foam.

explosion (*n*) a sudden expansion (p.38) of gases, produced by rapid combustion, which exerts a very strong force when shut in a small space. **explode** (*v*), **explosive** (*adj*).

inflammable (*adj*) can burn easily and readily when a flame is applied.

spontaneous (*adj*) caused by itself, e.g. spontaneous combustion has no outside cause, a substance starts to burn due to causes inside itself.

空氣（名） 包圍地球形成地球大氣層的氣體混合物（第 95 頁）。乾燥空氣含氮約 78％，氧 21％，二氧化碳 0.03％，其餘爲氦、氖、氬之類稀有氣體（也稱爲惰性氣體或不活潑氣體）。

燃燒（名） 物質和空氣中的氧所發生的化學反應，反應中放出熱並可能有火焰，即着了火。“速燃”時，在高溫下發出熱量，“緩燃”時，是在較低溫度下發出熱量，而且絕無火焰。（形容詞爲 combustible）

火焰（名） 熾熱致發光的氣體，其熱量是由燃燒產生的。

chemical reaction between hydrogen and oxygen
氫和氧之間的化學反應

a simple flame
簡單的火焰

hot gas gives out light
熾熱氣體發出光

complete combustion
完全燃燒

unburnt gas
不燃的氣體

hydrogen 氫

flame 火焰

滅火器（名） 一種提供滅火流體的裝置。例如二氧化碳的泡沫滅火器。

爆炸（名） 快速燃燒所致的氣體突然膨脹（第 38 頁），如果此氣體被封閉在一個小空間，就會產生猛烈的爆炸力。（動詞爲 explode，形容詞爲 explosive）

易燃的（形） 指用火焰點燃時易迅速燃燒。

自發的（形） 自然引起的。例如，自發燃燒沒有外因，只是物質由於內在原因而起燃。

water cycle 水循環
clouds 雲
rain 雨
water vapour 水汽
evaporation 蒸發
water vapour 水汽
lake 湖
spring 春天
sea 海
animals 動物
river 河
plants 植物

water-cycle the changes in the state of water: (1) evaporation (p.42) from rivers, lakes and the sea and respiration (p.191) of plants and animals producing water vapour; (2) water vapour condensing to form clouds; (3) rain falling from clouds; (4) rain water passing through earth to rivers and lakes and on to the sea. Each event follows the one before it, so the changes form a continuous process.

hard water water containing salts (p.115) of calcium, magnesium or both metals. With soap, hard water does not form a **lather** (lots of bubbles) immediately; instead it forms a **scum** (particles of solid on the water surface) first and a lather later. Hard water which can be softened (↓) by boiling is called temporary hard water; hard water which can be softened by adding sodium carbonate is called permanent hard water.

soft water water not containing salts of calcium or magnesium. With soap, soft water forms a lather (↑) immediately. **to soften water**.

水循環 水的狀態變化：(1)河、湖和海水蒸發(第 42 頁)以及動、植物呼吸(第 191 頁)產生水汽；(2)水汽冷凝而形成雲；(3)雲下降成雨；(4)雨水通過地面流向河流和湖泊，最後流向海洋。上述的每一現象接踵發生而形成一個持續不斷的過程。

硬水 含有鈣鹽(第 115 頁)和鎂鹽或兩者兼有的水。肥皂和硬水不能立即形成**泡沫**(大量氣泡)；而是首先形成**浮渣**(水面上的固體微粒)。而後才形成肥皂沫。經煮沸可軟化(↓)的硬水稱爲暫硬水；加入碳酸鈉可軟化的硬水稱爲永硬水。

軟水 不含鈣鹽或鎂鹽的水。肥皂和軟水可立即形成泡沫(↑)。(**to soften water** 意爲使水軟化。)

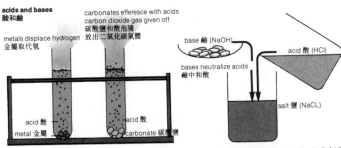

acids and bases
酸和鹼

metals displace hydrogen
金屬取代氫

carbonates effervesce with acids
carbon dioxide gas given off
碳酸鹽和酸泡騰
放出二氧化碳氣體

base 鹼 (NaOH)

acid 酸 (HCl)

bases neutralize acids
鹼中和酸

salt 鹽 (NaCL)

acid 酸

acid 酸

metal 金屬

carbonate 碳酸鹽

acid (n) a substance which (1) forms hydrogen ions (p.101) in solution; (2) contains hydrogen which can be replaced by a metal to form a salt (↓). Some acids are corrosive (p.99) and most acids change an indicator (p.116). **acidic** (adj), **acidify** (v).

base (n) a substance which reacts with an acid to form a salt and water only, generally an oxide or a hydroxide of a metal. **basic** (adj).

alkali (n) a soluble base (↑) which forms hydroxyl (↓) ions (p.101) in solution.

酸（名）　（1）指在溶液中能形成氫離子（第 101 頁）的一種物質或（2）指含有的氫能被金屬置換成鹽（↓）的物質。有些酸有腐蝕性（第 99 頁），大多數酸能使指示劑（第 116 頁）變色。（形容詞爲 acidic，動詞爲 acidify）

鹼；醶（名）　一種和酸反應只形成鹽和水的物質，通常是金屬的氧化物或氫氧化物。（形容詞為 basic）

強鹼；鹼（名）　在溶液中可形成氫氧根（↓）離子（第 101 頁）的可溶性鹼（↑）。

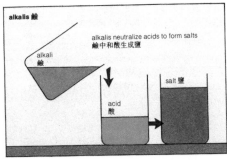

alkalis 鹼

alkalis neutralize acids to form salts
鹼中和酸生成鹽

alkali
鹼

salt 鹽

acid
酸

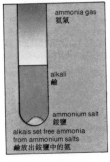

ammonia gas
氨氣

alkali
鹼

ammonium salt
銨鹽

alkais set free ammonia from ammonium salts
鹼放出銨鹽中的氨

hydroxyl (n) the group of atoms —OH; the hydroxyl ion (OH⁻) has a negative charge and an electrovalency of 1.

neutralization (n) the reaction between an acid and a base, in which both lose their properties, and a salt is formed. **neutralize** (v), **neutral** (adj), e.g. hydrochloric acid neutralizes sodium hydroxide and sodium chloride is formed.

氫氧根（名）　—OH 原子團；氫氧根離子（OH⁻），帶負電，電價爲 1。

中和作用（名）　指酸和鹼的反應，反應中二者都失去本身的性質而形成鹽。例如，鹽酸中和氫氧化鈉生成氯化鈉。（動詞爲 neutralize，形容詞爲 neutral）

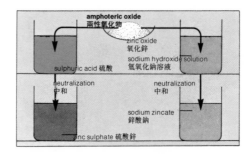

amphoteric (*adj*) describes a substance which can neutralize an acid and also neutralize a base, e.g. zinc oxide neutralizes acids to form zinc salts and it neutralizes bases to form zincates.

salt (*n*) a compound formed when the hydrogen of an acid is replaced by a metal. The salt is named from the metal and the acid; if soluble, it produces ions (p.101) in solution, a cation from the metal and an anion from the acid.

acid salt a salt formed when there is not enough acid to neutralize (↑) a base completely. Acid salts are formed only with dibasic (↓) acids, as only part of the hydrogen is replaced by a metal, e.g. sodium hydrogen sulphate.

basic salt a salt formed when there is not enough base to neutralize (↑) an acid completely; it consists of a neutral salt with a definite proportion of the base, e.g. basic lead carbonate $2PbCO_3$, $Pb(OH)_2$.

basicity (*n*) (1) the number of hydrogen atoms in a molecule of an acid which can be replaced by a metal, e.g. ethanoic (acetic) acid has the formula CH_3COOH but only one hydrogen atom can be replaced, so its basicity is 1. (2) the number of moles (p.105) of hydrogen ions formed from one mole of an acid, e.g. 1 mole of sulphuric acid forms 2 moles of hydrogen ions, its basicity is 2.

monobasic (*adj*) of an acid, having a basicity of 1, e.g. hydrochloric acid.

dibasic (*adj*) of an acid, having a basicity of 2, e.g. sulphuric acid. Dibasic acids can form acid salts.

兩性的(形) 描述一種能中和酸也能中和鹼的物質。例如，氧化鋅能中和酸生成鋅鹽，也能中和鹼生成鋅酸鹽。

鹽(名) 酸中的氫爲金屬置換所形成的化合物；鹽是根據此金屬和酸命名。如爲可溶性鹽，則在溶液中產生離子(第 101 頁)，金屬產生陽離子酸產生陰離子。

酸式鹽 當酸量不足，不能完全中和(↑)鹼所形成的鹽。酸式鹽只由二元(↓)酸形成，因爲只有二元酸才有部分氫可被金屬取代。例如硫酸氫鈉。

鹼式鹽 當鹼量不足，不能完全中和(↑)酸所形成的鹽；它含有中性鹽和一定比例的鹼。例如鹼式碳酸鉛 $2PbCO_3 \cdot Pb(OH)_2$。

鹼度(名) (1)酸分子中能被金屬所置換的氫原子數。例如，乙酸(醋酸)的化學式爲 CH_3COOH，但只有一個氫原子能被置換，故其鹼度爲 1；(2)由 1 摩爾酸形成的氫離子摩爾(第 105 頁)數。例如，1 摩爾硫酸形成 2 摩爾氫離子，其鹼度爲 2。

一元的(形) 指具有鹼度爲 1 的酸，例如鹽酸。

二元的(形) 指具有鹼度爲二的酸，例如硫酸。二元酸可形成酸式鹽。

indicator (*n*) a substance that changes colour when put in acids and alkalis; it is used to determine when neutralization has been completed, e.g. litmus (↓). **indicate** (*v*).

litmus (*n*) a soluble purple substance obtained from plants. In acids it turns red and in alkalis it turns blue. In neutralization, litmus is usually put in some alkali (it turns blue) and acid added until the litmus just turns red, showing neutralization is complete.

pH value a value on a numerical scale from 0–14 indicating the concentration of hydrogen ions (p.101) and hydroxyl (p.114) ions. At a value of 7, the concentrations of both ions are equal and the solution is neutral. Between 0 and 7, a liquid is acidic, with the hydrogen ion concentration increasing as the pH value falls from 7 to 0. Between 7 and 14, a liquid is alkaline, with the hydroxyl ion concentration increasing as the pH value increases.

指示劑（名） 放入酸中和鹼中能改變顏色的物質；用以判定何時完成中和作用。例如石蕊（↓）。（動詞爲 indicate）

石蕊（名） 一種從植物中提取的可溶性紫色物質。它在酸中變紅，在鹼中變藍。在中和作用中，通常將石蕊放入某種鹼中（變成藍色），再加入酸至石蕊剛變紅色，即表示中和作用完成。

pH 值 數字範圍從 0-14，表示氫離子（第 101 頁）和氫氧根（第 114 頁）離子濃度的值。值爲 7 時，兩種離子的濃度相等，溶液呈中性。在 0 至 7 之間，液體呈酸性，氫離子濃度隨 pH 值從 7 降至 0 而增大。在 7 至 14 間，液體呈鹼性，氫氧根離子濃度也隨 pH 值增大而升高。

pH scale pH 值標度

acidic 酸性　　　　　neutral 中性　　　　alkaline 鹼性

0　　　　　　　　　　7　　　　　　　　　14

radical (*n*) a group of atoms which can take part, without change, in chemical reactions. It is present in compounds, or as an ion in solution, but it is not found free. An acid radical and displaceable hydrogen form a molecule of an acid. In an ionized acid, the radical has one or more negative charges, e.g. (a) the sulphate radical, SO_4, in calcium sulphate, $CaSO_4$; (b) the sulphate ion, SO_4^{2-}, present in solution.

inorganic (*adj*) from mineral (p.119) sources. Inorganic substances and materials include all elements and all their compounds except for compounds containing carbon; carbonates, however, are considered as inorganic compounds.

根（名） 能參與化學反應而不發生變化的原子團。根存在於化合物中或在溶液中成爲離子，但未見游離的根。酸根和可置換氫形成酸分子。在電離的酸中，根帶一個或多個負電荷。例如：（1）硫酸鈣 $CaSO_4$ 中的硫酸根 SO_4；（2）存在於溶液中的硫酸根離子 SO_4^{2-}。

無機的（形） 指來自礦物（第 119 頁）源的。無機物質和材料包括除含碳化合物之外的一切元素和其化合物，但碳酸鹽也被視爲無機化合物。

naming compounds
化合物的命名

prefix 詞頭	number 表示數目	example 例子	
mono-	1	carbon monoxide 一氧化碳	CO
di-	2	carbon dioxide 二氧化碳	CO_2
		sulphur dioxide 二氧化硫	SO_2
tri-	3	phosphorus trichloride 三氧化磷	PCl_3
		phosphorus trioxide 三氧化二磷	P_2O_3
tetra-	4	tetrachloromethane 四氯甲烷	CCl_4
penta-	5	phosphorus pentachloride 五氯化磷	PCl_5
hexa-	6	potassium hexacyanoferrate (III) 六氰合鐵(III)酸鉀	$K_3Fe(CN)_6$

oxide (*n*) a compound consisting of one element combined with oxygen, e.g. calcium oxide, CaO; nitrogen oxide, NO; iron (III) oxide, Fe_2O_3. The prefixes: mono-(one); di-(two); tri-(three); penta-(five) are used to show the covalency (p.109) of a non-metal, e.g. phosphorus pentoxide, P_2O_5.

peroxide (*n*) an oxide which forms hydrogen peroxide, formula H_2O_2, with a dilute acid instead of forming a salt and water, e.g. barium peroxide, BaO_2; the normal oxide is BaO.

hydroxide (*n*) a compound containing the group of atoms —OH; these compounds are bases, and, if soluble, form hydroxyl (p.114) ions in solution. When an alkali neutralizes an acid, the hydroxyl ions combine with hydrogen ions to form water. Examples are: sodium hydroxide, NaOH; calcium hydroxide, $Ca(OH)_2$.

chloride (*n*) the radical (↑) of hydrochloric acid (HCl). Chlorides are also considered as compounds consisting of one element combined with chlorine, e.g. phosphorus trichloride, PCl_3, sodium chloride, NaCl.

氧化物(名) 由一種元素與氧化合而成的化合物。例如氧化鈣(CaO)；氧化氮(NO)；氧化鐵 (III) (Fe_2O_3)。用詞頭 mono- (一)、di- (二)、tri-(三)、penta-(五)表示非金屬的共價數(第109頁)，例如五氧化二磷。

過氧化物(名) 與稀酸作用生成過氧化氫(化學式爲 H_2O_2)而不生成鹽和水的一種氧化物。例如過氧化鋇(BaO_2)，其正常氧化物爲氧化鋇 (BaO)。

氫氧化物(名) 含有 —OH 原子團的化合物。這些化合物是鹼，如爲可溶性鹼，則在溶液中形成氫氧根(第 114 頁)離子，酸鹼中和時，氫氧根離子和氫離子結合成水。例如：氫氧化鈉(NaOH)；氫氧化鈣($Ca(OH)_2$)。

氯根；氯化物(名) 鹽酸 (HCl) 根 (↑)。氯化物也被看成是由一種元素與氯化合而成的化合物。例如：三氯化磷(PCl_3)、氯化鈉 (NaCl)。

acid radicals and ions 酸根和離子

radical 根	formula 化學式	ion 離子
chloride 氯根	$-Cl$	Cl^-
bromide 溴根	$-Br$	Br^-
sulphate 硫酸根	$-SO_4$	SO_4^{2-}
sulphite 亞硫酸根	$-SO_3$	SO_3^{2-}
nitrate 硝酸根	$-NO_3$	NO_3^-
nitrite 亞硝酸根	$-NO_2$	NO_2^-
carbonate 碳酸根	$-CO_3$	CO_3^{2-}
sulphide 硫根	$-S$	S^{2-}
silcate 矽酸根	SiO_3	SiO_3^{2-}

sulphate (*n*) the radical (p.116) of sulphuric acid (H_2SO_4); any salt formed from a metal and sulphuric acid, e.g. sodium sulphate, Na_2SO_4.

sulphite (*n*) the radical (p.116) of sulphurous acid (H_2SO_3); any salt formed from a metal and sulphurous acid, e.g. sodium sulphite, Na_2SO_3.

nitrate (*n*) the radical (p.116) of nitric acid (HNO_3); any salt formed from a metal and nitric acid, e.g. sodium nitrate, $NaNO_3$.

nitrite (*n*) the radical (↑) having the group of atoms $-NO_2$; the acid, formula HNO_2, cannot be prepared. Salts include sodium nitrite, $NaNO_2$, potassium nitrite, KNO_2.

carbonate (*n*) the radical (↑) of carbonic acid (H_2CO_3); any salt formed by a metal and carbon dioxide in solution, e.g. sodium carbonate, Na_2CO_3.

硫酸根；硫酸鹽（名） 硫酸(H_2SO_4)的根（第 116 頁）；金屬和硫酸形成的任何鹽。例如硫酸鈉(Na_2SO_4)。

亞硫酸根；亞硫酸鹽（名） 亞硫酸(H_2SO_3)的根（第 116 頁）；金屬和亞硫酸形成的任何鹽。例如亞硫酸鈉(Na_2SO_3)。

硝酸根；硝酸鹽（名） 硝酸(HNO_3)的根（第 116 頁）；金屬和硝酸形成的任何鹽。例如硝酸鈉($NaNO_3$)。

亞硝酸根；亞硝酸鹽（名） 含有原子團 $-NO_2$ 的根（↑）；化學式爲 HNO_2 的酸無法製得。其鹽包括亞硝酸鈉($NaNO_2$)、亞硝酸鉀(KNO_2)。

碳酸根；碳酸鹽（名） 碳酸(H_2CO_3)的根（↑）；金屬和二氧化碳在溶液中形成的任何鹽。例如碳酸鈉(Na_2CO_3)。

rock (*n*) the solid material which forms the surface of the Earth. In some places, rock is covered by soil (p.230). There are different kinds of rock and each kind contains different minerals (↓). Examples of rocks are: granite (↓), chalk (p.128), coal (p.128), limestone (p.128). Rock can vary from very hard to soft; each kind of rock has its own particular hardness.

岩石(名) 形成地球表面的固體物質。有些地方的岩石被土壤(第 230 頁)覆蓋着。岩石有不同的種類，而每種岩石都含有不同的礦物(↓)。岩石的例子如：花崗石(↓)、白堊(第 128 頁)、煤(第 128 頁)和石灰石(第 128 頁)。岩石從很硬到很軟各不相同；每種岩石都有其特有的硬度。

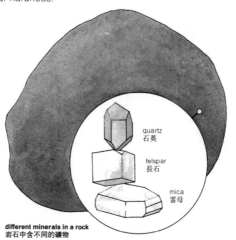

quartz
石英

felspar
長石

mica
雲母

different minerals in a rock
岩石中含不同的礦物

mineral (*n*) a substance found in the ground; it has a particular chemical composition (p.95); it possesses chemical and physical properties by which it can be recognized. Examples of minerals are: (a) native elements such as gold and silver; (b) rock salt, which is sodium chloride; (c) iron pyrites, a kind of iron sulphide; (d) quartz (p.120); (e) mica, a silicate of aluminium, potassium, magnesium and iron. **mineral** (*adj*).

ore (*n*) a mineral (↑) dug from the earth, from which a useful substance, generally a metal, is obtained, e.g. haematite is an ore, and iron is obtained from it. Some ores are used to obtain non-metals and their compounds.

granite (*n*) a very hard igneous (p.121) rock containing at least three minerals: quartz (p.120), felspar, and mica.

礦物(名) 在地底下發現的物質；礦物具有特定的化學成分(第 95 頁)，有可資識別的化學和物理性質。礦物的例子如：(a)金、銀之類天然元素；(b)岩鹽，即氯化鈉；(c)黃鐵礦，係一種鐵的硫化物；(d)石英(第 120 頁)；(e)云母係含鋁、鉀、鎂和鐵的矽酸鹽。(形容詞爲 mineral)

礦石(名) 從地下挖掘出的礦物(↑)。從礦石可以提取有用的物質，通常是金屬。例如，赤鐵礦是一種礦石，從這種礦石可提煉得鐵。有些礦石用來取得非金屬及其化合物。

花崗石(名) 一種很硬的火成(第 121 頁)岩，至少含三種礦物：石英(第 120 頁)、長石和雲母。

basalt (n) a dark-coloured or black glass-like igneous (↓) rock formed by volcanic (p.125) action. It is the commonest type of lava, and contains silicates of iron, magnesium and calcium, and oxides of lime.

quartz (n) a mineral with clear, uncoloured crystals in the shape of hexagonal prisms (p.110), consisting of silicon dioxide. The most common substance in rocks.

silica (n) a very hard white substance, it is silicon dioxide. Quartz and sand consist of silica. Silica also combines with basic substances to form silicates, compounds which are common in minerals.

weathering (n) the action of wind, rain, ice, water, frost, or chemical substances on rock. Weathering loosens and breaks up the surface of the rock, setting free small pieces. These pieces can be carried by wind or rivers to other places. **weathered** (adj).

erosion (n) the destruction of the Earth's surface by strong weathering agents followed by the removal of the products of weathering. In particular, it is the removal of soil (p.230), leaving rocks uncovered, so that no plants grow. **erode** (v).

leaching (n) (1) the washing away of soluble mineral salts from soil (p.230), by water, usually rain water, passing quickly through the soil. Leaching produces a poor soil, with few plants, so erosion follows quickly. (2) washing out a soluble constituent (p.96) from a mixture. **leach** (v).

detritus (n) loose solids caused by weathering, which are carried by wind or water. Detritus usually consists of gravel, sand and clay. When it is no longer carried by wind or water, it settles and forms a sediment. **detrital** (adj).

silt (n) a material formed from very small pieces of rock; it is like mud. Particles of different sizes are given different names, see diagram. The particles of silt are larger than those in clay (p.230), but smaller than those in sand (p.230).

sediment (n) solids which separate from water and are deposited by gravity.

玄武岩(名) 由火山(第 125 頁)作用形成的一種暗色或黑色玻璃狀火成(↓)岩。它是最普通的一類熔岩，含有鐵、鎂和鈣的矽酸鹽和石炭的氧化物。

石英(名) 一種六稜柱(第 110 頁)形的透明無色晶體礦物，由二氧化矽組成，是岩石中最常見的物質。

矽石；硅石(名) 一種極硬的白色物質，成分爲二氧化矽。石英和沙組成矽石。矽石也能和鹼性物質化合形成矽酸鹽，這是礦物中常見的化合物。

風化；風蝕(名) 風、雨、冰、水、霜或化學物質對岩石的作用。風化使岩石表面鬆潰破裂，碎成小塊。這些碎塊可被風吹或水沖到別處。(形容詞爲 weathered)

侵蝕(名) 地球表面經年累月受風吹雨打及化學作用而破壞，留下的碎石殘礫逐漸被移去。特別是它移走了土壤(第 230 頁)，留下無覆蓋的岩石，使植物無法生長。(動詞爲erode)

淋濾；瀝濾(名) (1)土壤(第 230 頁)中的可溶性礦物鹽被迅速流過的水(通常是雨水)沖走。淋濾導致土壤貧瘠，植物稀少，以致迅速受到侵蝕。(2)從混合物中沖掉可溶的組分(第 96 頁)。(動詞爲 leach)

岩屑(名) 風或水引致風化引成的疏鬆固體。岩屑通常是由砂礫、沙和粘土組成。在無風或水運載時，就沉積形成沉積物。(形容詞爲detrital)

粉沙(名) 由極小的岩石碎塊形成的一種物質，很像淤泥。大小不同的微粒具有不同的名稱(見圖)。粉沙的微粒比粘土(第 230 頁)的微粒大，但比砂(第 230 頁)的微粒小。

沉積物(名) 從水中分離並因重力而沉積的固體。

quartz crystal 石英晶體

a crystalline mineral
一種晶體狀礦物

weathering 風化

particle size 粒徑	name 名稱	diameter 直徑
	boulder 巨礫	> 200 mm
	cobble 大礫	> 200 - 50 mm
	pebble 卵石	50 - 10 mm
	gravel 砂礫	10-2 mm
	sand 砂	2-0.1 mm
	silt 粉沙	0.1-0.01 mm
	clay, dust 黏土、塵土	< 0.01 mm

sedimentary rock any rock formed from the settling of detritus (sediment) or from sediment formed in layers by chemical action or from the changing of plant and animal bodies into decayed (p.146) material. For example: (a) sand, brought by a river, settles, and in time changes into hard sandstone; (b) chalk (p.128) is a soft rock formed from seashells which settled on the seabed; (c) trees from forests on the Earth 200 million years ago, decayed and changed into coal. Sedimentary rocks are usually soft and easily weathered. **sediment** (n), **sedimentary** (adj).

metamorphic rock a rock formed from sedimentary rock by pressure and high temperature. Great movements of the Earth break up sedimentary rock and push it down into the Earth, where it is very hot. In the Earth, the pressure and heat form hard, crystalline, metamorphic rock, from the sedimentary rock, e.g. marble is formed from limestone by heat and pressure. Metamorphic rocks are the most common in Earth's crust (p.124). **metamorphism** (n), **metamorphic** (adj).

igneous rock a rock which has solidified from liquid magma (p.124). The magma comes near to the surface (a) through volcanoes (p.125), forming volcanic rocks or (b) from being forced between metamorphic rocks, forming plutonic rocks. Basalt (↑) is the commonest volcanic rock and granite (p.119) the commonest plutonic (p.127) rock.

igneous rock
火成岩

basalt
玄武岩

沉積岩；水成岩 指由岩屑（沉積物）沉降或因化學作用而在岩層中沉積的沉積物，或由植物和動物遺體腐敗（第 146 頁）所成的物質形成的任何岩石。例如：(a) 河水帶來的沙石沉積後隨時間推移而變成堅硬的沙岩；(b) 白堊（第 128 頁）是由沉積在海牀的海貝殼形成的一種軟岩石；(c) 兩億年前地球森林中的樹木腐敗後變成的煤。沉積岩通常鬆軟並容易風化。（名詞爲 sediment，形容詞爲 sedimentary）

變質岩 沉積岩因受壓力和高溫而形成的一種岩石。地殼的巨大運動破壞沉積岩並把它推入地球內部高熱之處。在地球內部，壓力和熱將沉積岩變成堅硬的、結晶的變質岩。例如大理石就是石灰石受熱和壓力形成的。變質岩是地殼（第 124 頁）內最普通的岩石。（名詞爲 metamorphism，形容詞爲 metamorphic）

火成岩 由液態岩漿（第 124 頁）凝固變硬而形成的岩石。岩漿接近地面時：(a) 通過火山（第 125 頁）而形成火山岩或 (b) 由於變質岩之間受到壓力而形成深成岩。玄武岩（↑）是最普通的火山岩，而花崗岩（第 119 頁）是最普通的深成岩（第 127 頁）。

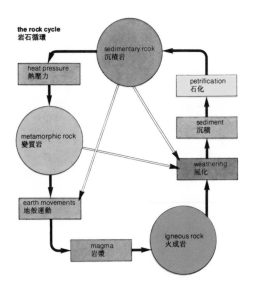

the rock cycle
岩石循環

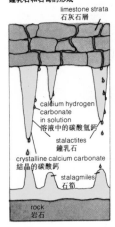

formation of stalactites
and stalagmites
鐘乳石和石筍的形成

rock cycle the changing of types of rock into other types. The changes form a continuous process, *see diagram*.

petrification (*n*) the changing of plant and animal bodies, and sediment into rock, brought about by pressure and increased temperature. **petrify** (*v*).

stalactite (*n*) a finger-like shape of crystalline calcium carbonate hanging down from the roof of a cave in limestone rock. Water, containing dissolved calcium hydrogen carbonate, passes through the limestone and drops evaporate on the stalactite, increasing its length with a deposit (p.102) of calcium carbonate.

stalagmite (*n*) a finger-like shape of crystalline calcium carbonate pointing upwards from the floor of a cave in limestone rock. Drops from a stalactite (↑) fall on the stalagmite and evaporate, increasing its length. A stalagmite is below a stalactite and they both grow, eventually to meet.

岩石循環 岩石類型變成其他類型。這些變化構成一個連續過程，（見圖）。

石化（名） 動、植物的遺體和沉積物因壓力和高溫而成岩石之變化。（動詞爲 petrify）

鐘乳石（名） 從石灰岩洞穴頂上垂掛下來的一種手指狀結晶碳酸鈣。溶有碳酸氫鈣的水穿過石灰石滴在鐘乳石上並蒸發，隨着碳酸鈣沉積物（第 102 頁）的增加而不斷增長鐘乳石的長度。

石筍（名） 從石灰岩洞穴地面指向上方的一種像手指狀結晶碳酸鈣。水滴從鐘乳石（↑）滴在石筍上並蒸發而不斷增加其長度。石筍在鐘乳石下面，二者均不斷增長直至會合。

structure of Earth
地球的結構

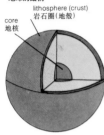

core
地核

lithosphere (crust)
岩石圈(地殼)

lithosphere (*n*) the outer, solid part of the Earth, which covers the hotter, inner parts. It is about 70 km thick. Metamorphic rocks (p.121) are the most common in the lithosphere, while sedimentary rocks are most common on its surface. The lithosphere is also known as the Earth's crust.

hydrosphere (*n*) the water portion of the Earth's crust as distinguished from the lithosphere, together with all the water vapour (p.41) in the atmosphere (p.51) around the Earth. The hydrosphere and the lithosphere together form the Earth's surface.

continent (*n*) a large, unbroken area of land. The lithosphere (↑) consists of the six great continental areas. **continental** (*adj*).

ocean (*n*) a large area of salt water, part of the hydrosphere (↑). The hydrosphere on the Earth's surface consists of four oceans. **oceanic** (*adj*).

sima (*n*) the lower layer of the lithosphere; it covers the whole of the Earth. The thickness of sima is not easily measured, *see diagram*. It has a density of 3000 kg/m³ and is mainly made up of basalt (p.120). **simatic** (*adj*).

sial (*n*) the part of the lithosphere which stands on sima (↑). Sial is the base on which the land surfaces of the Earth stand; it is not continuous, as different parts of the sial are separated by the oceans. The thickness varies from very thick below high mountains to thin below plains. Sial has a density of about 2700 kg/m³, and being less dense than sima, it floats on the sima. *See diagram*. **sialic** (*adj*).

岩石圈(名)　地球外層的固體部分，它包蓋着地球內部熾熱的部分，大約有 70 公里厚。變質岩(第 121 頁)在岩石圈中最常見，而沉積岩在地球表面最常見。岩石圈也叫做地殼。

水圈；水界(名)　地殼中有別於岩石圈的水域部分以及地球周圍大氣(第 51 頁)中的所有水汽(第 41 頁)合稱水圈。水圈和岩石圈合在一起形成地球的表層。

大陸(名)　一塊巨大的、完整的陸地區域。岩石圈(↑)由六塊巨大的大陸區域組成。(形容詞爲 continental)

海洋(名)　一片巨大的鹽水區域，爲水圈(↑)的一部分。地球表面的水圈由四大洋組成(形容詞爲 oceanic)

矽鎂層；矽鎂帶(名)　岩石圈的較低層；它覆蓋着整個地球。矽鎂層的厚度不易測量(見圖)。其密度爲 3000 kg/m³，主要由玄武岩(第 120 頁)構成。(形容詞爲 simatic)

矽鋁層；矽鋁帶(名)　矽鎂層(↑)之上的岩石圈部分。矽鋁層是地球陸地表面的基礎，它並非綿延不斷的，大洋將其各個不同部分隔開，其厚度從高山下很厚的部分到平原下薄的部分各不相同。矽鋁層的密度約爲 2700 kg/m³，比矽鎂層密度小，它浮在矽鎂層之上。(見圖)(形容詞爲 sialic)

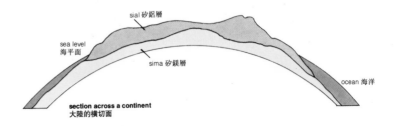

sial 矽鋁層

sea level
海平面

sima 矽鎂層

ocean 海洋

section across a continent
大陸的橫切面

magma (*n*) a hot, sticky liquid state of rock made by conditions of high temperature and great pressure. Rocks from the lithosphere are forced down below the lithosphere (p. 123) by earthquakes (p. 126) and other earth movements and are changed into magma by the conditions there. Magma has a temperature of about 1000°C, and contains floating crystals and pockets of gas. Volcanic action and plutonic (p. 127) movements push magma up into the lithosphere. Magma cools to form igneous rocks (p.121).

core² (*n*) the innermost part of the Earth with a radius of about 3500 km; it has a density of between 8000 and 10 000 kg/m³ and is much denser than the lithosphere. The core is a mixture of iron and nickel with an inner, solid core, and an outer, liquid core.

crust (*n*) (1) a hard, outer cover over an inner softer core. (2) the lithosphere (p.123) of the Earth.

stratum (*n*) (*strata*) (1) one of a number of layers, each on top of the other. (2) an approximately level, or formerly level, layer of sedimentary rock. **stratification** (*n*).

fault (*n*) a break in the stratification of rocks; large earth movements cause the break, *see diagram*, and the strata (↑) are displaced, so that they no longer form a continuous layer.

seam (*n*) a stratum (↑) of an ore, or other useful mineral, e.g. a seam of coal.

岩漿（名） 在高溫和巨大壓力條件下產生的熾熱黏液態岩石。地震（第 126 頁）和地殼其他運動的力量將岩石圈的岩石推到岩石圈（第 123 頁）下面，該處的條件使岩石變成岩漿。岩漿的溫度約爲 1000°C，其中含有漂浮的晶體和氣袋。火山作用和深成岩（第 127 頁）運動將岩漿向上推入岩石圈中。岩漿冷卻後形成火成岩（第 121 頁）。

地核（名） 地球最裏面的部分，半徑約爲 3 500 公里，其密度介於 8000 至 10 000 kg/m³ 之間，比岩石圈密度大得多。地核是鐵鎳混合物，內部爲固體的核，外部爲液態的核。

地殼（名） （1）地球內層較軟的地核外面包覆的堅硬殼；（2）地球的岩石圈（第 123 頁）。

地層（名） （1）層層相疊的許多岩層之一；（2）近於平坦的或曾經是平坦的沉積岩層。（名詞爲 stratification）

斷層（名） 岩石成層過程中發生的斷裂；地殼劇烈運動引起岩層斷裂（見圖），岩層（↑）發生位移，結果不再成爲綿延不斷的岩層。

礦層（名） 礦石或其他有用礦物的岩層（↑）。例如煤層。

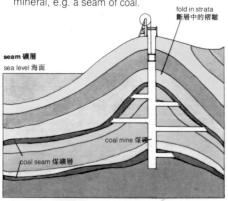

fold in strata 斷層中的褶皺

seam 礦層
sea level 海面
coal mine 煤礦
coal seam 煤礦層

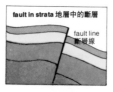

fault in strata 地層中的斷層
fault line 斷層線

volcano (*n*) a cone-shaped mountain formed when steam, lava (↓), gases, and rocks are forced out from inside the Earth by the pressure of gases and steam, usually with explosive force. **volcanic** (*adj*).

火山(名)　由於地球內部氣體和蒸汽的壓力，迫使蒸汽、熔岩(↓)、氣體和岩石從地球內部噴出所形成的錐形山，通常具有爆發力。(形容詞爲 volcanic)

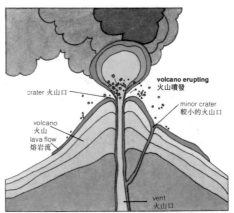

crater 火山口

volcano erupting
火山噴發

minor crater
較小的火山口

volcano
火山

lava flow
熔岩流

vent
火山口

crater (*n*) the hollow at the top of a volcano (↑). Minor craters may be formed on the sides of a volcano.

vent (*n*) (1) an opening or pipe to allow fluids to pass out. (2) the tube-like hollow in the middle of a volcano (↑), up which pass the steam, gases, lava and solid particles. When the volcano is no longer active, the vent is blocked by a *plug* of basalt (p.120).

lava (*n*) liquid magma (↑), usually white-hot, forced out of the vent (↑) of a volcano; and the solid magma formed when it cools.

pumice (*n*) a sponge-like (p.149), grey, glass-like solid formed as a crust on the top of lava when it cools. Many bubbles of gas in the lava form the sponge-like structure of pumice.

eruption (*n*) the explosive action of a volcano when gases, steam, and lava are thrown out of the vent (↑). **erupt** (*v*).

volcanic dust a black cloud of very small pieces of solid lava blown out during an eruption (↑).

火山口(名)　火山(↑)頂上的洞穴。在火山坡可能形成較小的火山口。

通氣口；噴發道(名)　(1)流體可外流的開口或管道；(2)火山(↑)中間的管狀洞穴，蒸汽、氣體、熔岩和固體粒可通過噴發道向上噴出。火山不活動時，噴發道被玄武岩(第 120 頁)堵塞。

熔岩(名)　液體岩漿(↑)，通常是白熱的，由噴發道(↑)噴出；熔岩冷卻後成爲固體岩漿。

浮石(名)　一種海綿狀的(第 149 頁)、灰色、玻璃狀固體，它是在冷卻的熔岩上形成的硬殼。熔岩中的許多氣泡形成浮石的海綿狀結構。

噴發(名)　當氣體、蒸汽和熔岩被噴出噴發道(↑)時產生的爆炸性活動。(動詞爲 erupt)

火山灰　在火山噴發(↑)過程中，由噴出的固體熔岩細小碎塊所形成的黑雲。

earthquake (*n*) the shaking of the Earth's crust caused mostly by displacement along a fault (p. 124), or by volcanic action. Earthquakes only occur where such movements take place. They produce shock waves (↓) felt as vibrations of the Earth's surface and may destroy buildings. The place of maximum displacement is the focus of the earthquake. Of the many thousand earthquakes each year, only a few are destructive or even noticed by man.

地震(名)　主要是沿斷層(第124頁)位移或火山作用所致的地殼搖動。地震只出現於這些運動發生之處。地震產生震波(↓)，令人感到地面振動，並可能摧毀建築物。位移最大之處即爲震源。每年發生的數千次地震中，僅少數地震有破壞性或是人們可以感覺得到的。

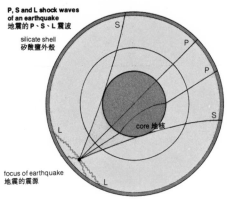

P, S and L shock waves of an earthquake
地震的 P、S、L 震波

silicate shell
矽酸鹽外殼

S

P

P

S

core 地核

L

focus of earthquake
地震的震源

L

shock wave a wave motion (p.65) of compression or deformation (p.27) transmitted by rock structures in the Earth. The greater the rigidity (p.28) of a rock, the faster the waves travel through it. There are three kinds of shock waves, called P, S, and L waves. P waves are the fastest, travelling at approximately 650 km/hour; they pass through the Earth's core and are refracted by the different structural layers of the Earth, *see diagram*. S waves, slightly slower, follow the P waves, but do not pass through the Earth's core. L waves travel more slowly, round the Earth's surface, and have a much greater amplitude than the other two waves. The focus of an earthquake, from which shock waves start, is usually only a few kilometres below the Earth's surface. The point directly above the focus is the epicentre.

震波　地球內部岩石結構所傳遞的擠壓或變形(第27頁)波動(第65頁)。岩石的剛硬度(第28頁)越大，則震波通過它傳遞得越快。震波有三種，即：P波、S波和L波。P波是最快的波，傳遞速度約650公里/小時；P波通過地核，被地球的不同的結構層所折射(見圖)。S波稍慢一些，跟在P波之後但不通過地核。L波更慢，它繞地球表面傳遞，其振幅比另兩種波大得多。震波起始的震源通常僅離地下幾公里處。正對震源上方的點即爲震中。

tremor (*n*) a vibration of the Earth's surface, caused by shock waves; a minor earthquake. Tremors are caused by P and S shock waves (↑) and are also called preliminary tremors. **tremble** (*v*).

secondary tremor a vibration of the Earth's surface caused by shock waves of large amplitude (L waves). These tremors cause almost all the damage from earthquakes.

seismograph (*n*) an instrument for recording tremors (↑); it produces a **seismogram** showing the amplitude of the tremors, *see diagram*.

地動；小震（名） 震波引起的地球表面振動；係一種較小的地震。地動係由 P 震波和 S 波（↑）引致，亦稱初期微震。（動詞爲 tremble）

次地動 由大振幅震波（L 波）引起的地球表面的振動。地震造成的一切破壞幾乎都可歸咎於這類地動。

地震儀（名） 一種記錄地動（↑）的儀器；它產生一種表明地動振幅的**地震波曲線圖**（見圖）。

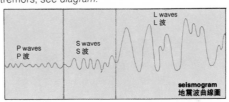

seismogram
地震波曲線圖

tectonic (*adj*) formed in the Earth's crust, e.g. earthquakes are tectonic movements.

plutonic (*adj*) formed, or coming from, deep inside the Earth, below the Earth's crust, e.g. movements of magma are plutonic.

tide (*n*) the regular rise and fall of water in the sea or ocean, taking place twice in a lunar (p.62) day; it is caused by the attraction of the moon for the water, *see diagram*. A *high tide* is produced at a place opposite the moon, and also at a place on the far side of the Earth. A *low tide* is produced between the places of high tides on the Earth's surface. **tidal** (*adj*).

(地殼)構造(上)的（形） 在地殼中形成的。例如地震就是構造運動。

深成的（形） 形成或來源於地殼下，地球內部深處的。例如岩漿運動就是深成的。

潮汐（名） 海或海洋中水的定期漲落。每一個太陰（第 62 頁）日發生兩次；月球對水的引力引起潮夕（見圖）。高潮發生於正對月球之處，也產生於遠離地球之處。低潮產生於地球表面上兩處高潮之間。（形容詞為 tidal）

tides 潮汐

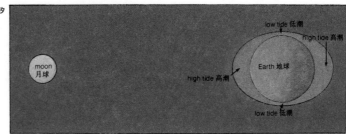

chalk (*n*) a soft, white or grey variety of limestone (↓) formed from the shells of very small sea animals deposited on the sea bed. Chalk is mainly used industrially (↓) to produce lime (calcium oxide), the cheapest available alkali. **chalky** (*adj*).

limestone (*n*) a medium hard, grey, sedimentary (p. 121) rock, of calcium carbonate, formed from sediment on shallow sea beds. Limestone is used for buildings, for making cement and glass and also for making lime.

coal (*n*) a soft, black, sedimentary (p.121) rock formed by the petrification (p.122) of plant material. The plants grew in the Carboniferous period, over 200 million years ago. Coal consists of carbon and other compounds; it burns slowly, producing heat. Its chemical products are obtained by destructive distillation (p.135).

petroleum (*n*) a liquid mineral containing many different hydrocarbons (p.131); formed in pockets in folds of rock strata (p.124); usually found in layers of porous rock. About 260 litres occur in a cubic metre of rock. It is thought to have been formed from decaying plants and animals, in a way similar to coal. The composition varies from place to place. Petroleum is usually under pressure from natural gas (↓) and when a pipe is forced through the earth into a pocket of petroleum, the pressure forces the petroleum up the pipe. The chemical products of petroleum are obtained by fractional distillation (p.134). They include petrol, kerosene, lubricating oils and asphalt.

白堊（名） 一種鬆軟白色或灰色的石灰石（↓），由微小海生動物的甲殼沉積在海底所形成。白堊主要用於工業上（↓）生產石灰（氧化鈣），提供最廉宜的鹼。（形容詞爲 chalky）

石灰石（名） 一種硬度適中的灰色沉積（第 121 頁）岩石，成分爲碳酸鈣，係由淺海底的沉積物所形成。石灰石用於建築業，製造水泥和玻璃，也用於生產石灰。

煤（名） 一種鬆軟、黑色的沉積（第 121 頁）岩石，由植物石化（第 122 頁）形成。這些植物生長於兩億多年前的石炭紀。煤含有碳和其他化合物；它燃燒緩慢，產生熱量。通過乾餾（第 135 頁）方法，可從煤製得其化學產品。

石油（名） 一種含有多種不同烴（第 131 頁）的液態礦物；形成於岩層（第 124 頁）褶層的礦穴裏；通常存在於多孔的岩層中。一立方米的岩石中約含 260 升石油。一般認爲石油是由腐爛的動、植物形成的，形成過程與煤相似。石油的成分因地而異，通常處於天然氣（↓）的壓力下。當一條管道通過地面深入到石油礦穴時，天然氣的壓力將石油通過管道壓上地面。經分餾（第 134 頁）可製得石油的化學產品，包括汽油、煤油、潤滑油和瀝青。

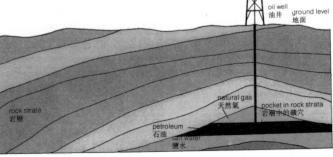

obtaining petroleum
石油的取得

oil well 油井
ground level 地面

rock strata 岩層

natural gas 天然氣

pocket in rock strata 岩層中的礦穴

petroleum 石油

salt water 鹽水

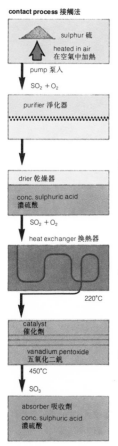

contact process 接觸法

sulphur 硫

heated in air
在空氣中加熱

pump 泵入

$SO_2 + O_2$

purifier 淨化器

drier 乾燥器

conc. sulphuric acid
濃硫酸

$SO_2 + O_2$

heat exchanger 換熱器

220°C

catalyst
催化劑

vanadium pentoxide
五氧化二釩

450°C

SO_3

absorber 吸收劑
conc. sulphuric acid
濃硫酸

natural gas gas formed in pockets above petroleum (↑); also gas sent out through volcanoes. In some places it consists mainly of methane (formula CH_4), in other places it consists of mixed hydrocarbons (p.131). It is used as a fuel (↓).

fuel (n) a material which can be burned to give out heat or to provide chemical energy. Most fuels are carbon compounds, e.g. wood, coal, petroleum.

raw materials materials obtained from natural sources (p.52) for use in making chemical substances, or other materials needed in industry (↓), e.g. (a) ores of iron are the raw material for making iron; (b) sea water is a raw material for making common salt; (c) sugar-cane is a raw material for making sugar.

industry (n) the making of materials, substances, articles in large amounts for selling to people, e.g. the clothing industry, the chemical industry, the steel industry. **industrial** (adj).

process[1] (n) (1) a set of events, following one after the other, all concerned with the same activity, e.g. the process of lighting a fire. (2) in chemistry, an industrial (↑) method of making a desired material or substance generally involving many operations (↓), e.g. (a) the Bessemer process for making steel; (b) the contact process for making sulphuric acid.

operate (v) to make a process (↑) or a machine work, e.g. (a) to operate the contact process so that sulphuric acid is produced; (b) to operate a hydraulic press. Simple machines, e.g. a lever, and simple processes, e.g. lighting a fire, are used, not operated. **operation** (n), **operator** (n), **operative** (adj).

by-product (n) a product other than the product for which a process (↑) is intended, e.g. a process intended to make sodium hydroxide also produces chlorine; the sodium hydroxide is the main product, while chlorine is the by-product.

waste product a substance made in a process (↑) but which has no industrial use.

天然氣　礦穴中石油(↑)上方形成的氣體；它也是通過火山發出的氣體。有些地方的天然氣主要是由甲烷(化學式爲 CH_4)組成，在另一些地方則由混合的碳氫化合物(第131頁)組成。天然氣可用作燃料(↓)。

燃料(名)　一種能夠燃燒發出熱量或提供化學能的物質。大多數燃料是碳的化合物。例如木材、煤、石油。

原料　從各種天然源(第52頁)取得的材料，用於製造化學物質，或工業(↓)上所需要的其他材料。例如：(a)鐵礦石是煉鐵的原料；(b)海水是提取食鹽的原料；(c)甘蔗是製糖的原料。

工業(名)　大量製造材料、物質、商品供應市場。例如服裝工業、化學工業、鋼鐵工業。(形容詞爲 industrial)

過程；方法(名)　(1)一個接一個連貫進行的一系列事件，所有這些事件都與同一活動有關。例如點火的過程；(2)化學上指製造所需材料或物質的工業(↑)方法，通常包括許多操作(↓)。例如(a)柏塞麥煉鋼法；(b)製備硫酸的接觸法。

操作；施行(動)　過程(↑)或機械運轉。例如：(a)運用接觸法生產硫酸；(b)操作水壓機。簡單機械如槓桿和簡單過程如點火，不說操作而只說使用。(名詞爲 operation，operator，形容詞爲 operative)

副產物(名)　生產過程(↑)所預定產品以外的產物。例如：預定生產氫氧化鈉的過程，同時也生產出氯氣；其中氫氧化鈉是主要產品，而氯氣是副產品。

廢產物　生產過程(↑)所產生的無工業用途的物質。

smelt (*v*) to heat the ore of a metal with a suitable substance, usually coke, in order to obtain the metal from its ore, e.g. to smelt iron ore in a blast furnace (↓) to obtain iron. **smelting** (*n*).

furnace (*n*) a device to obtain strong heating effects using a fuel or electric current, e.g. (a) a furnace to heat a boiler to produce steam; (b) a furnace to smelt metals.

blast furnace a furnace which uses a forced current of air (a blast) to smelt ores of iron.

熔煉（動）　用適當的物質（通常是焦碳）加熱金屬礦石以從礦石中取得金屬。例如，在高爐（↓）中熔煉鐵礦石，製得鐵。（名詞爲 smelting）

爐；熔爐（名）　用燃料或電流獲得強熱效果的設備。例如：(a)將鍋爐加熱以產生蒸汽的爐子；(b)熔煉金屬的熔爐。

鼓風爐；高爐　使用強力氣流（鼓風）熔煉鐵礦石的熔爐。

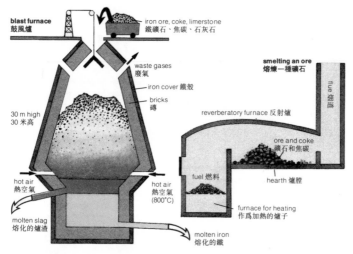

reverberatory furnace a furnace in which a material is heated by flames passing above it, and being directed down on to the material by the roof of the furnace, *see diagram*. It is used for smelting ores of metals.

slag (*n*) waste material, mainly non-metallic, formed during the smelting of ores, and floating on the molten metal. It contains silicates and other mineral material.

refine (*v*) to make as pure as possible, e.g. (a) sugar is refined to remove unwanted plant material; (b) metals are refined to remove impurities. **refinement** (*n*), **refinery** (*n*).

反射爐　火焰在材料上方通過並由爐頂返回到材料上以加熱材料的熔爐（見圖）。用於熔化金屬礦石。

爐渣；熔渣（名）　熔煉礦石過程中形成的廢料，它飄浮在熔化的金屬上，主要是非金屬的廢料。含有矽酸鹽和其他礦物質。

精煉；精製（動）　使盡可能純。例如：(a)精製糖以除去無用的植物質；(b)精煉金屬以除去其中含的雜質。（名詞爲 refinement，refinery）

carbon (*n*) an element, atomic number 6, relative atomic mass 12.01; it is a non-metal with three allotropic (p.103) forms. It has a covalency of 4 and the ability to combine its atoms in long chains and in rings; this allows a great variety of compounds to be formed, many of which are found in plants and animals.

organic (*adj*) describes materials, substances and compounds which contain carbon, other than the oxides of carbon and the carbonates. All organisms (p.147) consist of organic compounds.

hydrocarbon (*n*) a compound consisting of carbon and hydrogen only. The simplest hydrocarbon is methane, formula CH_4 (p. 109). Hydrocarbons can consist of long chains of carbon atoms, with hydrogen combined by covalent bonds (p.109), or they can consist of rings of carbon atoms with hydrogen combined, e.g. benzene (↓).

碳（名） 原子序數爲 6，相對原子質量爲 12.01 的一種元素，具有三個同素異形（第 103 頁）體的非金屬元素。其共價爲 4，能將原子結合成長鏈和環狀；這使之可以形成各種不同的化合物，其中許多存在於植物和動物體內。

有機的（形） 描述含碳的材料、物質和化合物，但不包括碳的氧化物和碳酸鹽。一切有機體（第 147 頁）都是有機化合物組成的。

碳氫化合物；烴（名） 只含碳和氫的化合物。最簡單的碳氫化合物是甲烷，化學式爲 CH_4（第 109 頁）。碳氫化合物可由碳原子長鏈上的碳原子和氫原子通過共價鍵（第 109 頁）結合而成，也能由碳原子環鏈和氫原子結合而成。例如苯（↓）。

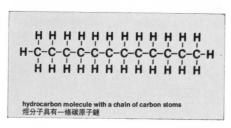

hydrocarbon molecule with a chain of carbon stoms
烴分子具有一條碳原子鏈

benzene ring the structure of a molecule of benzene in which six carbon atoms are joined by covalent bonds in the shape of a ring. The benzene ring is also seen in other compounds, e.g. in naphthalene and anthracene.

苯環 苯的分子結構中，六個碳原子由共價鍵連接成環狀。其他化合物中也有苯環。例如萘和蒽。

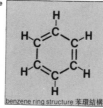

molecule of benzene
苯的分子

benzene ring structure 苯環結構

symbol for a benzene ring
苯環的符號

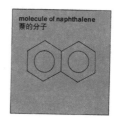

molecule of naphthalene
萘的分子

isomer (n) one of two compounds with the same number of atoms of each element contained in the molecule, but with different physical or chemical properties, e.g. ethanol and dimethyl ether both possess a molecular formula of C_2H_6O, but their structure differs, *see diagram*, and their physical and chemical properties are different. **isomeric** (*adj*), **isomerism** (*n*).

同分異構物（名） 兩種化合物分子中所含每種元素的原子數目相同，但其物理或化學性質不同，這兩種化合物爲同分異構物。例如：乙醇和二甲醚的分子式都是 C_2H_6O，但它們的結構卻不同（見圖），物理性質和化學性質也各異。（形容詞爲 isomeric，名詞爲 isomerism

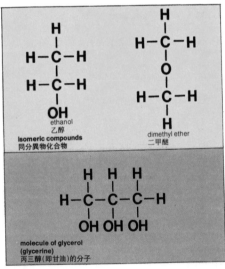

ethanol
乙醇
isomeric compounds
同分異物化合物

dimethyl ether
二甲醚

molecule of glycerol
(glycerine)
丙三醇（即甘油）的分子

alcohol (n) a substance with the structure of a hydrocarbon but with one, or more, hydrogen atoms replaced by hydroxyl groups (p.114), e.g. ethanol is an alcohol, glycerol is an alcohol, *see diagram*. **alcoholic** (*adj*).

ester (n) a compound formed from an alcohol and an acid, usually an organic acid. The chemical reaction is: alcohol + acid → ester + water, e.g. ethanol and ethanoic (acetic) acid react to form ethyl ethanoate. **esterification** (*n*).

soap (n) a salt of a metal and a fatty acid (p.175) e.g. stearic, oleic, and palmitic acids. Soap is made by the action of alkalis on fats (p.175).

醇；酒精（名） 具有烴的結構，但其中一個或多個氫原子爲羥基（第114頁）所取代的物質。例如：乙醇是醇，丙三醇也是醇（見圖）。（形容詞爲 alcoholic）

酯（名） 由醇和酸（通常爲有機酸）反應形成的化合物。其化學反應是：醇＋酸→酯＋水。例如：乙醇和乙酸（即醋酸）反應生成乙酸乙脂。（名詞爲 esterification）

肥皂（名） 金屬和脂肪酸（第175頁），例如硬脂酸、油酸和棕櫚酸所形成的鹽。肥皂是由鹼作用於脂肪（第175頁）而製成的。

polymer (*n*) a molecule with a high relative molecular mass (p. 104) formed by the chemical linking of many simpler molecules, each of the same substance, called monomers, e.g. ethene (ethylene) is a simple molecule, formula $CH_2{:}CH_2$; many molecules of ethene combine together to form polyethylene, formula: . . . $CH_2{\cdot}CH_2{\cdot}CH_2{\cdot}CH_2$. . .; polyethylene is a polymer of ethene known as polythene. **polymerize** (*v*), **polymeric** (*adj*).

聚合物（名）　由許多較簡單分子靠化學聯接形成的一種分子，此分子具有很高的相對分子質量（第 104 頁），每一個簡單分子都是同一種物質，稱爲單體。例如乙烯是一種簡單的分子，結構式爲 $CH_2{:}CH_2$；許多乙烯分子結合一起就形成聚乙烯，聚乙烯分子式：$\cdots CH_2{\cdot}CH_2{\cdot}CH_2{\cdot}CH_2\cdots$；聚乙烯是乙烯的聚合物。（動詞爲 polymerize，形容詞爲 polymeric）

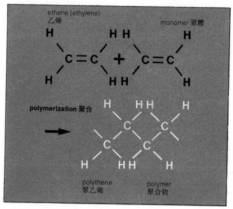

polymerization (*n*) the forming of a polymer from a simpler compound called a monomer. **polymerized** (*adj*).

聚合作用（名）　由稱爲簡單化合物的單體形成聚合物的作用。（形容詞爲 polymerized）

thermoplastic (*n*) a polymerized (↑) substance which becomes plastic (p.28) on heating and can be shaped by pressure and heat, without changing its chemical properties; the process can be repeated on further heating, e.g. polythene is a thermoplastic.

熱塑塑膠（名）　一種聚合的（↑）物質，在加熱時有可塑性（第 28 頁），並能藉壓力和加熱成形，而不改變其化學性質；這一過程可以再次加熱重複進行。例如，聚乙烯就是一種熱塑塑膠。

thermosetting (*adj*) describes a polymerized (↑) substance which first becomes plastic (p.28) on heating and then becomes hard because of a chemical change. A different substance is formed which cannot be made plastic by reheating. Thermoplastic and thermosetting substances are called **plastics**. They are shaped by heat and pressure, and keep their shape when cooled.

熱固性的（形）　描述一種聚合的（↑）物質，它在加熱時首先變成塑性（第 28 頁），然後由於化學變化而變硬。所形成的不同物質不能通過再加熱製成塑膠。熱塑性的和熱固性的物質都稱爲塑膠。它們可藉加熱和加壓成形，並在冷卻後繼續保持其形狀。

oil (*n*) any one of a number of greasy, combustible substances obtained from plants, animals or minerals. Oils and fats (p.175) from living organisms are esters (p.132) of glycerol and organic acids; Oils are liquid, and fats are solid, at room temperature. Mineral oil consists of hydrocarbons and is obtained from petroleum (p. 128). **oily** (*adj*).

油(名) 從植物、動物或礦物中取得的多種油脂性可燃物質之一。從活有機體取得的油和脂肪(第 175 頁)都是丙三醇和有機酸形成的酯(第 132 頁)。在室溫下油是液體,脂肪是固體。礦物油碳氫化合物,是從石油(第 128 頁)提取的。(形容詞爲 oily)

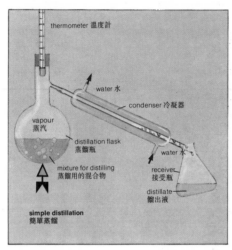

thermometer 溫度計
water 水
condenser 冷凝器
vapour 蒸汽
distillation flask 蒸餾瓶
mixture for distilling 蒸餾用的混合物
water 水
receiver 接受瓶
distillate 餾出液
simple distillation 簡單蒸餾

distillation (*n*) a physical process for separating (p.91) volatile (p.98) liquids from mixtures. The mixture is heated, the liquid is vaporized (p. 41); the vapour is then condensed (p.42) and collected. Liquids with different boiling points can be separated by the process. **distil** (*v*), **distillate** (*n*).

fractional distillation a distillation (↑) process for separating liquids with boiling points close together. A **fractionating column** is used, and vapour is taken out at different levels of the vertical column; the vapour is condensed and the different liquids are *fractionated* by the process. Fractionating is used to separate (p. 91) the different substances in petroleum (p.128).

fractionate (*v*) to split into fractions.

蒸餾作用(名) 將揮發性(第 98 頁)液體從混合物中分離(第 91 頁)出來的物理方法。這種混合物被加熱後,液體蒸發(第 41 頁);隨後蒸氣冷凝(第 42 頁)並收集之。可分離出具有不同沸點的液體。(動詞爲 distil,名詞爲 distilate)

分餾作用 用於分離沸點相近的液體的蒸餾(↑)方法。使用**分餾柱**,在立柱的不同高度排出蒸氣;蒸氣冷凝後就分餾出不同的液體,分餾用來分離(第 91 頁)出石油(第 128 頁)中的不同物質。

分餾(動) 分成幾個分餾成分。

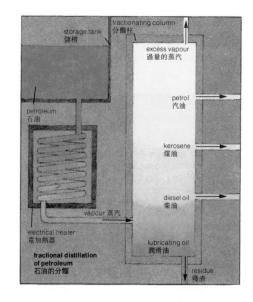

fractional distillation
of petroleum
石油的分餾

refluxing a mixture of organic liquids
有機液體混合液的回流加熱

vapour condenses and falls back into flask
蒸汽冷凝后落回燒瓶中

water
水

water
水

condenser
冷凝器

vapour
蒸汽

flask
燒瓶

destructive distillation the distillation (↑) of volatile (p.98) substances from a solid, which is decomposed by the process, e.g. coal is destructively distilled leaving a solid residue, **coke**, while **coal-tar** and **pitch** are collected as a distillate, and coal-gas is collected in a gasometer.

still (*n*) an apparatus for distillation, especially for distilling alcohol or petroleum; also the vessel in which a mixture is heated for distillation.

reflux (*v*) to boil a liquid with a condenser above the liquid so that the vapour is condensed and falls back into the boiling liquid. Refluxing is a process for making chemical reactions of volatile (p.98) organic compounds take place. **reflux** (*adj*).

cracking (*n*) a kind of pyrolysis (p.99) used with mineral oils (↑) of high boiling point. The oils are passed through a red hot tube, decomposition takes place and mineral oils of lower boiling point, e.g. petrol, are produced.

分解蒸餾 蒸餾(↑)出固體的揮發性(第98頁)物質，此過程使固體分解。例如，煤乾餾後留下固體殘渣**焦炭**，**煤焦油**和**瀝青**則作爲餾出物收集起來，煤氣也收集在煤氣罐裏。

蒸餾器(名) 一種蒸餾器具，特別用於蒸餾酒精或石油；也是一種可在其中加熱混合物進行蒸餾的器皿。

回流；回流加熱(動) 煮沸液體並在該液體上方置一個冷凝器，使蒸氣冷凝並落回煮沸的液體中。回流是一種使揮發性(第98頁)有機化合物發生化學反應的過程。(形容詞爲reflux)

裂化(名) 用於熱解(第99頁)高沸點礦物油(↑)的方法。油通過赤熱的導管被分解，產生低沸點的礦物油，例如汽油。

fermentation (*n*) a chemical change caused by yeasts (p.146) and bacteria (p.145), in which a carbohydrate (p.173), usually a sugar, is changed to ethanol (ethyl alcohol) and carbon dioxide. Enzymes (p.167) catalyse the change. **ferment** (*v*).

saponification (*n*) the hydrolysis (↓) of an ester (p.132) using an alkali. An alcohol and a salt of the organic acid, or acids present, are formed. **saponify** (*v*).

hydrolysis (*n*) the chemical decomposition (p.95) of a substance by water, e.g. (a) some salts, such as iron (III) chloride, are partly hydrolyzed by water to form an acidic, and not a neutral, solution; (b) esters are hydrolyzed by water into an alcohol and an acid; the chemical reaction is slow, so an acid or an alkali is added to make the reaction go faster. **hydrolyze** (*v*), **hydrolytic** (*adj*).

substitution (*n*) the replacing (p.97) of an atom in a molecule of an organic compound by an atom of a different element or by a group of atoms, e.g. to replace a hydrogen atom in methane (CH_4) by a chlorine atom, forming chloromethane (CH_3Cl). **substitute** (*v*).

vulcanization (*n*) the process of making natural rubber harder and less elastic, so that it keeps its shape when in use. Rubber is heated with sulphur to form vulcanized rubber, as used in tyres. **vulcanize** (*v*).

synthesis (*n*) the combining (p.94) of elements to form a simple compound, or of simple compounds to form a compound with many atoms in its molecules, e.g. the synthesis of explosives (p.112) such as *dynamite*, which is made from glycerol and nitric acid. **synthesize** (*v*).

synthetic (*adj*) made by chemical processes from simpler compounds, not obtained from natural products, e.g. synthetic threads, such as nylon (↓) and artificial silk, used in making clothes.

nylon (*n*) a polymer (p.133) with a long chain of carbon atoms to which *amide groups* ($-CONH_2$) are combined at intervals. There are many different polymers, with different physical properties, all called nylon. A common kind is the nylon used for thread.

發酵（名） 由酵母（第 146 頁）和細菌（第 145 頁）引致的一種化學變化，其中碳水化合物（第 173 頁）（通常是糖）分解成乙醇和二氧化碳。酶（第 167 頁）對此變化起催化作用。（動詞為 ferment）

皂化作用（名） 用鹼使酯（第 132 頁）起水解作用（↓）。結果形成乙醇和有機酸鹽或其中存在的其他酸的鹽。（動詞為 saponify）

水解作用（名） 用水使物質發生化學分解（第 95 頁）。例如：（a）某些鹽如氯化鐵（III）被部分水解形成酸性溶液，而不是中性溶液；（b）酯被水解成為乙醇和酸；這種化學反應很慢，所以要加酸或鹼以加速反應。（動詞為 hydrolyze，形容詞為 hydrolytic）

取代作用（名） 有機化合物分子中的一個原子被不同元素的一個原子或原子團所置換（第 97 頁），例如：甲烷（CH_4）中的一個氫原子被一個氯原子所置換，形成氯代甲烷（CH_3Cl）。（動詞為 substitute）

硫化（名） 使天然橡膠更硬及彈性較低的方法。硫化後的橡膠在使用時能保持其形狀。橡膠和硫磺一起加熱形成硫化橡膠，如在輪胎所用的硫化橡膠。（動詞為 vulcanize）

合成（名） 許多元素結合（第 94 頁）成簡單的化合物，或許多簡單化合物形成其分子中含有許多原子的化合物。例如：炸藥（第 112 頁）的合成。達納炸藥是由甘油和硝酸製成的。（動詞為 synthesize）

合成的（形） 以各種化學方法由較簡單的化合物製成產物，而不是得自天然產物。例如，用於製衣的合成纖維如尼龍（↓）和人造絲。

尼龍（名） 一種具有碳原子長鏈的聚合物（第 133 頁），長鏈上間隔結合有醯胺基（$-CONH_2$）。有多種不同的聚合物，具有不同的性質，總稱為尼龍。普通的一種是用來做線的尼龍。

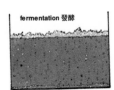
fermentation 發酵

Biology 生 物 學

cell (*n*) the smallest part of a plant or animal; some organisms (p.147) consist of one cell only, e.g. amoebae and bacteria, while others, e.g. trees, humans, contain many millions of cells. A cell consists of protoplasm (↓) with a membrane (↓) around it. It has the ability to take in chemical substances and use them to make various substances it needs for the process of living. Many cells have the ability to divide and form two new cells. **cellular** (*adj*).

protoplasm (*n*) a grey, jelly-like material containing many organic (p.131) chemical compounds, and organelles (p.141). Protoplasm is divided into cytoplasm (↓) and a nucleus (↓). **protoplasmic** (*adj*).

membrane (*n*) (1) a thin skin-like piece of material, covering or supporting a part of a plant or animal. (2) a cell-membrane is a very thin membrane (about 10 nm thick) composed of fat (p.175) and protein (p.172); it covers a cell and allows some substances to pass through into the cell, but not others. If the cell-membrane is damaged, the cell is destroyed. **membranous** (*adj*).

cell wall a wall of cellulose (↓) round the cell of a plant. It is formed by the protoplasm and is tight against the cell-membrane, *see diagram*. The cell wall is rigid (p.28) and gives a plant its strength to stand upright.

cellulose (*n*) a long carbon-chain polymer (p.133) of glucose (p.174). Cellulose molecules bind together to form a strong structure.

cytoplasm (*n*) all the protoplasm (↑) in a cell, except the nucleus (↓). It contains many different organic (p.131) chemical compounds and organelles (p.141). It is always in a state of movement, with a continuous change of compounds taking place. It is divided into ectoplasm and endoplasm. **cytoplasmic** (*adj*).

ectoplasm (*n*) (1) in plants, another name for cell-membrane. (2) in animal cells, it is the outer layer of cytoplasm. It is lighter in colour, clearer, and with fewer granules (↓) than endoplasm.

endoplasm (*n*) the inner layer of cytoplasm (↑); it is more liquid than ectoplasm (↑) and contains many granules (↓).

細胞（名） 動物或植物體的最小部分。某些生物體（第 147 頁）僅由一個細胞組成，例如變形蟲和細菌，而另一些生物體如樹、人則包含數以百萬計的細胞。細胞由細胞質（↓）和外包着的細胞膜（↓）組成。細胞能吸收化學物質並利用化學物質製造生命過程中所需的各種物質。很多細胞都能分裂成兩個新細胞。（形容詞爲 cellular）

原生質（名） 一種含有多種有機（第 131 頁）化合物和細胞器（第 141 頁）的灰色膠狀物質。原生質可分爲細胞質（↓）和細胞核（↓）。（形容詞爲 protoplasmic）

膜（名） (1) 呈皮膚狀的薄片物質，包覆或支撐動、植物的組成部分；(2) 細胞膜是一層極薄的膜（約 10 nm 厚），由脂肪（第 175 頁）和蛋白質（第 172 頁）組成；它包着細胞，有些物質可穿過膜進入細胞，而另一些物質則不能穿過。如果細胞膜受損傷，則細胞也就毀壞。（形容詞爲 membranous）

細胞壁（名） 包覆植物細胞的纖維素（↓）壁，它由原生質組成，並緊貼着細胞膜（見圖）。細胞壁堅韌（第 28 頁），它使植物有力挺立。

纖維素（名） 葡萄糖（第 174 頁）長碳鏈聚合物（第 133 頁）。纖維素分子結合在一起形成堅固的結構。

細胞質（名） 細胞中除核（↓）以外的全部原生質（↑）。它含有各種不同的有機（第 131 頁）化合物和細胞器（第 141 頁）。它由於化合物不斷發生變化而經常處於運動狀態。細胞質分爲外質和內質。（形容詞爲 cytoplasmic）

外質（名） (1) 植物細胞膜的另一名稱；(2) 動物細胞中的外層細胞質。顏色比內質淡而透明，且含的顆粒（↓）較少。

內質（名） 內層細胞質（↑）；它比外質（↑）有更多液體，且含有許多顆粒（↓）。

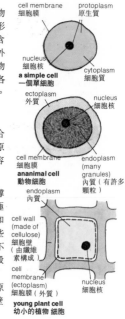

cell membrane 細胞膜　protoplasm 原生質
nucleus 細胞核
cytoplasm 細胞質
a simple cell 一個單細胞

ectoplasm 外質
nucleus 細胞核
cell membrane 細胞膜
an animal cell 動物細胞
endoplasm (many granules) 內質（有許多顆粒）
endoplasm 內質

cell wall (made of cellulose) 細胞壁（由纖維素構成）
cell membrane (ectoplasm) 細胞膜（外質）
nucleus 細胞核
young plant cell 幼小的植物細胞

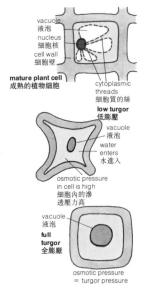

vacuole
液泡
nucleus
細胞核
cell wall
細胞壁

mature plant cell
成熟的植物細胞

cytoplasmic
threads
細胞質的絲

low turgor
低膨壓

vacuole
液泡

water
enters
水進入

osmotic pressure
in cell is high
細胞內的滲
透壓力高

vacuole
液泡

full
turgor
全膨壓

osmotic pressure
= turgor pressure
滲透壓力 = 緊漲壓力

cell
membrane
細胞膜

cytoplasm
細胞質

cell
wall
細胞壁

osmotic pressure 溶液的滲透壓力
of solution is 高於細胞的滲透壓力
greater than
osmotic pressure
of cell

plasmolysis 質壁分離

nucleus[2] (n) a small body, of dense material, covered with a membrane, in the cytoplasm of a cell. It controls all the activities of the cell, and without it, the cell dies. The nucleus contains nuclear sap (a liquid) and chromosomes (p.142). **nuclear** (adj).

granule (n) a small piece of material, usually hard, separate from other granules, and from other material. **granular** (adj).

vacuole (n) a space in cytoplasm (↑) enclosed by a membrane and filled with liquid. Many plant cells have a single vacuole which takes up most of the volume of the cell; it is filled with cell sap (a liquid) which has the same osmotic pressure (↓) as the cytoplasm around it. Also a minute space in any tissue. **vacuolar** (adj).

osmotic pressure the pressure arising from a solute (p.89) in a solution, which causes water to pass through a membrane (↑) to dilute the solution, provided the solute itself cannot pass through the membrane. The osmotic pressure is the pressure which has to be applied to the solution to stop the movement of water through the membrane. Water passes through a membrane from a solution with a lower osmotic pressure to one with a higher osmotic pressure, until both pressures are equal. The greater the concentration of a solution, the greater the osmotic pressure. (If a **solute** passes through a membrane then diffusion takes place.)

turgor (n) the state of a plant cell when fully expanded because of water absorbed (p.163) by its cytoplasm and vacuole; this produces a turgor pressure which keeps the cell wall (↑) rigid. When turgor pressure is equal to the osmotic pressure (↑) of the cytoplasm, no more water enters the cell through the cell membrane (↑). **turgid** (adj).

plasmolysis (n) the shrinking of the cytoplasm (↑) of a plant cell away from the cell wall (↑) when the cell is put in a solution with a higher osmotic pressure (↑) than that of the cytoplasm. Water passes through the cell membrane (↑) from the vacuole (↑) to the solution and turgor (↑) is lost. **plasmolyse** (v).

細胞核(名) 細胞質內有致密物質，外包一層膜的小物體。它控制細胞的全部活動。沒有細胞核，細胞就會死亡。細胞核內含有核液（一種液體）和染色體(第 142 頁)。(形容詞爲 nuclear)

顆粒(名) 小粒的物質，通常是堅硬的，每顆粒都與其他顆粒及其他物質分開。(形容詞爲 granular)

液泡(名) 外有一層膜包覆着的細胞質(↑)內充滿液體的空腔。很多植物細胞都有一個單一的，佔有細胞大部分體積的液泡；液泡內充滿細胞液(一種液體)。泡內細胞液的滲透壓力(↓)與泡外細胞液的滲透壓力相同。此外液泡還指任何組織的微小空間。(形容詞爲 vacuolar)

滲透壓力 指溶液中溶質(第 89 頁)所產生的壓力。如溶質本身不能透過膜(↑)，滲透壓力可使水透過膜而使溶液稀釋。滲透壓力是一種施於溶液以阻止水透過膜的壓力。水從滲透壓力低的溶液透過膜流向滲透壓力高的溶液，直到兩邊壓力相等爲止。溶液的濃度愈大，滲透壓力也愈大。(如果**溶質**透過膜，則產生滲濾現象)。

膨壓現象(名) 由於細胞質和液泡吸收(第 163 頁)水分，植物細胞充分膨脹的狀態，結果產生膨壓使細胞壁(↑)保持堅韌。當膨壓與細胞質的滲透壓力(↑)相等時，水再也不能透過細胞膜(↑)進入細胞。(形容詞爲 turgid)

質壁分離(名) 當植物細胞放入滲透壓力(↑)高於細胞壁(↑)滲透壓力的一種溶液中時，植物細胞的細胞質(↑)收縮離開細胞壁(↑)的一種現象。水從液泡(↑)透過細胞膜(↑)進入溶液，於是膨壓現象(↑)消失。(動詞爲 plasmolyse)

function (*n*) the work done by any part of a plant or animal, and also the purpose of that work, e.g. (a) a function of the nose is to take in air; (b) a function of the blood is to provide the body tissues (↓) with oxygen. **function** (*v*).

inhibit (*v*) in animals, to stop or slow down any function (↑) or action as a result of control by nerves. **inhibition** (*n*).

physiological (*adj*) concerning the function (↑) of any part of an organism (p.147) as opposed to its structure, e.g. a nerve poison has a physiological effect on the nerves of an animal. **physiology** (*n*).

biological (*adj*) to do with biology.

功能（名）　動物或植物體的任何一部分所作的工作及其目的。例如：(a)鼻的功能是吸入空氣；(b)血液的功能是給身體組織(↓)提供氧氣。（動詞爲 function）

抑制（動）　對動物而言，指其受神經控制而停止或減緩體內的某種功能(↑)或活動。（名詞爲 inhibition）

生理學的（形）　與生物體（第 147 頁）任何一部分的功能(↑)有關，而與它的結構無關的。例如動物神經中毒對其神經有生理影響。（名詞爲 physiology）

生物學的（形）　與生物學有關的。

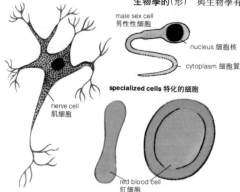

male sex cell 男性性細胞
nucleus 細胞核
cytoplasm 細胞質
specialized cells 特化的細胞
nerve cell 肌細胞
red blood cell 紅細胞

specialization (*n*) the change in an organism (p.147), part of an organism, or a cell (p.138), in order to perform a particular function or to improve a function, e.g. (a) the specialization of cells as seen in nerve cells and muscle cells; (b) specialization in animals as seen in the growth of feathers by birds to help in flying. **specialized** (*adj*).

tissue (*n*) a mass of cells and the intercellular material surrounding them all of which perform the same function, e.g. muscle tissue has the ability to contract (p.38). The tissue consists mainly of cells of the same function, e.g. muscle cells in muscle tissue, but other cells are needed to bind the tissue together.

特化作用（名）　指生物體（第 147 頁）、生物體的部分、或細胞（第 138 頁）爲執行某種特殊功能或爲改善某種功能而發生的變化。例如：(a)神經細胞和肌細胞的特化；(b)動物的特化，如鳥類生長羽毛有助於飛行。（形容詞爲 specialized）

組織（名）　執行同一功能的細胞群和細胞周圍的細胞間質。例如肌組織有收縮（第 38 頁）的能力。組織主要由具有同一功能的許多細胞組成（例如肌組織中的肌細胞），但也需要有其他細胞將肌肉組織聯結在一起。

organelle (*n*) a specialized part or unit of a cell with a particular function, (↑), e.g. mitochondria (↓), ribosomes (↓) are organelles.

mitochondrion (*n*) (*mitochondria*) a very small granular or rod-shaped body in the cytoplasm (p.138) of all cells. It uses oxygen and substances in the cytoplasm to set free energy for many of the cell's functions.

ribosome (*n*) a very small granular body in the cytoplasm (p.138) of all cells. Ribosomes synthesize (p.136) proteins (p.172) from amino acids (p.172).

plastid (*n*) a very small body, of various shapes, in the cytoplasm (p.138) of plant cells; some cells have few, some cells have many. Their function is to provide colour in plants; some produce starch (p.174), some protein (p.172) and some store starch and protein.

細胞器(名) 有特殊功能(↑)的細胞的特化部分或單位。例如粒線體(↓)、核糖體(↓)都是細胞器。

粒線體(名) 一切細胞的細胞質(第138頁)中含有的一種小的粒狀體或桿狀體。它利用氧和細胞質中的物質放出能量供細胞完成多種功能。

核糖體(名) 一切細胞的細胞質(第138頁)中存在的一種極小的粒狀體。核糖體將氨基酸(第172頁)合成(第136頁)爲蛋白質(第172頁)。

質體(名) 植物細胞的細胞質(第138頁)中的一種微小物體,有各種不同形狀;有些細胞只有少幾個質體,而另一些細胞則有很多質體。其功能是使植物有顏色;有一些是製造澱粉(第174頁),有一些是製造蛋白質(第172頁),還有一些是貯藏澱粉和蛋白質。

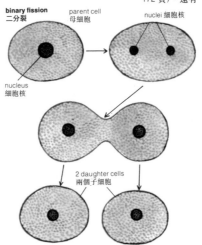

binary fission the division of one cell into two daughter cells of equal size. The nucleus (p.139) divides into two, the two nuclei separate, and the cytoplasm then divides into two and the two cells are formed.

二分裂 一個細胞分裂爲兩個大小相等的子細胞。細胞核(第139頁)一分爲二,兩個核分開,然後細胞質也一分爲二,形成兩個細胞。

centriole (*n*) a very small granule just outside the nuclear membrane (p.138). When binary fission (p.141) is about to take place, the centriole divides in two and the process starts of the nucleus dividing in two.

chromosome (*n*) a thread-like body in the nucleus (p.139) of a cell. Each animal and each plant has a particular number of chromosomes, e.g. human beings have 46 chromosomes. All chromosomes are in pairs, so a human being has 23 pairs in the nucleus of every cell in his body, except for sex cells which have half that number through meiosis (↓). The chromosomes in the nuclei determine the particular characteristics of an organism, e.g. height, appearance, colour of hair and eyes, etc. **Sex chromosomes** a pair of chromosomes in the nucleus of a cell which determine the sex of an animal. One sex has a pair of identical (p.94) chromosomes, called X-chromosomes. The other sex has one X-chromosome and either a different chromosone, called a Y-chromosome, or no other chromosome. In mammals (p.150), a Y-chromosome produces a male, i.e. the sex chromosomes are XY; two X-chromosomes produce a female; the sex chromosomes are XX.

meiosis (*n*) the process by which a nucleus (p.139) divides to form sex-cells. The chromosomes of a chromosome pair separate, one going to each side of the nucleus under the control of the centriole (↑). This forms two nuclei, each with half the number of chromosomes, hence the name **reduction division**. The chromosomes then each form two chromosomes and a further division, which is exactly the same as in mitosis (↓), takes place. This results in four nuclei each containing half the number of chromosomes that the nucleus had at the start of meiosis. The cytoplasm divides to form four sex cells with the four new nuclei. **meiotic** (*adj*).

mitosis (*n*) the process by which a nucleus (p.139) divides in two during binary fission (p.141). Each chromosome first forms two chromosomes; these two chromosomes separate, one going to each side of the nucleus under the control of the

中心粒(名) 緊貼核膜(第138頁)外側的一個極小的顆粒。二分裂(第141頁)即將發生之際中心粒一分為二，開始細胞核一分為二的過程。

染色體(名) 細胞核(第139頁)中的絲狀體。動物或植物個體都有一定數目的染色體，例如人體有46個染色體。所有的染色體都是成對的，所以人體每個細胞的細胞核中都有23對染色體。性細胞除外，因為它經過成熟分裂(↓)後只有此數一半的染色體。細胞核中的染色體決定生物體的個別特徵，例如高度、外貌、毛髮和眼睛的顏色等。**性染色體**是細胞核中決定動物性別的一對染色體。其中一種性染色體有一對相同的(第94頁)染色體，稱為X染色體。而另一種性染色體有一個X染色體和另一個不同的染色體(Y染色體)或者沒有別的染色體。哺乳動物(第150頁)的Y染色體產生雄性，即其性染色體是XY；兩個X染色體產生雌性；性染色體是XX。

成熟分裂(名) 細胞核(第139頁)分裂成性細胞的過程。一對染色體分開時，每個染色體都受中心粒(↑)控制，各自移向細胞核的一邊形成兩個核，每個核帶有半數染色體，故亦稱**減數分裂**。然後這些染色體分開，形成兩個染色體，跟著又再分裂。這和在有絲分裂(↓)中發生的情況一樣。結果產生四個細胞核，每個核含有細胞核開始成熟分裂時的半數染色體。細胞質分開形成四個性細胞和四個新核。(形容詞為 meiotic)

有絲分裂(名) 二分裂時(第141頁)細胞核(第139頁)一分為二的過程。每個染色體首先形成兩個染色體，這兩個染色體分開，各在中心粒(↑)的控制下移向細胞核的一邊，同時核膜(第138頁)消失。在每組染色體的周圍

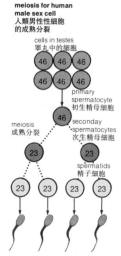

meiosis for human male sex cell
人類男性性細胞的成熟分裂

cells in testes
睾丸中的細胞

primary spermatocyte
初生精母細胞

meiosis
成熟分裂

secondary spermatocytes
次生精母細胞

spermatids
精子細胞

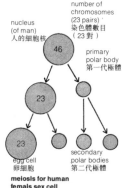

number of chromosomes (23 pairs)
染色體數目(23對)

nucleus (of man)
人的細胞核

primary polar body
第一代極體

egg cell
卵細胞

secondary polar bodies
第二代極體

meiosis for human female sex cell
人類女性性細胞的成熟分裂

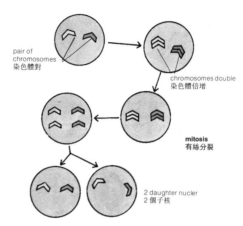

pair of chromosomes
染色體對

chromosomes double
染色體倍增

mitosis
有絲分裂

2 daughter nuclei
2 個子核

centriole (↑), while the nuclear membrane (p.138) disappears. A new nuclear membrane is formed round each group of chromosomes and two new nuclei are formed. The process usually takes between 30 and 180 minutes. **mitotic** (*adj*).

polar body a very small cell containing a nucleus, but almost no cytoplasm (p.138) at all, formed during the formation of a female sex cell. In the first division of meiosis, one egg-cell (a female cell) and a primary polar body are formed. Each of these two cells then divides mitotically (↑) forming one egg-cell and three polar bodies, *see diagram*. The polar bodies die leaving the egg-cell.

Protozoa (*n*) a group of animals which are different from all other animals because they consist of only one cell, and thus are microscopically (p.61) small. Some protozoa show a relation with simple plants. Protozoa live under almost all conditions found in nature, and are important in ecological (p.227) relations.

Amoeba (*n*) a group of protozoa (↑) of irregular shape. An amoeba continually changes its shape by movement of its cytoplasm. It feeds by closing pseudopodia (p.144) round a food particle and absorbing the food inside a vacuole (p.139). **amoebic** (*adj*), **amoeboid** (*adj*).

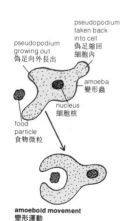

pseudopodium taken back into cell
偽足縮回細胞內

pseudopodium growing out
偽足向外長出

amoeba
變形蟲

nucleus
細胞核

food particle
食物微粒

amoeboid movement
變形運動

形成新的核膜並形成兩個新核。這一過程通常需 30 至 180 分鐘。(形容詞為 mitotic)

極體 一種極小的細胞，它有一個細胞核，但幾乎全無細胞質(第 138 頁)。極體是在雌性細胞形成的過程中產生的。第一次成熟分裂時產生一個卵細胞(雌性細胞)和一個第一代極體。之後這兩個細胞各自有絲分裂(↑)形成一卵細胞和三個極體(見圖)。三個極體死亡而留下一個卵細胞。

原生動物(名) 和所有其他動物不同的一類動物，這類動物很細小，只由一個細胞構成，需用顯微鏡(第 61 頁)才能看見。有些原生動物與低等植物有關係。原生動物幾乎在自然界的任何環境下都能生存，它在生態(第 227 頁)關係中極為重要。

變形蟲(名) 一種形狀不規則的原生動物(↑)。變形蟲靠其細胞質的流動而不斷改變形狀。其攝食的方法是用偽足(第 144 頁)包圍食物微粒，再將之吸入液泡(第 139 頁)。(形容詞為 amoebic，amoeboid)

pseudopodium (n) (*pseudopodia*) a part of a cell (in some protozoa) formed by the cytoplasm (p.138) pushing out temporarily an arm of irregular shape. Pseudopodia are used in locomotion (p.194) and in feeding; they are formed, are used for these purposes, and then are taken back into the cell. This kind of locomotion is called **amoeboid movement**.

structure (n) (1) a part of an organism which is complete in itself, but which has no shape that can be described, i.e. it is not a surface, a cavity or a vessel. Examples of structures are: a tooth, an ear, a tail. (2) the arrangement of cells and tissues (p.140) in a part of an organism, e.g. the structure of a muscle. **structural** (adj).

cilium (n) (*cilia*) a thread-like structure (↑) which together with many others grows from the surface of a cell. Cilia beat, one after the other, in a regular way; they either push water past a cell, or push the cell through water, *see diagram*. **ciliary** (adj).

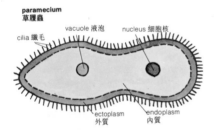

paramecium
草履蟲

cilia 纖毛　　vacuole 液泡　　nucleus 細胞核

ectoplasm 外質　　endoplasm 內質

Paramecium (n) a group of protozoa (p.143) of a shoe-like shape. The surface of a Paramecium is covered with cilia (↑) and ciliary beating drives the animal through the water.

flagellum (n) (*flagella*) a long, thread-like structure, longer than a cilium (↑), with a wave-like movement for beating in water; it is used for locomotion (p.194). One or two may be found on some unicellular organisms but bacteria (↓) can have groups of flagella called **tufts** (n). Organisms with flagella are described as **flagellate** (adj).

偽足（名）　（有些原生動物)細胞的一部分是由細胞質(第 138 頁)暫時向外突出形成一形狀不規則的肢體。偽足用於運動(第 194 頁)和攝食；為運動和攝食而形成偽足，然後再縮回細胞內，這種移動稱爲**變形運動**。

結構（名）　（1)指生物體中本身是完整的一個部分，但沒有可描述的形狀，亦即它不是表皮、空腔或血管。結構的例子如：牙、耳和尾；(2)生物體內某一部分細胞和組織(第 140 頁)的排列。例如肌肉的結構。(形容詞爲 structural)

纖毛（名）　在細胞表膜上長出的許多絲狀結構(↑)。纖毛一根接一根地有規則地擺動，它們或推動水流過細胞，或推動細胞在水中流動(見圖)。(形容詞爲 ciliary)

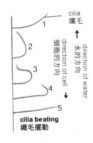

cilia 纖毛

1
2
3
4
5

direction of cell 細胞的方向

direction of water 水的方向

cilia beating
纖毛擺動

草履蟲（名）　一種形似草鞋的原生動物(第 143 頁)。其表面滿佈一層纖毛(↑)，纖毛擺動推動身體在水中運動。

鞭毛（名）　一種較纖毛(↑)長的長絲狀結構。在水中呈波浪形擺動；鞭毛用於運動(第 194 頁)。一些單細胞生物體上可有一兩根鞭毛，但細菌(↓)上有一片密密的鞭毛，稱爲**鞭毛叢**。具鞭毛的生物被描述爲 **長鞭毛的**。

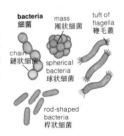

bacteria
細菌

mass 團狀細菌

tuft of flagella 鞭毛叢

chain 鏈狀細菌

spherical bacteria 球狀細菌

rod-shaped bacteria 桿狀細菌

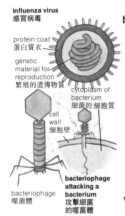

influenza virus
感冒病毒

protein coat
蛋白質衣

genetic
material for
reproduction
繁殖的遺傳物質

cytoplasm of
bacterium
細菌的細胞質

cell
wall
細胞壁

bacteriophage
噬菌體

**bacteriophage
attacking a
bacterium**
攻擊細菌
的噬菌體

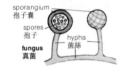

sporangium
孢子囊

spores
孢子

fungus
真菌

hypha
菌絲

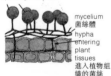

sporangia 孢子囊

mycelium
菌絲體

hypha
entering
plant
tissues
進入植物組
織的菌絲

**mushrrom
(a fungus)**
蘑菇（一種真菌）

spores
孢子

mycelium
菌絲體

ground
level 地面

bacterium (n) (*bacteria*) an organism (p.147) with a single cell, having no separate nucleus; it is neither plant nor animal. In shape some are spherical, some rod-like, and some spiral (p.219). Spherical bacteria have no flagella, and are not able to move themselves. The other bacteria usually possess flagella, and are able to carry out locomotion. Bacteria are found in very large numbers everywhere, and are concerned with (a) the decay (p.146) of plant and animal tissues (p.140), (b) chemical changes of inorganic salts, e.g. nitrites to nitrates, (c) the spread of disease. Bacteria are 0.5 to 2.0 μm in length or breadth; they reproduce by binary fission (p.141). **bacterial** (*adj*).

virus (n) an agent of disease which is not usually considered to be living, as some viruses can be crystallized (p.110), but as all viruses can reproduce inside a living cell, they can be considered to be alive. Bacteria can be grown on suitable substrates (p.146) outside the body, but viruses cannot. The diameter of a virus is between 0.5 μm and 0.03 μm, i.e. much smaller than that of bacteria. **viral** (*adj*).

bacteriophage (n) a virus that attacks and destroys bacteria.

fungus (n) (*fungi*) a plant containing no chlorophyll (p.159). It consists either of a single cell, or of cellular tube-like threads; it feeds on dead or living organisms (p.147) and is an agent of decay (p.146) or disease. Fungi reproduce by spores (p.146). **fungal** (*adj*).

hypha (n) a thread-like part of a fungus (↑), e.g. as seen growing on old bread. It grows at the end, or by branches; it is tube-like, and filled with cytoplasm (p.138), which is not separated by cell walls or membranes; nuclei are present in the cytoplasm. **hyphal** (*adj*).

mycelium (n) (*mycelia*) the mass of hyphae (↑) which forms all parts of the fungus, other than the reproductive organs, *see diagram.*

sporangium (n) (*sporangia*) a structure, covered by a wall, containing spores (p.146). It is formed at the end of a hypha (↑) growing up from a mycelium (↑).

細菌（名） 無單獨細胞核的單細胞生物體（第 147 頁）；它既不是植物，也不是動物。有些是球狀，有些是桿狀，還有些是螺旋狀（第 219 頁）。球狀細菌無鞭毛，自己不能行動。其他細菌通常都有鞭毛，能行動。到處都可發現大量細菌，並與下列情況有關：(a) 動、植物組織（第 140 頁）腐爛（第 146 頁）；(b) 無機鹽的化學變化，例如亞硝酸鹽變爲硝酸鹽；(c) 疾病的傳播。細菌長或寬爲 0.5 至 2 μm；細菌是藉二分裂（第 141 頁）繁殖的。（形容詞爲 bacterial）

病毒（名） 一種傳染疾病的媒介。因爲有些病毒能形成結晶（第 110 頁），所以一般認爲病毒是無生命的，但因爲一切病毒都能在活的細胞中繁殖，所以也被認爲是有生命的。細菌能在體外適宜的基質（第 146 頁）上生長，病毒則不能。病毒的直徑介於 0.5 至 0.03 μm 之間，比細菌小得多。（形容詞爲 viral）

噬菌體（名） 係一種能攻擊和破壞細菌的病毒。

真菌（名） 一種不含葉綠素（第 159 頁）的植物。或由單細胞，或由多孔的管狀菌絲構成；它從死的或活的生物體（第 147 頁）中吸取營養，並且是腐爛（第 146 頁）或傳播疾病的媒介。真菌靠孢子（第 146 頁）繁殖。（形容詞爲 fungal）

菌絲（名） 真菌（↑）的絲狀部分，例如在發霉的麵包上可見到菌絲；它在端部發展或分叉；呈管形，內充滿不被細胞壁或膜分開的細胞質（第 138 頁）；其細胞核在細胞質中。（形容詞爲 hyphal）

菌絲體（名） 組成真菌除生殖器官之外的其他部分的大量菌絲（↑）（見圖）。

孢子囊（名） 上有壁覆蓋，內含孢子（第 146 頁）的結構。它形成於菌絲（↑）的端部，由菌絲體（↑）發育而成。

spore (*n*) a reproductive cell, or a group of cells, with a wall around it, formed by certain plants, particularly ferns and fungi; also formed by some bacteria and some protozoa (p.143). Plant spores are produced by vegetative reproduction (p.213) in very large numbers, and from each spore, a new plant can grow under favourable conditions; the spores differ from seeds in not having an embryo (p.154). Spores of bacteria and protozoa are formed from a single organism (↓) when conditions are unfavourable.

cyst (*n*) a bag-like cover formed round a resting protozoan, e.g. an amoeba forms a cyst when conditions are unfavourable. **cystic** (*adj*).

yeast (*n*) a fungus consisting of single cells; it multiples by budding, i.e. small cells grow out from the mother cell, *see diagram*. Yeasts produce enzymes (p.167) which cause the decomposition of starch and sugars to alcohol (ethanol) and carbon dioxide. There are many different kinds of yeasts, each acting on a different substrate (↓).

mould (*n*) a grey or white growth of a fungus (p.145) on the surface of a living, or specially a dead, organism (↓); it causes decay of the organism, e.g. mould growing on old bread. **mouldy** (*adj*).

substrate (*n*) (1) the ground, or other object, on which animals live; (2) the substance on which microscopic organisms (↓) live and feed; (3) the particular substance on which an enzyme (p.167) acts.

decay (*n*) the chemical processes, caused by bacteria (p.145) and fungi (p.145), which take place in animal and plant substances after death. Decay breaks down the tissues and chemical compounds until only animal bones are left, while plants completely disappear. **decay** (*v*).

Penicillium (*n*) a mould (↑) from which penicillin (↓) is produced.

penicillin (*n*) a substance which prevents the growth of bacteria; it is an antibiotic (p.240). Fungi (p.145) produce substances which prevent other organisms growing near them, and penicillin is such a substance.

孢子(名)　某些植物，特別是蕨類和真菌所產生的生殖細胞或一群細胞其外有細胞壁覆蓋，某些細菌和原生動物(第143頁)也產生孢子。植物孢子藉營養繁殖(第213頁)大量產生，在有利的環境下每個孢子長出一新株。孢子不同於種子，它沒有胚(第154頁)。環境不利時，細菌或原生動物的孢子由單個的生物體(↓)產生。

包囊(名)　休眠的原生動物周圍所形成的袋狀外層，例如當環境不利時，變形蟲產生包囊。(形容詞爲 cystic)

酵母菌(名)　單細胞組成的真菌；酵母菌藉出芽生殖成倍繁殖，例如由母細胞長出小細胞(見圖)。酵母菌產生酶(第167頁)使澱粉和糖分解爲酒精(乙醇)和二氧化碳。酵母菌種類繁多，各對不同的基質(↓)起作用。

黴菌(名)　長在活的或特別是死的生物體(↓)表面的灰色或白色的真菌(第145頁)；它使生物體腐爛。例如在不新鮮麵包上長着的霉。(形容詞爲 mouldy)

基質；被酶作用物；受質(名)　(1)動物賴以生存的地面或其他物體；(2)微生物(↓)在上面生存或攝食的物質；(3)被酶(第167頁)作用的特殊物質。

腐爛(名)　在死亡的動、植物體上由細菌(第145頁)和真菌(第145頁)引起的化學過程。腐爛分解組織和化合物直到剩下動物的骨骼爲止，而植物則完全消失。(動詞爲 decay)

青黴菌屬(名)　產生青黴素(↓)的黴菌(↑)。

青黴素(名)　一種阻止細菌生長的物質；是一種抗生素(第240頁)。黴菌(第145頁)產生一種阻止其他生物體在它附近生長的物質，而青黴素即爲此類物質。

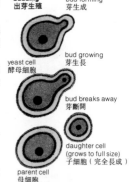

budding 出芽生殖
bud forming 芽生成
yeast cell 酵母細胞
bud growing 芽生長
bud breaks away 芽斷開
daughter cell (grows to full size) 子細胞 (完全長成)
parent cell 母細胞

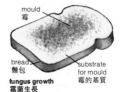

mould 霉
bread 麵包
substrate for mould 霉的基質
fungus growth 霉菌生長

yeast 酵母
sugar solution 糖液
substrate for yeast 霉的基質
fermentation of substrate 基質的醱酵

vertebrates
脊椎動物

mammal
哺乳動物
hair 毛
mammary 有乳
glands 腺體
yong 幼獸一出世
born alive 就能活動
warm
blooded
溫血的

bird 鳥
feathers 羽毛
beak 喙
wings 翼
warm-
blooded
溫血的

reptile 爬蟲類

scales cold-blooded
有鱗、冷血的

organism (n) a plant, an animal, anything that lives, such as a tree, a man, a bacterium.

nature (n) (1) all objects, organisms and forms of energy considered as a system (p.162) but excepting objects made by man, e.g. mountains, rivers, the atmosphere, plants, animals, are all parts of nature. (2) all the properties (p.27) and characteristics (↓) which determine the whole behaviour of an object, organism, or form of energy, e.g. (a) the nature of a metal depends on its properties; (b) the nature of a fish depends on its characteristics.

characteristic (n) any part, shape, way of behaving, by which an organism, or group of organisms, can be recognized, e.g. (a) hair on the body is a characteristic of mammals; (b) feathers are a characteristic of birds; (c) green-coloured leaves are a characteristic of most plants; (d) walking on two legs is a characteristic of man. Objects can have their characteristics, e.g. the conduction of heat is a characteristic of metals. **characteristic** (adj).

classify (v) to put objects or organisms into named groups, chosen by examining the properties (p.27) or characteristics (↑) of the object or organism and of the group, e.g. (a) an animal has a beak, feathers, two legs and two wings, and these are the characteristics of a bird, so the organism is classified as a bird; (b) a substance is hard, is a good conductor of heat and electric current, reacts with acids to form a salt, and these are the characteristics of a metal, so the substance is classified as a metal. **classification** (n).

生物體；有機體(名) 植物、動物、任何有生命的東西。例如樹、人、細菌。

自然界；本性(名) (1)所有被認爲屬一個體系(第 162 頁)的物體、生物體和各種形式的能，但人造的物體除外。例如山、河、大氣、植物、動物都是自然界的組成部分；(2)決定物體、生物體或一種形式的能的整個行爲的全部屬性(第 27 頁)和特徵(↓)。例如：(a)金屬的本性決定於它的性質；(b)魚的本性決定於它的特徵。

特徵(名) 任何一個部位、形狀和行爲方式，人們可以藉此辨認一種生物體或一類生物體。例如(a)體上長毛是哺乳動物的特徵；(b)長有羽毛是鳥類的特徵；(c)綠葉是大部分植物的特徵；(d)而用兩腿走路是人類的特徵。物體也有其特徵，例如能導熱是金屬的特徵。(形容詞爲 characteristic)

分類(名) 將物體或生物體歸入已命名的各類別，類別的選擇是通過審查物體、生物體或它們這一類的屬性(第 27 頁)或特徵(↑)。例如，一個動物有喙、羽毛、兩腿和兩翼，這是鳥的特徵，所以這一生物體歸入鳥類；(b)一物質是堅硬的，是熱和電的良導體，與酸起作用形成鹽，這些是金屬的特徵，所以這一物質歸入金屬類。(名詞爲 classification)

fish 魚
scales 有鱗
fins 有鰭
cold-blooded 冷血的

amphibian
兩棲動物

smooth skin 皮膚光滑
four legs 四條腿
cold-blooded 冷血的

species (*n*) (*species*) the smallest group in the classification (p.147) of organisms (p.147). Members of a species can be parents of young, who can in turn be parents. Pairs from different species usually cannot produce young; if they do, the young cannot themselves reproduce (p.209), e.g. two horses are members of the same species and can reproduce young horses; a horse and a donkey can produce a mule, but mules cannot reproduce, as horses and donkeys belong to different species.

animal (*n*) a living being which cannot make its food from simple inorganic (p.116) substances, but obtains its food from plants or other animals. Most animals use locomotion (p.194) to help them find food.

vertebrate (*n*) an animal possessing a line of bones, a backbone, which supports and protects its body; each bone is called a vertebra (p.193). The main groups of vertebrates are: fishes, amphibians (↓), reptiles (p.150), birds and mammals (p.150).

invertebrate (*n*) an animal possessing no backbone; some are completely soft, e.g. worms, and some have a hard outer covering, e.g. insects, crabs.

物種 (名) 生物 (第 147 頁) 分類 (第 147 頁) 中最小的分類群。類群中的成員可以是幼體的親體,幼體也能依次成爲親體。不同物種的一對通常不能生殖幼體,如果能生殖幼體,則幼體不能繁殖 (第 209 頁)。例如兩匹同屬一個種的馬,能生殖幼馬,馬和驢能生出騾,但騾不能再生殖,因爲馬和驢屬不同的畜種。

動物 (名) 不能以純無機 (第 116 頁) 物爲食物,但能以植物或其他動物爲食物的生物。大部分動物都要爲覓食而四處活動 (第 194 頁)。

脊椎動物 (名) 有一系列骨骼 (即脊骨) 支持和保護身體的動物,每塊骨骼稱爲脊椎骨 (第 193 頁)。脊椎動物的主要種類有魚類、兩棲動物 (↓)、爬蟲動物 (第 150 頁)、鳥類和哺乳動物 (第 150 頁)。

無脊椎動物 (名) 沒有脊骨的動物;有些完全是軟體動物,例如蠕蟲,還有些有一層硬的外殼。例如昆蟲、蟹類。

crab 蟹

insect 昆蟲

earthworm 蚯蚓

shellfish 甲殼動物

jellyfish 水母

invertebrates 無脊椎動物

amoeba 變形蟲

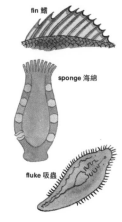

fin 鰭

sponge 海綿

fluke 吸蟲

earthworm 蚯蚓

spider 蜘蛛

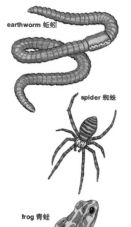

frog 青蛙

appendage (n) a relatively large structure attached to the main body of an animal, e.g. a leg, an arm, a tail, are all appendages. **Paired appendages**, with one on each side of the body, are common in many animals, e.g. a pair of arms. **appendicular** (adj).

fin (n) a thin, flat structure consisting usually of many small bones joined by a continuous piece of skin; found on all fish and some other animals which live in water. Fins are used to control the direction of motion, and also are used in locomotion (p.194).

sponge (n) an animal, living in water, consisting of many cells, but having no nerves, and not able to move. It has tissues, but no organs. Water is drawn in through small holes by the movement of flagella (p.144) and passed out through larger holes; food particles are caught from the water.

fluke (n) a kind of very small worm with a flat body; some flukes can be as long as 1 cm. Fully grown flukes live in the liver, gut, lung or blood vessels of vertebrates (↑), with different kinds of fluke in each place. Flukes can cause serious diseases, especially in man.

worm (n) a general name for many different kinds of invertebrates (↑) having a long, thin, soft body, and no appendages (↑). It is not a biological name for an animal.

vermiform (adj) describes a structure with a shape like a worm.

arthropod (n) one of a group of animals which possess (a) a hard outer covering to their body instead of bones to support the body, (b) paired legs consisting of jointed tube-like structures. Some live in water, e.g. crabs; others live on land, e.g. centipedes, spiders, insects (p.151).

amphibian (n) one of a group of animals which are vertebrates (↑) and possess (a) four legs, each with five digits (fingers or toes), (b) a moist, smooth skin with no scales; their eggs are not protected by a shell. Amphibians live partly in water and partly on land. Examples of amphibians are frogs and toads. **amphibious** (adj).

附肢(名)　附屬於動物軀體上的一種相當大的結構。例如腿、臂、尾㪗是附肢。很多動物身上都有成對的附肢，一側一個，這是很普遍的現象。例如一雙臂。(形容詞爲 appendicular)。

鰭(名)　通常由一片皮膚連結許多小骨組成的薄而平的結構，所有魚類和某些水生動物身上都有鰭。鰭用於控制運動方向，也用於游動(第194頁)。

海綿(名)　一種由許多細胞組成，無神經也不能行動的水生動物。它有組織，但無器官。其攝食方法是利用鞭毛(第144頁)運動，從小孔吸入水並讓其從大孔流出，從中攝取微粒食物。

吸蟲(名)　一種極小、身體扁平的蠕蟲；有些吸蟲可長達1cm。完全發育成熟的吸蟲生活在脊椎動物(↑)的肝、腸、肺或血管裏，長在各個部位的吸蟲，其種類各不相同。吸蟲能引致嚴重疾病，尤其是對人。

蠕蟲(名)　多種有細長軟體、無附肢(↑)的無脊椎動物(↑)的總稱。蠕蟲並非一種動物的生物學名稱。

蠕蟲狀的(形)　描述形狀像蠕蟲的結構。

節肢動物(名)　動物的一個類群，具有以下特徵：(a)用堅硬的外殼代替骨骼支持身體；(b)由節管狀結構組成的成對的腿。有些節肢動物生活在水中，例如蟹類；另一些生活在陸上，例如蜈蚣、蜘蛛、昆蟲(第151頁)。

兩棲動物(名)　脊椎動物(↑)的一個類群，具有以下特徵：(a)有四條腿，每腿有五趾(手指或腳趾)；(b)皮膚濕潤而光滑，無鱗；其卵無外殼保護。兩棲動物有時生活在水中，有時生活在陸上。青蛙和蟾蜍是兩棲動物的例子。(形容詞爲 amphibious)

reptile (*n*) one of a group of animals which are vertebrates and possess (a) horny skins or skin covered with horny plates, (b) four legs, or no legs; their eggs (p. 217) are protected by a hard, horny shell. Some reptiles live in water, e.g. turtles and crocodiles, but breathe air, other reptiles live on land, e.g. lizards, tortoises and snakes. **reptilian** (*adj*).

bird (*n*) one of a group of animals which are warm-blooded (p.188) vertebrates and possess (a) skin covered in feathers, (b) legs covered in scales, (c) beaks without teeth; their young are hatched from eggs (p.217) with a large yolk and a hard shell, e.g. doves, sparrows, eagles.

mammal (*n*) one of a group of animals which are warm-blooded (p.188) vertebrates and possess (a) skin covered in hair, (b) on females, mammary glands producing milk; their young are born alive. Some mammals live in the sea, e.g. whales and seals, others live on land, e.g. horses, lions, monkeys, men. **mammalian** (*adj*).

爬蟲動物(名) 脊椎動物的一個類群,具有以下特徵:(a) 有角質皮膚或披着角質板的皮膚;(b) 有四條腿,或無腿;其卵(第 217 頁)有堅硬的角質外殼保護。一些爬蟲動物生活在水中,例如海龜、鰐魚,但呼吸空氣;另一些爬蟲動物生活在陸上,例如蜥蜴、烏龜和蛇。(形容詞爲 reptilian)

鳥(名) 動物的一個類群,溫血(第 188 頁)脊椎動物,其特徵是:(a) 皮膚長滿毛;(b) 腿長滿鱗;(c) 喙內無牙;幼鳥是從一個大蛋黃和硬殼的卵(第 217 頁)中孵化出來的。例如鴿子、麻雀和鷹。

哺乳動物(名) 動物中的一類群,爲溫血(第 188 頁)脊椎動物,其特徵是:(a) 皮膚長滿毛;(b) 雌性身上有產生乳汁的乳腺;幼子出世即能活動。有些哺乳動物生活在海裏。例如鯨和海豹,另一些生活在陸上。例如馬、獅、猴、人。(形容詞爲 mammalian)

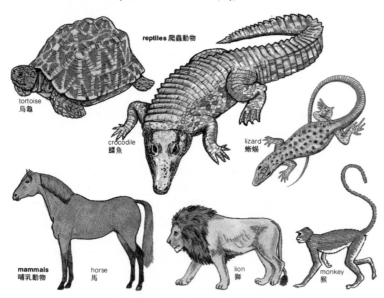

reptiles 爬蟲動物

tortoise
烏龜

crocodile
鰐魚

lizard
蜥蜴

mammals
哺乳動物

horse
馬

lion
獅

monkey
猴

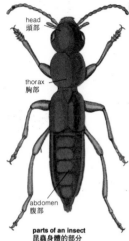

parts of an insect
昆蟲身體的部分

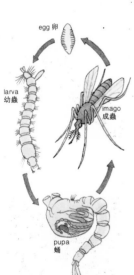

life cycle of a mosquito
蚊的生活週期

insect (*n*) a fully grown arthropod (p.149) with its body divided into three separate parts, head, thorax and abdomen, *see diagram*. It possesses (a) three pairs of jointed legs, (b) one pair of antennae, (c) two pairs of wings (in many, but not all, insect groups). It has a life cycle (↓) in which it changes its appearance and way of life, e.g. flies, bees, moths.

larva (*n*) (*larvae*) one part of the life cycle (↓) of an insect, and of other invertebrates (p. 148); it comes after an egg (p. 217). An egg hatches (p.217) and a larva comes out. A larva is quite different in appearance from the fully grown insect, usually being worm-like, *see diagram*; it cannot reproduce. It changes into a pupa or a fully grown insect or other invertebrate. **larval** (*adj*).

pupa (*n*) (*pupae*) a resting stage (p.152) in the life of some insects; it no longer moves or feeds. It is enclosed in a case, and great changes take place in the structure of its body, e.g. a caterpillar changes into a butterfly. **pupate** (*v*), **pupal** (*adj*).

cocoon (*n*) a protective cover made by a larva (↑) for the pupa (↑); also a protective cover for eggs made by some invertebrates, e.g. earthworms make a cocoon for their eggs. The cocoon of the silkworm is the source of silk.

imago (*n*) a fully grown insect into which a pupa changes. It can reproduce and the female forms eggs, which starts the life cycle (↓) of the insect.

metamorphosis (*n*) the change of form from a larva (↑)to an imago (↑). *Complete metamorphosis* is the change of an insect from larva to pupa to imago; *incomplete metamorphosis* is the change of an insect from larva to imago, with the larva being similar in appearance to the imago, e.g. a cockroach undergoes incomplete metamorphosis.

life cycle the changes, one after another, through which a plant or animal passes, e.g. the changes which take place from the production of an egg to the death of the organism. For many insects this is egg-larva-pupa-imago-death. It is represented in the diagram with the imago producing an egg for the next generation.

昆蟲(名) 發育成熟的節肢動物(第149頁),蟲體分爲頭、胸和腹三個部分〔見圖〕。昆蟲有:(a) 三對有關節的腿;(b) 一對觸角;(c) (很多昆蟲,但不是所有的昆蟲)有兩對翅膀。昆蟲有生活週期(↓),在此期間改變其外形及生活方式。例如蒼蠅、蜜蜂和蛾有生活週期。

幼蟲(名) 指昆蟲或其他無脊椎動物(第148頁)生活週期(↓)的一個階段;幼蟲在卵(第217頁)之後出現,卵孵化(第217頁)後幼蟲出來。幼蟲的外形與發育成熟的昆蟲很不同,一般呈蠕蟲狀(見圖);幼蟲不能生殖,只能化蛹或發育成熟的昆蟲或其他無脊椎動物。(形容詞爲 larval)

蛹(名) 一些昆蟲生活過程中的一個休眠階段(第152頁);在此階段,昆蟲不再活動,不進食,蜷伏在一個囊內,身體結構發生很大變化。例如毛蟲變成蝴蝶。(動詞爲 pupate,形容詞爲 pupal)

繭(名) 幼蟲(↑)爲蛹(↑)作的一種保護性的覆蓋物;也是一些無脊椎動物爲卵作的保護性的外殼,例如蚯蚓爲卵作繭,蠶繭是絲的來源。

成蟲(名) 由蛹變成的一隻發育成熟的昆蟲。它能繁殖,雌蟲產卵,昆蟲的生活週期(↓)從卵開始。

變態(名) 從幼蟲(↑)的形狀變爲成蟲(↑)。完全變態是昆蟲從幼蟲變爲蛹,再變爲成蟲;不完全的變態是昆蟲從幼蟲變爲成蟲,幼蟲的外形與成蟲相似,例如蟑螂就經歷不完全變態。

生活週期 動、植物經過的一個接一個的變化。例如生物體從產卵到死亡所發生的變化。對很多昆蟲來說,這就是卵——幼蟲——蛹——成蟲——死亡。圖示產出下一代卵的成蟲的這一變化過程。

moult (*v*) to lose an outer covering such as hair, feathers or skin at regular times, e.g. a cat moults and loses hair in hot weather. In larvae (p.151) which undergo incomplete metamorphosis (p.151), the hard outer cover splits and the larva draws itself out, with a new, soft cover which hardens in the air.

life history the changes which take place in an organism, from the production of an egg to death. Also, the changes in one stage (↓) of a life cycle (p.151) of an organism, e.g. the life history of a larva. The life cycle of a mammal is the same as its life history, but the life cycle of an insect consists of the life histories of each stage.

stage (*n*) one part, out of two or more parts, of a course of work, or a particular process or the life of an organism (p.147), e.g. (a) the stages of school work; (b) the stages in a process of distillation, i.e. boiling the mixture – collecting the distillate; (c) the stages in the life of a mosquito: egg-larva-pupa-imago.

maggot (*n*) a worm-like larva of an insect, e.g. the maggot of a house-fly.

caterpillar (*n*) the larva of a butterfly (↓) or moth. Its body is soft, long and round, and is divided into fourteen parts, called segments. The first segment is the head, the next three are thoracic (p.189) segments and the last ten are abdominal (p.162) segments. There are a pair of legs on each thoracic segment and a pair of structures on each abdominal segment called prolegs. The body is usually covered in fine hairs. A caterpillar eats leaves and plants, and some caterpillars are serious pests (p.231).

chrysalis (*n*) the pupa of a butterfly (↓), a moth and of some other insects.

butterfly (*n*) an insect (p.151) with a short, fat body divided into fourteen segments, the same as in a caterpillar (↑). Two pairs of large wings are fixed to the 2nd and 3rd thoracic segments, and a pair of jointed legs to each thoracic segment. The wings and body are covered with small scales. Butterflies fly by day and feed on nectar (p.213) from flowers. Moths are similar to butterflies, but fly by night.

蛻皮；換毛(動)　定時脫掉體外的覆蓋物如毛髮、羽毛或皮膚，例如貓在熱天時換毛。在經歷不完全變態(第 151 頁)的幼蟲(第 151 頁)體上，堅硬的外殼破裂，幼蟲脫殼而出，新的軟殼遇空氣變硬。

生活史　生物體從產卵到死亡所發生的變化。也指生物生活週期(第 151 頁)中一個時期(↓)的變化，例如幼蟲的生活史。哺乳動物的生活週期即其生活史，但昆蟲的生活週期是由每個時期的生活史組成的。

階段；時期(名)　兩個或多個部分中的一個部分；工作進程中一個部分；特殊過程中的某個階段，或生物體(第 147 頁)中的某個時期；例如 (a) 學校工作中的各個階段；(b) 蒸餾過程中的各個階段，即：從使混合物沸騰到收集餾出物；(c) 蚊蟲一生的各時期：卵 — 幼蟲 — 蛹 — 成蟲。

蛆(名)　昆蟲的蠕蟲狀的幼蟲，例如家蠅的蛆。

毛蟲(名)　蝴蝶(↓)或蛾的幼蟲。其軀體柔軟、長而圓，分為十四段，稱為十四節。第一節是頭部，後三節是胸(第 189 頁)節，最後十節是腹(第 162 頁)節。每一胸節上有一對腿，每一腹節上有一對腹足結構。軀體上通常長滿細毛。毛蟲食樹葉和植物，而有些毛蟲是危害極大的害蟲(第 231 頁)。

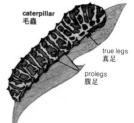

caterpillar
毛蟲

true legs
真足

prolegs
腹足

蠋蛹(名)　蝴蝶(↓)、蛾和一些其他昆蟲的蛹。

蝴蝶(名)　蟲體肥而短，與毛蟲(↑)一樣分為十四節的昆蟲(第 151 頁)。兩對大翅膀長在第二、第三胸節上，每一胸節上長有一雙有關節的腿。翅膀和軀體上密佈細鱗。蝴蝶白天飛行，以花蜜(第 213 頁)為食。蛾與蝴蝶相似，但在夜間飛行。

antenna
觸角

butterfly
蝴蝶

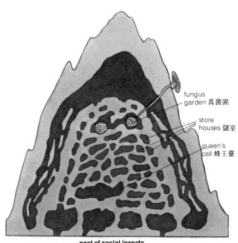

nest of social insects
群居昆蟲的巢

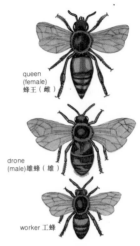

queen
(female)
蜂王（雌）

drone
(male)雄蜂（雄）

worker 工蜂

**three castes of
honey bee**
蜜蜂的三個職級

social insect an insect which lives in a group with a division of work between the different castes (↓), e.g. bees and termites are social insects.

society (*n*) a group of social insects.

solitary insect (*n*) an insect that lives alone and does not belong to a society.

caste (*n*) a particular kind of social insect (↑) with a structure and a function for a special kind of work, e.g. honey bees have three castes for the division of labour in a hive (↓): queens (↓), workers (↓) and drones (↓).

queen (*n*) a female social insect (↑) that produces eggs. There is only one queen in a hive (↓) of honey bees; she is larger than the members of the other two castes and has a pointed abdomen.

drone (*n*) a male bee which does no work; it mates with a queen.

worker (*n*) a female social insect not able to produce eggs; it is smaller than either a queen or a drone and does all the work in a hive (↓).

hive (*n*) the place in which bees live, a kind of very large nest; also the bees gathered together in that place.

fungus garden 真菌園
store houses 儲室
queen's cell 蜂王臺

群居昆蟲；社會性昆蟲 一種成群生活的昆蟲，不同職級(↓)的昆蟲間有分工。例如蜜蜂和白蟻都是群居昆蟲。

群居(名) 一群群居昆蟲。

獨居昆蟲(名) 一種單獨生活的昆蟲，不屬於群居昆蟲。

職級(名) 群居昆蟲(↑)中的特殊一類。其身體結構和功能適合於特種工作。例如蜜蜂在蜂群(↓)中按分工的不同，有三種職級：蜂后(↓)、工蜂(↓)和雄蜂(↓)。

蜂王(名) 擔負產卵的雌性羣居昆蟲(↑)。一個蜂羣(↓)中只有一隻蜂王；其身體比其他兩職級蜂大，並有一尖形腹部。

雄蜂(名) 雄蜂不作工，只與蜂王交配。

工蜂(名) 不能產卵的雌性羣居昆蟲，其身體比蜂王和雄蜂都小，擔負蜂窩(↓)中的全部工作。

蜂群；蜂窩(名) 蜜蜂的住所，是一種非常大的巢，也指群集於此的蜜蜂。

seed (*n*) the small body, produced by flowering plants, from which a new plant grows. It is a product of sexual reproduction (p.209) from the union of male and female parts of flowers (p.211) and consists of several different structures.

testa (*n*) (*testae*) the hard, outer covering of a seed. It does not let water pass through; when dry it does not let oxygen pass, but when it is wet, oxygen passes through.

micropyle (*n*) a very small hole in the testa (↑) of a seed. Water enters a seed through its micropyle, then the seed begins to germinate (p.156).

hilum (*n*) a scar, i.e. a mark, on the testa (↑) of a seed; it shows where the seed was fixed to the parent plant.

embryo[1] (*n*) a young plant in a seed; it is formed from an ovule (p.211) in a flower. An embryo is the part of a seed which grows into a new plant. **embryonic** (*adj*).

plumule (*n*) the part of an embryo which grows to form the main stem of a new plant.

radicle (*n*) the part of an embryo which grows to form the main root of a new plant.

shoot (*n*) the stem of a young plant which grows from the plumule (↑); leaves grow from the shoot.

endosperm (*n*) food material for the use of an embryo (↑) in a seed. Not all seeds have an endosperm, but where they do, the endosperm surrounds the embryo. **endospermous** (*adj*).

cotyledon (*n*) a simple leaf, usually lacking chlorophyll (p.159), forming part of the embryo of a seed. Flowering plants have one or two cotyledons. In some plants, e.g. legumes (↓), the cotyledons have a food store for the growing embryo; in other plants, e.g. maize, the cotyledon takes in food from the endosperm and passes it to the embryo. In many plants, the cotyledons appear above ground, produce chlorophyll, and make food for the plant. **cotyledonous** (*adj*).

dormant (*adj*) describes any organism that is resting and not growing. Seeds are dormant in soil until the temperature is high enough.

種子(名) 開花植物長出的小果實。新株由它發芽而來。種子是花(第 211 頁)的雄蕊和雌蕊有性生殖(第 209 頁)的產物。種子由幾個不同的結構組成。

外種皮(名) 種子堅硬的外殼。可阻止水流入，乾燥時，不讓氧氣透入，但種子受潮時，氧氣可透入。

珠孔(名) 種子外種皮(↑)上的一個極小的洞孔，水通過珠孔進入種子，於是種子開始發芽(第 156 頁)。

種臍(名) 一個瘢疤，即種子外種皮(↑)上的一個疤痕，顯示種子與母株相連之處。

胚胎(名) 種子內的幼株；係由花的胚珠(第 211 頁)所形成。胚是種子中發育成新株的部分。(形容詞爲 embryonic)

胚芽(名) 胚中發育成新株主莖(第 156 頁)的部分。

胚根(名) 胚中發育成新株主根的部分。

幼芽(名) 從胚芽(↑)發育成幼株的莖；葉子從幼芽上長出。

胚乳(名) 種子中供胚胎(↑)使用的營養物質。並非所有種子都有胚乳，但有胚乳的種子，胚乳總是包着胚。(形容詞爲 endospermous)

子葉(名) 通常是一片缺乏葉綠素(第 159 頁)的單葉，形成種子胚的一部分。開花植物有一、兩片子葉，有些植物中，例如豆科植物(↓)，子葉有一供胚發育的營養庫；其他植物，例如玉蜀黍，子葉從胚乳攝取營養，並輸送給胚。很多植物的子葉露在土面上製造葉綠素，並爲植物製造營養。(形容詞爲 cotyledonous)

休眠的(形) 描述正在休眠和停止生長的任何生物體。溫度不夠高時種子就在土中休眠。

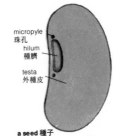

micropyle 珠孔
hilum 種臍
testa 外種皮

a seed 種子

embryo 胚

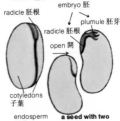

radicle 胚根
radicle 胚根
open 開
plumule 胚芽
cotyledons 子葉

endosperm 胚乳

a seed with two cotyledons and no endosperm 有兩片子葉而無胚乳的種子

endosperm 胚乳
plumule 胚芽
radicle 胚根
cotyledon 子葉

a seed with two cotyledons and an endosperm 有兩片子葉和一個胚乳的種子

cotyledon 子葉

seed with one cotyledon and endosperm 有一片子葉和胚乳的種子

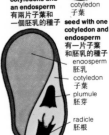

endosperm 胚乳
cotyledon 子葉
plumule 胚芽
radicle 胚根

monocotyledons
單子葉植物

dicotyledons
雙子葉植物

arallel venation of leaves
的平行葉脈序

net ventation
of leaves 葉的網狀葉脈序

irregular
arragement
of vascular
bundles in stem
莖內維管束
的不規則排列

regular
arrangement
of vascular
burdles in stem
莖內維管束的
規則安排

tomato 蕃茄

skin 皮
fleshy wall
肉質壁
membrane 膜
seeds 種子

berries 漿果

membrane 膜
skin 皮
fleshy
wall
肉質壁
seeds
種子
succulent
hairs 多汁的毛狀物

orange 橙

plum 李

hard stony
wall
梗如石的壁
skin 皮
single
seed
單種子

drupes 核果

mango 芒果

pore 孔

seeds 種子

poppy 罌粟
a capsule
一種朔果

seeds 種子

pea 豌豆
a legume
莢果

pod
豆莢

monocotyledon (n) a flowering plant with one cotyledon in its seed. All monocotyledons have (a) parallel venation (p.158); (b) flower parts (p.211) in groups of threes; (c) vascular bundles (p.160) irregularly arranged. Most monocotyledons are small plants, but a few are large, e.g. bananas, palm trees.
monocotyledonous (adj).

dicotyledon (n) a flowering plant with two cotyledons in its seed. All dicotyledons have (a) net venation (p.158); (b) flower parts (p.211) in groups of four or five; (c) vascular bundles (p.160) in the form of a ring round the centre of a stem. All trees except palms are dicotyledons; there are many more species of dicotyledons than of monocotyledons. **dicotyledonous** (adj).

dispersal (n) the process by which seeds get to places away from the parent plant. The means of dispersal are wind, water and animals.
disperse (v).

fruit (n) a body, formed from the ovary (p.211) of a plant, containing and protecting seeds. Common kinds of fruits include berries, drupes, capsules, legumes (↓). When describing biological structures, fruit has a different meaning from that in everyday use.

berry (n) a succulent (↓) fruit containing many seeds, with an outer skin, a thick fleshy wall, and an inner membrane round the inside part which contains seeds, e.g. tomato, orange, guava.

drupe (n) a succulent (↓) fruit containing a single seed with an outer skin, a thick fleshy wall, and a hard woody layer round the seed, e.g. plum, mango, cherry.

capsule (n) a dry fruit which splits open to set free seeds, e.g. poppy.

legume (n) (1) a dry fruit, also known as a pod, with a long narrow skin consisting of two halves with seeds inside. The skin splits along the join of the two halves and the seeds fall out, e.g. pea, bean. (2) any plant belonging to the pea or bean family.

succulent (adj) soft, thick and containing a lot of water, e.g. cacti and many fruits that are eaten are succulent.

單子葉植物（名） 種子內有一片子葉的開花植物。所有的單子葉植物都有 (a) 平行的葉脈序（第 158 頁）；(b) 一簇三瓣的花部（第 211 頁）；(c) 排列不規則的維管束（第 160 頁）。單子葉植物大部分是小型植物，但也有些是大型植物。例如香蕉、棕櫚樹。（形容詞爲 monocotyledonous）

雙子葉植物（名） 種子內有兩片子葉的開花植物。所有的雙子葉植物都有 (a) 網狀葉脈序（第 158 頁）；(b) 一簇四瓣或五瓣的花部（第 211 頁）；(c) 圍繞著莖中心的環狀維管束（第 160 頁）。除了棕櫚樹，所有的樹都是雙子葉植物。雙子葉植物的種較單子葉植物的種多得多。（形容詞爲 dicotyledonous）

散播（名） 種子離開母株到別處的過程。傳播的媒介有風、水和動物。（動詞爲 disperse）

果實（名） 果實由植物的子房（第 211 頁）發育而成，果實內含種子並保護種子。常見的果實包括漿果、核果、蒴果和莢果（↓）。在描述生物的結構時，果實的含意與日常生活中所説的水果有所不同。

漿果（名） 一種多汁的（↓）果實，內含許多種子，有外皮，肉質肥厚，內果膜包着有種子的內心部分，例如番茄、橙和番石榴。

核果（名） 一種多汁的（↓）果實，內含許多種子，有外皮，肉質肥厚，種子外包有一堅硬的木質層，例如李、芒果、櫻桃。

蒴果（名） 一種裂開發散出種子的乾果。例如罌粟。

莢果（豆科植物）（名） (1) 一種稱為豆莢的乾果，果皮狹長，由兩個半片組成，內含種子。果皮沿兩個半片的連結處裂開，種子掉出。例如豌豆、蠶豆；(2) 任何屬於豌豆或蠶豆科的植物。

多汁的 軟而肥厚的，且含有大量水分的。例如仙人掌和很多食用果實都是多汁的。

prolific (*adj*) producing seeds, spores, eggs, young, in large numbers.

germinate (*v*) to start to grow, of seeds and spores. The growth produces a new plant, called a seedling (↓). In order for a seed, or spore, to germinate, the conditions round it must be suitable. **germination** (*n*).

seedling (*n*) a newly formed organism growing from a seed. The cotyledons (p.154) help to provide food during this stage of growth; they die when the seedling becomes a plant.

epigeal (*adj*) having growth above ground level; epigeal seedlings have their cotyledons (p.154) above ground during growth. Most dicotyledons (p.155) are epigeal.

hypogeal (*adj*) having growth below ground level; hypogeal seedlings have their cotyledons (p.154) below ground level during growth. Most monocotyledons (p.155) are hypogeal.

epicotyl (*n*) the part of a seedling above the cotyledons (p.154) and below the first foliage (p.158) leaves. It grows bent in a half circle to protect the plumule (p.154) during germination; when the plumule is above ground, it straightens and becomes the first part of the stem.

hypocotyl (*n*) the part of a seedling below the cotyledons (p.154) and above the root. It is the stem of the embryo seed plant.

permeable (*adj*) allows molecules (p.103) and ions (p.101) to pass through, e.g. a permeable membrane allows molecules or ions, but does not allow fluids, to pass through; such a membrane is *impervious* to fluids but permeable to solvents, solutes and ions.

impermeable (*adj*) does not allow molecules or ions to pass, i.e. it is not permeable (↑).

impervious (*adj*) does not allow fluids to pass, but may be permeable (↑).

plant (*n*) a living object that makes its own food from simple inorganic substances which are gases or in solution; it usually has no power of locomotion (p.194).

stem (*n*) the part of a plant that bears leaves and buds (p.213). Most stems grow above ground, but some stems grow below ground level.

多育的(形) 能生產大量種子、孢子、卵、幼體的。

萌發；發芽(動) 指種子或孢子開始發育。發育成新株稱爲苗(↓)，種子或孢子發芽必須有適宜的環境。(名詞爲 germination)

籽苗(名) 從種子發育成的新生物體。在發育期間，子葉(第 154 頁)幫助提供營養，苗長成植株後，子葉即死亡。

出土的(形) 指長出地面上的；出土的籽苗在生長期間，其子葉(第 154 頁)處於地面之上。大部分雙子葉植物(第 155 頁)是出土的。

留土的(形) 指長在地面下的；留土的籽苗在生長期間，其子葉(第 154 頁)處於地面之下。大部分單子葉植物(第 155 頁)是留土的。

上胚軸(名) 子葉(第 154 頁)以上和第一簇營養葉(第 158 頁)以下的幼苗部分。出芽期間胚軸發育成彎曲的半圓形以保護胚芽(第 154 頁)。胚芽出土後，它伸直成爲莖的第一節。

下胚軸(名) 子葉(第 154 頁)以下根以上的幼苗部分，它是胚種子植物的莖。

可滲透的(形) 允許分子(第 103 頁)和離子(第 101 頁)通過的，例如可滲透性膜允許分子或離子通過，但不讓流體通過；此種膜對流體是不能滲透的，但對溶劑、溶質和離子則是可滲透的。

不滲透的(形) 不允許分子或離子通過，即它是不可滲透的(↑)。

不可流過的(形) 不允許流體通過，但可能是可滲透的(↑)。

植物(名) 能從純無機物質即氣體或溶於水的純無機物質中自己製造養料的生物；植物通常不能行動(第 194 頁)。

莖(名) 植物上長着葉子或蓓蕾(第 213 頁)的部分。大部分莖長在地面上，但有些莖長在地面下。

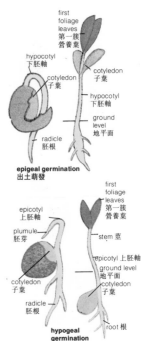

epigeal germination
出土萌發

hypogeal germination
留土式萌發

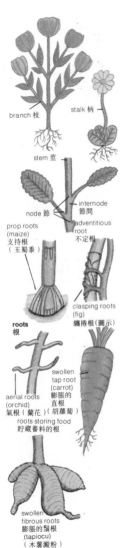

branch 枝

stalk 柄

stem 莖

node 節

internode 節間

adventitious root 不定根

prop roots (maize) 支持根（玉蜀黍）

clasping roots (fig) 纏捲根（圖示）

roots 根

swollen tap root (carrot) 膨脹的直根（胡蘿蔔）

aerial roots (orchid) 氣根（蘭花）

roots storing food 貯藏養料的根

swollen fibrous roots (tapiocu) （木薯澱粉） (cassava) （木薯根）

stalk (*n*) a structure without branches, supporting one or several objects at the top. It may be found in objects, or in plants, e.g. a straight, unbranched stem with a flower or flowers, or in animals, e.g. the eye-stalk of a crab.

node² (*n*) the part of a stem (↑) from which leaves (p.158) grow. Adventitious roots (↓) may also grow from nodes. **nodal** (*adj*).

internode (*n*) the part of a stem (↑) between two nodes (↑); no leaves grow on it.

lenticel (*n*) a small, raised hole, usually elliptical, formed in woody (p.161) stems; it allows gases to enter and leave the stem.

root (*n*) the part of a plant that grows downwards into the earth; it differs from a stem in not having leaves or buds. Its functions are: (1) to hold the plant firmly in the ground; (2) to absorb water from the earth; (3) to absorb inorganic salts (p.115) from the earth. A root-cap protects the end of the root.

tap root the main root, growing straight downwards, of dicotyledon (p.155) plant; other roots branch from it. It is formed from the radicle (p.154) of the seed. The tap roots of some plants become swollen with stored food, e.g. carrot.

adventitious root a root growing from any node (↑) on a stem. There are several kinds of adventitious root, e.g. a prop root (↓) is an adventitious root. Adventitious roots also grow from bulbs.

prop root a root growing from a node (↑), on a stem, down into the earth; it supports the stem.

fibrous root one of many branching roots growing from the bottom of a stem, usually with no tap root in the system of roots. Monocotyledons (p.155) have fibrous roots, and no tap root.

柄；梗（名） 沒有分枝，頂端托着一個或幾個物體的結構。物體或植物中都有此結構。例如挺直無枝有花或花簇的莖，動物體上亦有此結構，例如蟹的眼柄。

節（名） 莖（↑）上長葉子（第 158 頁）的部位。不定根（↓）也可從節長出。（形容詞爲 nodal）

節間（名） 莖（↑）上兩個莖節（↑）間的部位；其上不長葉子。

皮孔（名） 在木質的（第 161 頁）樹莖上形成的隆起小洞，通常爲橢圓形的；爲莖上氣體的進出口。

根（名） 植物向地下生長的部分；它與莖的區別是根沒有葉或蓓蕾。其功能是：(1) 使植物牢固定在土裏；(2) 從土中吸收水分；(3) 從土中吸收無機鹽（第 115 頁）。根冠保護根尖。

直根 雙子葉植物（第 155 頁）向地下直生的根，直根上長出其他側根。直根是由種子中的胚根（第 154 頁）形成的。有些植物的直根因儲藏着養料而膨大。例如胡蘿蔔。

不定根 在莖上從任一節（↑）長出的根。不定根分幾種，例如支持根（↓）是一種不定根。鱗莖上也可長不定根。

支持根 在莖上從節（↑）長出向下伸入土中的根；它支撐着莖。

鬚根 從莖基部長出的許多側根之一，根系中通常無直根。單子葉植物（第 155 頁）有鬚根，無直根。

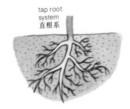

tap root system 直根系

fibrous root system 鬚根系

root hair a small, hair-like growth from a root, with a very thin cell wall. Large numbers of root hairs grow near the end of the root; their function is to take in water and inorganic salts from the earth. They have a short life, as they die when the root grows downwards.

piliferous layer the part of the surface of a root bearing root hairs; it covers only part of the root, near the end. A layer of cells near the outside of the root forms the piliferous layer.

leaf (*n*) (*leaves*) a flat structure, usually green, growing on a stalk from the node (p.157) of a stem or branch of a plant. The important functions of a leaf are photosynthesis (↓) and transpiration (↓). **leaflet** (*n*) a small leaf.

margin (*n*) (1) the edge of a leaf, which can have various shapes depending on the species of plant, *see diagram.* (2) the edge, and the flat space near the edge, of an object, e.g. the margin of an insect's wing. **marginal** (*adj*).

petiole (*n*) the stalk of a leaf; it grows from a node (p.157).

venation (*n*) the arrangement of small tubes seen in a leaf; the tubes are called *veins*, and the veins act as a support and conduct water and nutrients (p.171) for the leaf. There are two kinds of venation, parallel and net (see p.158).

foliage (*n*) the mass of leaves on a plant. A foliage leaf is the common kind of leaf, different from special leaves such as cotyledons (p.154) and other leaf-like parts of flowers.

axil (*n*) the angle between the upper side of a petiole and the stem of the plant. **axillary** (*adj*).

modified (*adj*) changed in shape and function. A modified leaf has a different shape from a common leaf, and has a different function, e.g. (1) leaf tendril, in which the leaf is modified to form a long, thin, thread-like shape (tendril) which twists round an object in order to support the plant, (2) leaf spine, such as a spine on a cactus which is a small, sharp-pointed, modified leaf.

根毛 根上長出的細毛狀物，有極薄的細胞壁，大量根毛長在根尖附近；根毛的功能是從土壤中吸收水分和無機鹽。根毛的壽命很短，在根往下生長時，根毛即脫落死去。

根毛層 長有根毛的根部表面部分；它只覆蓋接近根尖的根部，靠近根外部的一層細胞組成根毛層。

葉片(名) 通常是綠色的扁平組織，長在從植物莖或枝的莖節(第 157 頁)上伸出的葉柄上。葉的重要功能是光合作用(↓)和蒸騰作用(↓)。(名詞 leaflet 意爲小葉片)

葉緣；緣(名) (1) 葉的邊緣，其形狀各不相同，取決於植物的種，(見圖)；(2) 一個物體的邊，接近邊緣的平坦的空間；例如昆蟲翅膀的邊。(形容詞爲 marginal)

葉柄(名) 葉子的柄；它從莖節(第 157 頁)長出。

葉脈序(名) 葉內可見的諸多小管排列；這些管稱爲葉脈，葉脈起支撐作用並給葉子輸送水分和養料(第 171 頁)。葉脈序有兩種，平行脈序和網狀脈序(見本頁)。

簇葉(名) 植物上密聚的葉子。簇葉是一種普通的葉子，不同於子葉(第 154 頁)。

葉腋(名) 葉柄上面的邊和植物莖間的夾角。(形容詞爲 axillary)

變態的(形) 指形狀和功能上的改變。變態葉與普通葉的形狀和功能都不同，例如 (1) 葉捲鬚，葉子變態形成長而細的絲狀(捲鬚)，纏繞在物體上以支持植物；(2) 葉刺，例如仙人掌上的刺是小而尖的變態葉。

cell of root 根的細胞

root hair 根毛

cell of piliferous layer 根毛層的細胞

root hairs 根毛

simple leaf 單葉

root hairs 根毛

leaf blade 葉片

leaflet 小葉

petiole (stalk) 葉柄(柄)

compound leaf 複葉

leaflet 小葉

leaf blade 葉片

different kinds of leaves 不同種類的葉

margin 葉緣

margins of leaves 葉緣

axil 葉腋

leaf 葉

petiole 葉柄

stem 莖

leaf 葉

vein 葉脈

venation 葉脈序

tendril for clasping 纏繞的捲鬚

leaf tendrils (gloriosa) 葉捲鬚

deciduous (*adj*) losing all leaves at a certain season of the year, e.g. many deciduous trees lose their leaves in autumn.

evergreen (*adj*) bearing leaves at all times, e.g. pine trees are evergreen as they never lose all their leaves at one time.

chlorophyll (*n*) a green substance which gives leaves their colour. Chlorophyll takes in energy from sunlight, and a plant uses this energy to make food for itself (the process known as photosynthesis (↓)).

photosynthesis (*n*) in green plants, sunlight is absorbed by chlorophyll (↑) contained in plastids (p.141). This energy is used to synthesize (p.136) organic (p.131) compounds from carbon dioxide and water. These carbohydrates (p.173) are mainly stored as starch. A small part is used as the substrate (p.146) for respiration.

photosynthetic (*adj*)

stoma (*n*) (*stomata*) a very small hole in the surface of a leaf; it has two guard cells (↓) round it. Oxygen and carbon dioxide from the air enter through the stomata; oxygen, carbon dioxide and water vapour leave through the stomata. See transpiration (↓) and respiration (p.191).

guard cell one of a pair of bean-shaped cells round a stoma (↑). The cell-wall facing the stoma is thick. When the turgor (p.139) pressure is high, the stoma is open as the guard cell wall is curved, but when the turgor pressure is low, the stoma is closed.

root pressure the pressure which forces water up from a root into the stem of a plant. The osmotic pressure of sap (p.160) in the cells of the root causes root pressure.

transpiration (*n*) the loss of water vapour by plants, mainly through their stomata (↑). This action helps root pressure by drawing water up the stem, and a plant thus increases the amount of water passing up its stem. **transpire** (*v*), **transpiratory** (*adj*).

potometer (*n*) a piece of apparatus (p.88) for measuring the rate of transpiration (↑), or the rate of movement of the transpiration stream in the stem of a plant.

guard cells and stomata
保衞細胞和氣孔

guard cell (open) 保衞細胞 (開)

stoma 氣孔

guard cell (closed) 保衞細胞 (閉)

water vapour leaves 水蒸汽離開

photosynthesis 光合作用

plant food material 植物的養料

transpiration 水蒸汽離開

pull 拉

plant processes 植物的過程

root pressure 根壓

落葉的(形) 在每年的一定季節葉子落盡的。例如很多落葉樹在秋季落葉。

常綠的(形) 任何時侯都長有葉子。例如松樹不在同一個時候落盡葉子是常綠樹。

葉綠素(名) 賦予葉子顏色的綠色物質。葉綠素從陽光吸取能量,植物利用這種能量爲自己創造養料(這一過程稱爲光合作用(↓))。

光合作用(名) 綠色植物質體(第 141 頁)內所含的葉綠素(↑)吸收陽光。這種能量用於從二氧化碳和水合成(第 136 頁)有機(第 131 頁)化合物。這些碳水化合物(第 173 頁)主要作爲澱粉儲藏起來。小部分用作呼吸的基質(第 146 頁)。(形容詞爲 photosynthetic)

氣孔(名) 葉面上極小的孔;其周圍有兩個保衞細胞(↓)。氧和二氧化碳從空氣中進入氣孔;氧、二氧化碳和水蒸汽經氣孔出去。參見蒸騰作用(↓)和呼吸作用(第 191 頁)。

保衞細胞 氣孔(↑)周圍的一對豆狀細胞之一。對着氣孔一面的細胞壁是厚的。膨壓(第 139 頁)非常高時,隨保衞細胞壁彎曲,氣孔打開;當膨壓低時,氣孔閉合。

根壓 迫使水從根部上升到植物莖部的壓力。根細胞中的胞液(第 160 頁)的滲透壓產生根壓。

蒸散作用;蒸騰作用(名) 植物主要通過氣孔(↑)散逸水蒸汽。這種作用有助於根細胞將水汲升到莖部,植物因此增加了汲入莖部的水量。(動詞爲 transpire,形容詞爲 transpiratory)

蒸騰計(名) 一種儀器裝置(第 88 頁),供測量蒸騰(↑)速度或植物莖中水分蒸騰運動的速度。

xylem (n) the part of the vascular system (p.178) containing pipe-like conducting vessels and wood (p.161). The vessels conduct water from the root to all parts of the plant. The wood contains dead cells and provides support for the plant stem. Water passes up the xylem vessels pushed by root pressure (p.159) and pulled by transpiration (p.159).

phloem (n) the part of the vascular system (p.178) containing tube-like conducting vessels in which dissolved food material passes from the leaves to all parts of a plant.

木質部(名) 含有管狀輸導管和木質(第 161 頁)的維管系統(第 178 頁)部分。輸導管將水從根部輸送到植物各部分。木質含有死細胞並支撐植物莖部。藉根壓(第 159 頁)的推力和蒸騰作用(第 159 頁)的拉力將水上升到木質導管。

韌皮部(名) 含有管狀輸導管的維管系統(第 178 頁)。將溶解的養料從葉子輸送到植物各部分。

stem of young dicotyledon plant
年幼雙子葉植物的莖

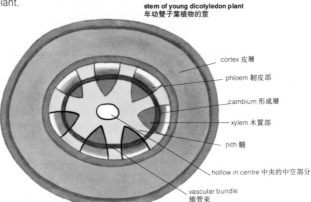

cortex 皮層
phloem 韌皮部
cambium 形成層
xylem 木質部
pith 髓
hollow in centre 中央的中空部分
vascular bundle 維管束

vascular bundle a bundle of xylem (↑) and phloem (↑) vessels passing from the end of a root to the end of a stem. The arrangement of vascular bundles in monocotyledons (p.155) is different from that in dicotyledons (p.155). Vascular bundles pass into leaves, where they form the veins, see venation (p.158).

sap (n) the liquid in a plant; it can be seen when a stem or a root is cut.

cortex (n) (1) the outer layer of a plant, surrounding the vascular system (p.178), *see diagram*. The cortex is covered by the cells of the outside surface of a plant. (2) the outer part of any structure, e.g. the cortex of a kidney, **cortical** (adj).

pith (n) the central sponge-like part of the stem of a plant; it helps in the storage of plant food.

維管束 從根尖通向莖端的一束木質(↑)和韌皮(↑)導管。單子葉植物(第 155 頁)的維管束排列與雙子葉植物(第 155 頁)不同,維管束通入葉子形成葉脈,(參見"葉脈"(第 158 頁))。

汁液(名) 植物的液體,割開莖或根時即可見。

皮層(名) (1) 包圍在維管系統(第 178 頁)的植物外皮層,(見圖)。皮層上佈滿植物的外皮細胞;(2) 任何結構的外面部分,例如腎的皮層。(形容詞為 cortical)

髓部(名) 植物莖中心部位的海綿狀部分,它有助於儲藏植物的養料。

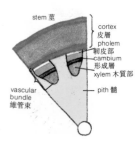

stem 莖
cortex 皮層
phloem 韌皮部
cambium 形成層
xylem 木質部
pith 髓
vascular bundle 維管束

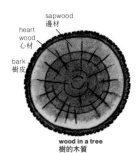

sapwood
邊材
heart
wood
心材
bark
樹皮

wood in a tree
樹的木質

cambium (*n*) a layer of cells between the xylem and the phloem of dicotyledons (p.155). The cells are actively dividing by binary fission (p.141) forming xylem on one side and phloem on the other; their action makes the stem thicker as the plant grows. Monocotyledons do not possess a cambium.

wood (*n*) the hard part of a stem formed from plant cells whose cellulose walls have been made stronger by a deposit of **lignin.** As the cells get older, they lose all cytoplasm, and only act as a support for the plant. **Heartwood** contains old cells which no longer conduct water or have any cytoplasm. **Sapwood** contains xylem vessels and is not as strong as heartwood. **woody** (*adj*).

tropism (*n*) the tendency of a plant to have curved growth under the effect of its outside conditions, e.g. light, water, gravity. **tropic** (*adj*).

phototropism of stems
莖的向光性

light
光線

phototropism (*n*) a tropism (↑) caused by light. Plant stems become curved so that the plant grows towards the light; this is a *positive tropism.* Some roots grow away from the direction of light; they are *negatively phototropic.* Leaves are always positively phototropic.
phototropic (*adj*).

hydrotropism (*n*) a tropism (↑) caused by the presence of water. Roots tend to grow towards water in a soil; they show positive hydrotropism.
hydrotropic (*adj*).

geotropism (*n*) a tropism (↑) caused by gravity. The main stems of a plant show negative geotropism as they grow upwards. The main roots of a plant are positively geotropic as they grow downwards. **geotropic** (*adj*).

形成層（名） 雙子葉植物（第 155 頁）木質部和韌皮部之間的一層細胞。這些細胞積極進行二分裂（第 141 頁），在一面形成木質層，在另一面形成韌皮層；形成層的作用是使植物生長時莖變粗。單子葉植物無形成層。

木質（名） 莖上由植物細胞組成的堅硬部分，這些細胞的纖維素壁由於木質素的沉積而變爲較堅硬。這些細胞隨着年老而喪失全部細胞質，以致僅起支持植物的作用。心材含有不能再輸送水分亦即沒有任何細胞質的老細胞。邊材含有木質輸導管，但不如心材那樣堅硬。（形容詞爲 woody）

向性（名） 在外界如光、水、重力的影響下植物有彎曲生長的傾向。（形容詞爲 tropic）

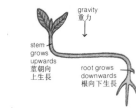

gravity
重力

stem
grows
upwards
莖朝向
上生長

root grows
downwards
根向下生長

geotropism of stem and root
莖和根的向地性

向光性（名） 指光引致的向性（↑）。植物爲了向着光而長成彎形；這是正向光性。有些根背光生長；這是負向光性的。葉都是正向光性的。（形容詞爲 phototropic）

向水性（名） 由於存在水而致的向性（↑）。根傾向於朝有水的土壤生長；表現正向水性。（形容詞爲 hydrotropic）

向地性（名） 重力引致的向性（↑）。植物的主莖往上長時表現負向地性。植物的主根是正向地性的，因爲它們向下生長。（形容詞爲 geotropic）

organ (*n*) a part of a plant or animal which has a definite structure and a particular function (p.140), e.g. (a) the stomach of an animal has a structure and a particular function to help in digesting (p.166) food; (b) the leaf of a plant has a definite structure and a particular function.

system² (*n*) all the organs (↑) and the tissues (p.140) concerned with a particular function of an organism, e.g. the nervous system, which includes nerves, brain and other structures, all concerned with the senses and body activities.

gland (*n*) an organ which takes water and substances from blood, and makes particular chemical substances needed by the organism (p.147). The compounds are passed to the outside of the gland in a solution, called a secretion (↓). **glandular** (*adj*).

secrete (*v*) to pass out from a cell a liquid containing substances made by the cell for use in other parts of the organism. In a gland (↑) special cells make the liquid. Secretion is both the process and the name of the liquid secreted. **secretion** (*n*), **secretory** (*adj*).

duct (*n*) a short pipe through which secretions (↑) leave a gland.

gut (*n*) the part inside an animal which changes food into chemical substances which are used by the body. A simple gut is seen in **hydra**, *see diagram*; it is a pipe with an opening at one end only. The gut in a human being has several organs and is a big system of pipes, glands and organs.

abdomen (*n*) a hollow part of the body of an animal which contains most of the gut (↑), e.g. the abdomen of an insect, or a man. **abdominal** (*adj*).

viscera (*n.pl.*) the organs in the abdomen, together with the heart and the lungs.

器官（名）　具有定形的結構並行使特定功能（第140頁）的動、植物體的一部分：例如 (a) 動物的胃具有一種有助消化（第166頁）食物的結構和特定功能；(b) 植物的葉有一定的結構和特定的功能。

系統（名）　與生物體的特定功能有關的全部器官（↑）和組織（第140頁），例如神經系統包括神經、腦和其他組織，所有這些都與感官和身體的活動有關。

腺體（名）　能從血液吸收水分及物質並製造生物體（第147頁）所需之特定化學物質的器官。這些化合物以一種溶液輸送到腺體外，稱爲分泌（↓）。（形容詞爲 glandular）

分泌（動）　指從細胞流出含有由細胞所製造的供生物體其他部分使用的某種物質的液體。腺體（↑）內的特殊細胞製造這種液體。分泌既指此過程，也指所分泌的液體。（名詞爲 secretion，形容詞爲 secretory）

導管（名）　指分泌物（↑）離開腺體所經的短管。

胃腸道（名）　動物將食物變成爲身體所用化學物質的身體的一部分。水螅體有簡單的腸，（見圖）；它只不過是一個在一端有開口的管，人體的腸有幾個器官，而且是由管、腺體和器官組成的一個大系統。

腹部（名）　動物體內爲大部分腸（↑）所佔的中空部分，例如昆蟲或人的腹部。（形容詞爲 abdominal）的。

內臟（名、複）　指腹部內的器官以及心和肺。

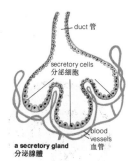

a secretory gland
分泌腺體

duct 管
secretory cells 分泌細胞
blood vessels 血管

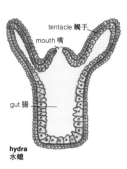

tentacle 觸手
mouth 嘴
gut 腸

hydra
水螅

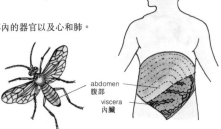

abdomen 腹部
viscera 內臟

absorb (v) to take in a liquid or a gas, e.g. (a) a brick absorbs water; (b) a plant absorbs dissolved mineral salts through its root hairs; (c) a leaf absorbs carbon dioxide through its stomata. The process by which the liquid or gas is absorbed, can be osmosis, diffusion, capillary attraction. **absorption** (n), **absorptive** (adj).

assimilate (v) to take in a material and to change it so that it becomes part of the thing it joins, e.g. when an organism assimilates food, the chemical substances are changed to become part of the body of the organism.

吸收 (動) 吸入液體或氣體，例如 (a) 磚吸收水；(b) 植物靠根毛吸收溶解的礦物鹽；(c) 葉靠其氣孔吸收二氧化碳。液體或氣體被吸收的過程可以是滲透、擴散或毛細管吸收。(名詞爲 absorption，形容詞爲 absorptive)

同化 (動) 攝入物質並將之轉變成身體自身的組成部分，例如生物同化營養時，化學物質轉變爲生物體自身的一部分。

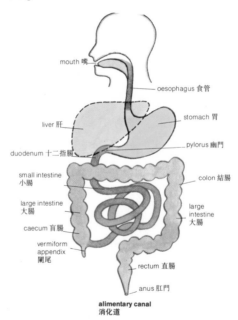

mouth 嘴
oesophagus 食管
stomach 胃
liver 肝
pylorus 幽門
duodenum 十二指腸
small intestine 小腸
colon 結腸
large intestine 大腸
large intestine 大腸
caecum 盲腸
vermiform appendix 闌尾
rectum 直腸
anus 肛門

alimentary canal
消化道

alimentary canal in all animals except very simple ones, a long tube, with an opening at each end, concerned with the digestion (p.166) and absorption (p.163) of food. Food enters at the mouth, passes through the alimentary canal, and the unassimilated material leaves the anus (p.165).

消化道 除極低級的動物外，其他一切動物都有一兩端開口的長管，其機能與消化（第 166 頁）和吸收（第 163 頁）食物有關。食物從口中進入，通過消化道，未被吸收的物質則由肛門（第 165 頁）排出體外。

oesophagus (*n*) the part of the alimentary canal between the throat and the stomach. Food is passed along it by peristalsis (p.169).

stomach (*n*) the part of the alimentary canal after the oesophagus, where the canal widens to form a bag-like structure. It has walls made of muscle, which press and turn the food into a liquid mass. A secretion (p.162) from the stomach walls acts on the food; it is called gastric juice (p.166).

pyiorus (*n*) the place where the stomach (↑) and the duodenum (↓) join; it is closed by a strong circular muscle, which opens and allows food to pass through after the food has been sufficiently mixed with gastric juice (p.166). **pyloric** (*adj*).

intestine (*n*) the part of the alimentary canal between the stomach and the anus (↓). In vertebrates (p.148) absorption of food takes place in the intestine. In reptiles, birds and mammals, the intestine is divided into a small intestine and a large intestine; the small intestine is smaller in diameter, but longer than the large intestine. **intestinal** (*adj*).

small intestine a long tube, with walls of muscle, between the stomach and the large intestine. The digestion (p.166) of food is completed and the absorption of food takes place in this part of the alimentary canal. The diameter of the small intestine is about 37 mm.

duodenum (*n*) the part of the small intestine, joined to the stomach (↑). It is joined to the liver and to the pancreas by ducts (p.162). Digestive (p.166) juices are very active in the duodenum. **duodenal** (*adj*).

ileum (*n*) in mammals, the part of the small intestine before the large intestine.

villus (*n*) (*villi*) a small, finger-like structure on the inside wall of the small intestine. In human beings, a villus is about 1 mm long and there are very large numbers of them; they increase the absorptive (p.163) surface of the intestine. Each villus contains blood vessels and the absorbed food material is carried away by the blood. The capillaries (p.179) surround a **lacteal**; the lacteal absorbs fats, which are carried away in lymph (p.182).

食道（名） 喉與胃之間的一部分消化道，食物蠕動（第 169 頁）通過食道。

胃（名） 食道之下的消化道部分，在此處食道下端變寬大形成袋狀結構。它有肌肉組成的壁，擠壓食物並將食物變爲液體物質。胃壁的分泌液（第 162 頁）作用於食物；此分泌液稱爲胃液（第 166 頁）。

幽門（名） 指胃（↑）和十二指腸（↓）的連接處；堅韌的環行肌封閉幽門，食物與胃液（第 166 頁）充分混合後，幽門張開，讓食物通過（形容詞爲 pyloric）

腸（名） 胃和肛門（↓）間的一段消化道，脊椎動物（第 148 頁）是在腸內吸收營養。爬蟲動物、鳥類和哺乳動物的腸分大腸和小腸；小腸的直徑比大腸細，但長度比大腸長。（形容詞爲 intestinal）

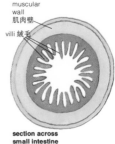

muscular wall 肌肉壁
villi 絨毛

section across small intestine
小腸的橫截面

小腸 胃和大腸之間一有肌壁的長管。食物在這段消化道完成消化（第 166 頁）和被吸收。小腸直徑約爲 37 mm。

十二指腸（名） 與胃（↑）連接的一段小腸。它藉導管（第 162 頁）與肝和胰相連。十二指腸中消化（第 166 頁）液起着積極作用。（形容詞爲 duodenal）

迴腸（名） 哺乳動物體內大腸之前的一段小腸。

絨毛（名） 小腸內壁上一種小的指狀結構。人體內的絨毛長約 1 mm，而且有無數的絨毛；絨毛擴大腸的吸收（第 163 頁）面積；每根絨毛內都有血管，被吸收的養料由血液送出。毛細血管（第 179 頁）包圍着乳糜管，乳糜管吸收脂肪，脂肪通過淋巴液（第 182 頁）輸送出去。

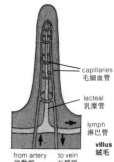

capillaries 毛細血管
lacteal 乳糜管
lymph 淋巴管
villus 絨毛

from artery 從動脈
to vein 至靜脈

large intestine a broad tube with muscular walls, connected to the end of the small intestine, and consisting of the colon (↓) and rectum (↓). It has the appearance of being divided into small bags. Its function is to absorb (p.163) water and mineral salts from the material passing through it, leaving faeces (p.169).

caecum (n) (caeca) a branch of the gut of an animal, closed at one end. Some vertebrates (p.148) have one or two caeca at the join of the small and large intestines; herbivorous (p.235) animals have a very large caecum which helps in the digestion of cellulose (p.138); carnivorous (p.235) animals have a very small caecum. In human beings, the caecum is a small bag at the start of the large intestine. **caecal** (adj).

vermiform appendix a small hollow finger-like structure leading from the caecum of some mammals; it contains lymphoid (p.182) tissue.

colon (n) the main part of the large intestine, concerned with the absorption of water and mineral salts. **colonic** (adj).

rectum (n) the end part of the large intestine; it has an opening at the end, either a cloaca (↓) or an anus (↓). Its function is to store faeces (p.169). **rectal** (adj).

anus (n) (ani) the opening at the end of the alimentary canal of mammals; through it passes the remains of undigested (p.166) food, i.e. faeces. **anal** (adj).

cloaca (n) the end of the alimentary canal of vertebrates, except mammals; tubes from the kidneys (p.185) and reproductive system (p.216) also enter into it. It has an opening to the exterior. Some invertebrates also have a cloaca. **cloacal** (adj).

大腸（名）　有肌肉壁的一段寬大導管，與小腸一端相連，由結腸（↓）和直腸（↓）組成。其外觀就像分成許多小袋似的。其功能是吸收（第163頁）通過大腸物質中的水分和礦物鹽，留下糞便（第169頁）。

盲腸（名）　動物腸的一個分支，其一端封閉。有些脊椎動物（第148頁）在大小腸連接處有一兩節盲腸，食草動物（第235頁）有一段很大的盲腸，以助消化纖維素（第138頁）；食肉動物（第235頁）的盲腸很小，人的盲腸是大腸起始段的一個小袋狀物。（形容詞爲caecal）

闌尾　一些哺乳動物由盲腸伸出的一個小的中空指狀結構，它含有淋巴（第182頁）組織。

結腸（名）　大腸的主要部分，其作用是吸收水分及礦物鹽。（形容詞爲colonic）

直腸（名）　大腸的末段；其末端有一開口，或爲泄殖腔（↓），或爲肛門（↓）。其作用是貯存糞便（第169頁）。（形容詞爲rectal）

肛門　哺乳動物消化道末端的開口；未消化（第166頁）的剩餘食物即糞便經此排出體外。（形容詞爲anal）

泄殖腔（名）　脊椎物動（哺乳動物除外）消化道的末端；腎（第185頁）和生殖系統（第216頁）都有導管與之相通，它有一開口通到外面。某些無脊椎動物也有泄殖腔。（形容詞爲cloacal）

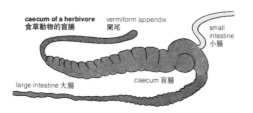

caecum of a herbivore
食草動物的盲腸

vermiform appendix
闌尾

small intestine
小腸

large intestine 大腸

caecum 盲腸

liver (*n*) a large gland (p.162) in many animals. In vertebrates it has many functions including (a) secretion of bile (p.168); (b) storage of glycogen (p.174); (c) deamination (p.186).

gall bladder a small vessel, in or near the liver; it stores bile (p.168) between meals. Bile leaves by the bile duct to enter the duodenum (p.164).

肝臟（名） 很多動物體內的一塊大腺體（第 162 頁）。脊椎動物的肝有多種功能，包括 (a) 分泌膽汁（第 168 頁）；(b) 貯存糖原（第 174 頁）；(c) 去氨基作用（第 186 頁）。

膽囊 在肝臟內或在肝臟附近的一個小器官；它在兩次進餐之間隔貯存膽汁（第 168 頁）。膽汁經由膽管進入十二指腸（第 164 頁）。

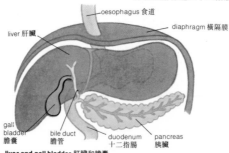

oesophagus 食道
liver 肝臟
diaphragm 橫隔膜
gall bladder 膽囊
bile duct 膽管
duodenum 十二指腸
pancreas 胰臟

liver and gall bladder 肝臟和膽囊

bile duct a tube from the liver to the duodenum (p.164); it conducts bile.

hepatic (*adj*) concerned with the liver, e.g. the hepatic artery takes blood to the liver.

pancreas (*n*) a gland in all vertebrates (p.148) except a few fishes; it produces an alkaline digestive (↓) juice, called pancreatic juice. Special groups of cells in the pancreas, called islets of Langerhans, produce insulin (p.209). The pancreatic duct, leads from the pancreas into the duodenum (p.164). **pancreatic** (*adj*).

digestion (*n*) the chemical decomposition (p.95) of food into substances which the body of an animal can absorb (p.163). The chemical change is brought about by digestive juices and, in most animals, takes place in the gut. **digestive** (*adj*), **digestible** (*adj*), **digest** (*v*).

juice (*n*) a liquid secretion (p.162) produced by animals for the purpose of digesting food.

saliva (*n*) a liquid secretion (p.162) produced by glands in the mouth. In land animals, saliva contains mucus (p.190), and provides lubrication (p.22) for food in the alimentary canal (p.163). **salivary** (*adj*).

膽管 肝通到十二指腸（第 164 頁）輸送膽汁的管道。

肝臟的（形） 有關肝的，例如肝動脈輸血入肝。

胰腺（名） 除少數魚類外，所有脊椎動物（第 148 頁）體內的都有的一個腺體；它產生鹼性的消化（↓）液，稱爲胰液。胰腺中有特殊的細胞團稱爲胰島，產生胰島素（第 209 頁）。胰腺管從胰腺通入十二指腸（第 164 頁）。（形容詞爲 pancreatic）

消化（名） 食物化學分解（第 95 頁）成爲動物體能吸收（第 163 頁）的物質。消化液引起化學變化，大部分動物中這種變化在腸內進行。（形容詞爲 digestive，digestible，動詞爲 digest）

胃液（名） 動物爲消化食物而產生的一種分泌（第 162 頁）液。

唾液（名） 口腔內腺體產生的一種分泌（第 162 頁）液。陸上動物的唾液內有粘液（第 190 頁），對消化道（第 163 頁）內的食物起着潤滑（第 22 頁）作用。（形容詞爲 salivary）

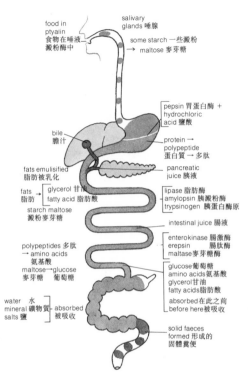

food in
ptyalin
食物在唾液
澱粉酶中

salivary
glands 唾腺

some starch 一些澱粉
→ maltose 麥芽糖

pepsin 胃蛋白酶 +
hydrochloric
acid 鹽酸

bile
膽汁

protein →
polypeptide
蛋白質 → 多肽

fats emulisified
脂肪被乳化

pancreatic
juice 胰液

fats
脂肪
─ glycerol 甘油
─ fatty acid 脂肪酸

lipase 脂肪酶
amylopsin 胰澱粉酶
trypsinogen 胰蛋白酶原

starch maltose
澱粉麥芽糖

intestinal juice 腸液

polypeptides 多肽
→ amino acids
氨基酸
maltose→glucose
麥芽糖 葡萄糖

enterokinase 腸激酶
erepsin 腸肽酶
maltase麥芽糖酶

glucose葡萄糖
amino acids氨基酸
glycerol甘油
fatty acids脂肪酸

water 水
mineral 礦物質
salts 鹽
─ absorbed
被吸收

absorbed在此之前
before here被吸收

solid faeces
formed 形成的
固體糞便

diagram of digestive processes
消化過程圖示

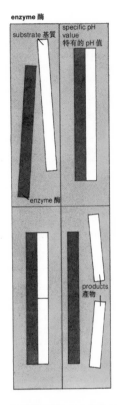

enzyme 酶

substrate 基質

specific pH
value
特有的 pH 值

enzyme 酶

products
產物

enzyme (*n*) any one of a large number of proteins (p.172) produced by all cells; an enzyme acts as a catalyst (p.99) on the chemical reactions which take place in all organisms (p.147). Most enzymes act on one substrate (p.146) only, so an organism produces a large number of enzymes. An enzyme is easily destroyed by heat, and by many chemical substances; it needs certain conditions, particularly a suitable pH value (p.116) before it will act. Most enzymes work inside cells; digestive (↑) enzymes are secreted into the gut and act on food there. **enzymatic** (*adj*).

酶(名) 一切細胞所產生的大量蛋白質(第 172 頁)中的任何一種;酶對所有生物體(第 147 頁)內發生的化學變化起催化(第 99 頁)作用,大部分酶只作用於一種受質(第 146 頁),所以生物體產生許多種酶。酶易被熱及多種化學物質所破壞;它需要一定的條件,特別是適當的 pH 值(第 116 頁)才能起作用。大部分酶在細胞內起作用;消化(↑)酶分泌入腸道,和腸內的食物起作用。(形容詞爲 enzymatic)

ptyalin (*n*) an enzyme (p.167) present in the saliva (p.166) of some mammals, including humans. It acts on cooked starch (p.174).

gastric (*adj*) concerned with the stomach, e.g. the walls of the stomach produce gastric juice (p.166).

pepsin (*n*) an enzyme (p.167) which decomposes proteins. It is secreted in gastric juice (↑) and needs the presence of acid in order to work. Gastric juice contains hydrochloric acid to provide an acidic solution for pepsin.

bile (*n*) a green, alkaline liquid produced by the liver of vertebrates (p.148); it passes through the bile-duct from the liver to the duodenum (p.164). The function of bile is to form an emulsion (p.111) with fat, so that enzymes (p.167) can act readily on fatty foods. **biliary** (*adj*), **bilious** (*adj*).

lipase (*n*) an enzyme (p.167) which decomposes fats (p.175) into alcohols (p.132) and organic (p.131) acids. It is produced by the pancreas (p.166) and is one of the enzymes in pancreatic juice. Lipase acts only in an alkaline solution; it is passed into the duodenum, where bile forms an alkaline solution.

invertase (*n*) an enzyme (p.167) which decomposes sucrose (p.174) into glucose (p.174) and fructose (p.174). Both plants and animals produce invertase. In human beings, invertase is produced in intestinal juice (↓).

amylases (*n.pl.*) a group of enzymes which decompose starch or glycogen (p.174); the products include maltose and glucose (p.174). Ptyalin is salivary (p.166) amylase.

amylopsin (*n*) amylase (↑) produced by the pancreas (p.166) and present in pancreatic juice (↓).

trypsinogen (*n*) an inactive form of the enzyme trypsin (↓); it is made active by an enzyme, enterokinase (↓).

trypsin (*n*) an enzyme (p.167) which further decomposes proteins after the action of pepsin (↑). It is produced from trypsinogen (↑) by the action of enterokinase (↓). Trypsin acts only in an alkaline solution.

enterokinase (*n*) an enzyme (p.167) produced by glands in the wall of the small intestine (p.164). It acts on trypsinogen (↑).

唾液澱粉酶（名） 一些哺乳動物，包括人的唾液（第 166 頁）中含有的酶（第 167 頁）。它對煮熟的澱粉（第 174 頁）起作用。

胃的（形） 有關胃的；例如胃壁分泌胃液（第 166 頁）。

胃蛋白酶（名） 一種分解蛋白質的酶（第 167 頁）。它分泌在胃液（↑）中並需要有酸才能起作用，胃液內含有的鹽酸，給胃蛋白酶提供酸性溶液。

膽汁（名） 脊椎動物（第 148 頁）的肝臟所產生的綠色鹼性液體；它從肝通過膽管流入十二指腸（第 164 頁）。膽汁的功能是與脂肪形成乳液（第 111 頁），使酶（第 167 頁）能與含脂肪食物起作用。（形容詞爲 biliary，bilious）

脂肪酶（名） 一種將脂肪（第 175 頁）分解爲醇（第 132 頁）和有機（第 131 頁）酸的酶（第 167 頁）。係由胰腺（第 166 頁）產生，是胰液中的酶之一。脂肪酶僅在鹼性溶液中起作用；它進入十二指腸與膽汁在此形成鹼性溶液。

轉化酶（名） 一種將蔗糖（第 174 頁）分解爲葡萄糖（第 174 頁）和果糖（第 174 頁）的酶（第 167 頁）。動、植物都能產生轉化酶，人體的轉化酶在腸液（↓）中產生。

澱粉酶（名、複） 分解澱粉或糖原（第 174 頁）的一組酶；分解產物包括麥牙糖和葡萄糖（第 174 頁）。唾液澱粉酶是唾液（第 166 頁）的澱粉酶。

胰澱粉酶（名） 胰腺（第 166 頁）產生並且存在於胰液（↓）中的澱粉酶（↑）。

胰蛋白酶原（名） 不活躍形式的胰蛋白酶（↓）；酶、腸致活酶（↓）可使這種酶活躍。

胰蛋白酶（名） 一種在胃蛋白酶（↑）起作用後，進一步分解蛋白質的酶（第 167 頁）。它是在腸致活酶（↓）的作用下，由胰蛋白酶原（↑）產生的。胰蛋白酶僅在鹼性溶液中起作用。

腸致活酶（名） 小腸（第 164 頁）壁內腺體產生的一種酶（第 167 頁）。它對蛋白酶原（↑）起作用。

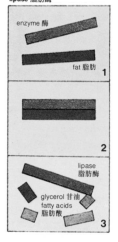

lipase 脂肪酶

enzyme 酶

fat 脂肪 **1**

2

lipase 脂肪酶

glycerol 甘油
fatty acids 脂肪酸 **3**

erepsin (*n*) a mixture of enzymes (p.167) which further decompose proteins after pepsin (↑) and trypsin (↑); it completes the decomposition of protein (p.172) to amino acids (p.172). Erepsin is secreted by glands in the wall of the small intestine (p.164).

maltase (*n*) an enzyme (p.167) which decomposes maltose (p.174) into glucose (p.174). It is secreted by glands in the wall of the small intestine (p.164).

pancreatic juice a secretion (p.162) of the pancreas; it contains lipase (↑), amylopsin (↑) and trypsinogen (↑). It passes to the duodenum (p.164) through the pancreatic duct.

intestinal juice a secretion (p.162) of glands in the walls of the small intestine (p.164). It contains invertase (↑), maltase (↑), enterokinase (↑), erepsin (↑) and completes the digestion of starch to glucose (p.174) and protein (p.172) to amino acids (p.172).

bolus (*n*) a round ball of food mixed with saliva ready to be passed down the oesophagus (p.164).

腸肽酶(名)　一種在胃蛋白酶(↑)和胰蛋白酶(↑)起作用後進一步分解蛋白質的混合酶(第167頁)；它能將蛋白質(第172頁)完全分解爲氨基酸(第172頁)。腸肽酶是由小腸(第164頁)壁內的腺體分泌的。

麥牙糖酶(名)　一種將麥牙糖(第174頁)分解爲葡萄糖(第174頁)的酶(第167頁)，是由小腸(第164頁)壁內的腺體分泌的。

胰液　胰腺的分泌液(第162頁)。它含有酯肪酶(↑)、胰澱粉酶(↑)和胰蛋白酶原(↑)。胰液經由胰腺管流入十二指腸(第164頁)。

腸液　小腸(第164頁)壁內腺體的分泌液(第162頁)。它含有轉化酶(↑)、麥芽糖酶(↑)、腸致活酶(↑)和腸肽酶(↑)，並完成將澱粉消化爲葡萄糖(第174頁)和蛋白質(第172頁)消化爲氨基酸(第172頁)的過程。

嚼過的食物團(名)　即將嚥入食道(第164頁)的唾液和食物的混合球團。

oesophagus
食道

contraction of muscle forces bolus down 肌肉收縮迫使食物團下移

food bolus
食物團

peristalsis
蠕動

wave of contraction passing down oesophagus
食道往下收縮的波形運動

peristalsis (*n*) a wave-like motion of muscles along a tube-shaped vessel. The muscles contract (p.38), in turn, along the vessel, so that an object in the vessel is pushed along it. Peristalsis in the alimentary canal pushes its contents along and also mixes the food with digestive juices. **peristaltic** (*adj*).

faeces (*n.pl.*) the remains of undigested food. Water is absorbed from this material in the colon (p.165) to form faeces. The faeces are stored in the rectum (p.165) and passed out through the anus (p.165) from time to time.

defaecation (*n*) the action of passing out faeces through the anus (p.165). **defaecate** (*v*).

蠕動(名)　肌肉沿着管狀器官作波浪形運動。肌肉沿管狀依次收縮(第38頁)，使物體沿管狀器官向前推進。消化道蠕動推動管內食物向前移動，並將食物和消化液混合。(形容詞爲 peristaltic)

糞便(名、複)　未消化食物的遺留物。在結腸中(第165頁)這些物質的水被吸收而形成糞便。糞便存於直腸(第165頁)中，通過肛門(第165頁)隨時排出體外。

排糞(名)　通過肛門(第165頁)排出糞便的動作。(動詞爲 defaecate)

tooth (n) (teeth) (1) in vertebrates (p.148) a hard, usually sharp, structure in the mouth used for biting food or for attacking and seizing other animals. In mammals the teeth vary in shape according to their use, and are fixed in hollows, called sockets, in the jaw bone. (2) any small pointed structure sticking out, such as the teeth in a cog-wheel, see diagram. (3) a small pointed part of a leaf margin (p.158).

permanent teeth the second set of teeth which mammals (p.150) grow. The first set grows in young mammals; it contains fewer teeth than the second set, having no molars.

enamel (n) a hard substance forming the outer cover of teeth in mammals.

dentine (n) a hard bone-like substance which forms the main part of a tooth.

pulp-cavity a hollow inside the dentine of a tooth. It contains nerves, blood vessels, tissues and cells producing dentine. A narrow opening in the root of the tooth allows nerves and blood vessels to enter the pulp-cavity.

crown (n) the outside part of a tooth which is seen in the mouth; it is covered in enamel (↑).

incisor (n) a tooth with a cutting edge in the front of the mouth of a mammal; incisors are used for biting.

canine (n) a tooth which is pointed and sharp and grows behind the incisors. Herbivorous (p.235) animals have very small canine teeth, or none; carnivorous (p.235) animals have large canine teeth.

molar (n) a tooth with a flat surface, usually with two or three roots, growing at the back of the mouth. Molars are used for grinding food. There are no molars in the first set of teeth of a mammal.

premolar (n) a tooth similar to a molar, but with one or two roots; premolars grow between molars and canine teeth, and are present in the first set of teeth of a mammal.

dental (adj) concerned with teeth, e.g. dental decay. Compare **dentate** which describes an animal possessing teeth, or a leaf with small pointed parts. See tooth (↑).

牙；齒(名)　(1) 脊椎動物(第 148 頁)口腔內有的一種堅硬的，通常是尖利的結構，用於咬嚼食物，攻擊和攫食其他動物。哺乳動物的牙齒形狀因其用途而不同，牙齒固定在頜骨的空腔裏，此空腔稱爲牙槽；(2) 任何向外突出的小牙或尖的結構，例如齒輪上的輪牙，(見圖)；(3) 葉緣(第 158 頁)上小而尖的部分。

恆牙　哺乳動物(第 150 頁)長出的第二套牙。幼小的哺乳動物長出第一套牙，數目比第二套牙少，且沒有臼牙。

琺瑯質(名)　哺乳動物牙齒外層的堅硬物質。

牙質(名)　構成牙齒主要部分的堅硬骨狀物質。

髓腔　在牙質內的空腔，內有神經、血管、組織和產生牙質的細胞。牙根部有一狹口，神經和血管經狹口進入牙髓腔。

牙冠(名)　口腔內可見的牙齒的外側部分，表面有琺瑯質(↑)。

切牙；門牙(名)　位於哺乳動物口腔前部有刃口的牙齒，切牙用於咬齧。

犬牙(名)　長在切牙之後的尖利的牙齒。食草動物(第 235 頁)的犬牙很小或者沒有，食肉動物(第 235 頁)的犬牙則很大。

臼牙(名)　長在口腔後部的牙齒，牙面平，通常有兩三個牙根。臼牙用於磨碾食物，哺乳動物的第一套牙沒有臼牙。

前臼牙(名)　與臼牙相似的牙，但只有一兩個牙根，前臼牙長在臼牙和犬牙之間，出現在哺乳動物的第一套牙中。

牙齒的(形)　與牙有關的；例如牙腐蝕。比較 **dentate**(齒狀的、有齒的)，這個詞描述有牙齒的動物，或有小而尖的部分的葉片。(見"牙；齒"(↑))。

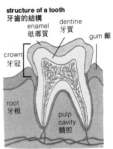

structure of a tooth
牙齒的結構
enamel 琺瑯質
dentine 牙質
gum 齦
crown 牙冠
root 牙根
pulp cavity 髓腔
capillaries and nerves 毛細血管和神經

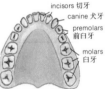

incisors 切牙
canine 犬牙
premolars 前臼牙
molars 臼牙
permanent teeth 恆牙

teeth 輪牙
a cog wheel 齒輪

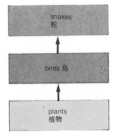

a simple food chain
簡單的食物鏈

snakes
蛇

birds 鳥

plants
植物

nutrition (n) the whole process of taking in food, digesting it, and using it to provide energy for living and materials for growth. The four stages of nutrition are: ingestion, digestion (p.166), absorption (p.163) and assimilation (p.163). **nutritious** (adj), **nutritional** (adj).

nutrient (n) a substance that can be used in the nutrition of an organism, e.g. (a) carbon dioxide is a nutrient for plants; (b) starch is a nutrient for human beings. **nutritive** (adj).

food chain a group of organisms (p.147) arranged in an order showing how each organism feeds on and obtains energy from the one before it, and is eaten and provides energy for the one after it. There are usually three or four organisms in a food chain. The first organism is a green plant, which obtains its food from inorganic compounds. The second organism is an animal which feeds on plants, a herbivorous animal. The third organism is an animal which eats herbivorous animals, a carnivorous animal. The fourth organism is a carnivorous animal which feeds on smaller carnivores. A typical food chain is: grass – insect – bird – snake.

營養 (名) 吞入和消化食物，為生存提供能量和為生長提供物質的整個過程。取得營養的四個步驟是：進食、消化 (第 166 頁)、吸收 (第 163 頁) 和同化 (第 163 頁)。(形容詞為 nutritious，nutritional)

營養素；養分 (名) 能用作生物體營養的物質，例如 (a) 二氧化碳是植物的營養素；(b) 澱粉是人體的營養素。(形容詞為 nutritive)

食物鏈 按某一順序排列的生物 (第 147 頁) 群落，以順序表示每種生物如何以排在其前的生物為食從而獲得能量，以及此生物如何為排在其後的生物吃掉並為之提供能量，一個食物鏈中通常有三或四種生物。第一種生物是從無機化合物中獲取養分的綠色植物，第二種生物是以植物為食的食草動物，第三種生物是以食草動物為食的食肉動物。第四種生物是以較小的食肉動物為食的食肉動物。典型的食物鏈是：草——昆蟲——鳥——蛇。

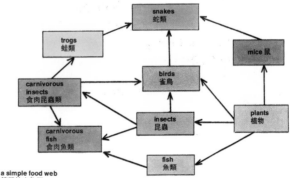

a simple food web
簡單的生物網

snakes
蛇類

trogs
蛙類

mice 鼠

carnivorous insects
食肉昆蟲類

birds
雀鳥

plants
植物

carnivorous fish
食肉魚類

insects
昆蟲

fish
魚類

food web an organism may feed on more than one organism *before* it in a food chain, and that organism may provide food for more than one organism *after* it in the food chain. Many food chains can thus be joined together to form a food web.

食物網 一種生物可吃食物鏈中排於其前的一種以上的生物，而此種生物又可為排在其後的一種以上的生物提供養料。很多食物鏈連結起來就形成食物網。

metabolism (*n*) the chemical reactions (p.96) which take place in an organism (p.147) or part of an organism; the reactions are controlled by enzymes (p.167). Metabolism includes the decomposition of organic compounds, as in digestion, and the synthesis (p.136) of new compounds, as in the production of digestive juices. **metabolic** (*adj*).

amino acid an organic (p.131) acid with an amino group of atoms ($-NH_2$) in the molecule. The formula of an amino acid is shown in the diagram.

新陳代謝（名）　在生物體（第 147 頁）或生物體的某部位發生的化學反應（第 96 頁），反應受酶（第 167 頁）控制。新陳代謝包括有機化合物的分解（（如消化過程）和新化合物的合成（第 136 頁）（如產生消化液）。（形容詞爲 metabolic）。

氨基酸　分子中有氨基原子團（$-NH_2$）的有機（第 131 頁）酸。氨基酸的結構式如圖示。

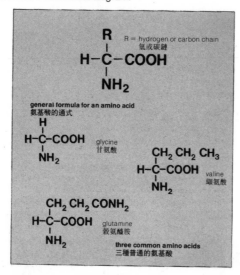

protein (*n*) a kind of compound formed by the chemical combination of many molecules of different amino acids. One protein molecule contains hundreds or thousands of amino acid molecules.

polypeptide (*n*) a compound formed by the chemical combination of molecules of amino acids, but containing fewer amino acids than a protein. Proteins are first decomposed by digestive juices to polypeptides. Polypeptides are then further decomposed to amino acids.

蛋白質（名）　由含不同氨基酸的許多分子化學結合形成的一類化合物。一個蛋白質分子含有數百個或數千個氨基酸分子。

多肽（名）　由氨基酸分子化學結合而成的一種化合物，但所含的氨基酸比蛋白質少。蛋白質先由消化液分解爲多肽。多肽再進一步分解爲氨基酸。

sources of carbohydrate in human diet
人類飲食中的碳水化合物來源

rice 米飯

bread 麵包

potato 馬鈴薯

carbohydrate (*n*) a compound consisting of carbon, hydrogen and oxygen with the atoms of hydrogen and oxygen in the same proportion as in water; the general formula of carbohydrate is $C_x(H_2O)_y$. Starch (p.174), cellulose (p.138) and all sugars (↓) are carbohydrates. Plants contain large amounts of carbohydrates, but animals contain much smaller amounts; in all organisms, carbohydrates are important in metabolism (p.172).

sugar (*n*) a crystalline (p.110) carbohydrate (↑) soluble in water and sweet to the taste. Plants produce various sugars in photosynthesis. A sugar is the simplest kind of carbohydrate.

monosaccharide (*n*) one of the simplest sugars (↑), containing either 5 or 6 carbon atoms. When a monosaccharide is decomposed, the products are no longer sugars, *see disaccharides*. The formula of a monosaccharide with 6 carbon atoms is $C_6H_{12}O_6$, and the atoms can be arranged in the molecule to form isomers (p.132). Glucose and fructose both have the same molecular formula $C_6H_{12}O_6$, but have different properties. Cell membranes are generally permeable to monosaccharides.

disaccharide (*n*) a sugar with a molecule made up of two monosaccharide (↑) molecules; the monosaccharide molecules can be the same, or different. Common disaccharides have 12 carbon atoms and a general formula of $C_{12}H_{22}O_{11}$. On hydrolysis (p.136) a disaccharide molecule is decomposed into two monosaccharide molecules. Isomers of disaccharides have different properties. Cell membranes are impermeable (p.156) to disaccharides and polysaccharides (↓).

polysaccharide (*n*) a carbohydrate with a molecule made up of many hundreds of monosaccharide molecules. The molecules are often fibrous. All plants store energy in polysaccharides; animals also use polysaccharides as an energy store, but to a lesser amount. On hydrolysis (p.136) polysaccharides are decomposed to disaccharides (↑) or monosaccharides (↑).

醣；碳水化合物（名） 指含碳、氫和氧的化合物，其中氫和氧原子的比例與水相同；碳水化合物的通式是 $C_x(H_2O)_y$。澱粉（第 174 頁）、纖維素（第 138 頁）和所有的糖類（↓）都是碳水化合物。植物含大量碳水化合物，但動物含的量卻少得多；在一切生物體中，碳水化合物在新陳代謝（第 172 頁）中都起重要作用。

糖（名） 一種晶狀（第 110 頁）碳水化合物（↑），可溶於水，味甜。植物在光合作用時產生不同的糖類。糖是最簡單的一種碳水化合物。

單醣；單糖（名） 含五個或六個碳原子的最簡單的糖（↑）。單醣分解後的產物不再是糖（參見雙醣）。含六個碳原子的單醣化學式爲 $C_6H_{12}O_6$，其原子可在分子中排列組成同分異構體（第 132 頁）。葡萄糖和果糖的分子式 $C_6H_{12}O_6$ 相同，但性質卻不同。單醣一般可滲透過細胞膜。

雙醣；雙糖（名） 由兩個單醣（↑）分子組成一個分子的糖；此兩個糖分子既可相同，也可不同。普通的雙醣有 12 個碳原子，通式爲 $C_{12}H_{22}O_{11}$。水解（第 136 頁）時一個雙醣分子分解爲兩個單醣分子。雙醣的同分異構體有不同的性質。雙醣和多醣（↓）都不能滲透過（第 156 頁）細胞膜。

多醣；多糖（名） 由數百個單醣分子組成爲一個分子的碳水化合物。其分子呈纖維狀。一切植物都將能量貯存在多醣中；動物也用多醣作貯存能量，但量很小。多醣水解（第 136 頁）時分解爲雙醣（↑）或單醣（↑）。

glucose (*n*) a monosaccharide (p.173) present in all plants and animals. It is produced in green plants by photosynthesis (p.159) from carbon dioxide and water, and stored as starch (↓). In animals, it is a last product of the digestion of carbohydrates and is stored as glycogen (↓). Its formula is $C_6H_{12}O_6$.

fructose (*n*) a monosaccharide (p.173) present in many plants. Its formula is $C_6H_{12}O_6$. Fructose can be absorbed by the gut (p.162).

maltose (*n*) a disaccharide, formula $C_{12}H_{22}O_{11}$, produced by the hydrolysis (p.136) of starch (↓). It is present in germinating (p.156) seeds, and is produced during the digestion (p.166) of starch. A molecule of maltose contains two molecules of glucose, chemically combined. Maltose, on hydrolysis, produces glucose.

sucrose (*n*) a disaccharide, formula $C_{12}H_{22}O_{11}$, present in many plants but not in animals. It is made from sugar cane. A molecule of sucrose contains one molecule of glucose (↑) combined with one molecule of fructose (↑). Sucrose is hydrolysed (p.136) to glucose and fructose (↑).

starch (*n*) a polysaccharide with a molecule formed from many molecules of glucose (↑) chemically combined. It is a white amorphous (p.110) substance, insoluble in water, and on hydrolysis (p.136) forms glucose. Plants store carbohydrates (p.173) as starch.

glycogen (*n*) a polysaccharide with a molecule formed from many molecules of glucose chemically combined; it is soluble in water. Animals and fungi (p.145) store carbohydrates (p.173) as glycogen. On hydrolysis (p.136) it forms glucose. In vertebrates it is present especially in the liver and in muscles.

葡萄糖（名） 存在於所有動、植物中的單醣（第173頁）。係由綠色植物通過光合作用（第159頁）以二氧化碳和水製成的。並作爲澱粉（↓）貯存起來。在動物體內它是碳水化合物消化的最後產物，並以醣原（↓）貯存。它的化學式是 $C_6H_{12}O_6$。

果糖（名） 存在於很多植物中的一種單醣（第173頁）。其化學式是 $C_6H_{12}O_6$。果糖能被腸（第162頁）吸收。

麥芽糖（名） 一種雙醣，化學式是 $C_{12}H_{22}O_{11}$，係由澱粉（↓）水解（第136頁）產生。它存在於發芽的（第156頁）種子中，也產生於澱粉消化（第166頁）的過程中。一個麥芽糖分子含兩個化學結合的葡萄糖分子。麥芽糖水解產生葡萄糖。

蔗糖（名） 一種雙醣，化學式是 $C_{12}H_{22}O_{11}$，存在於多種植物中，但不存在於動物中。可從甘蔗製得。蔗糖分子由一個葡萄糖（↑）分子和一個果糖（↑）分子組成。蔗糖水解（第136頁）成葡萄糖和果糖（↑）。

澱粉（名） 一種由許多葡萄糖（↑）分子化學結合爲一個分子的多醣。它是白色無定形（第110頁）的物質，不溶於水，水解（第136頁）時形成葡萄糖。植物以澱粉形式貯存碳水化合物（第173頁）。

醣原；肝醣（名） 一種由許多葡萄糖分子化學結合成一個分子的多醣；溶於水，動物和菌類（第145頁）以糖原形式貯存碳水化合物（第173頁）。醣原水解（第136頁）時形成葡萄糖。它存在於脊椎動物體內，特別是肝和肌肉中。

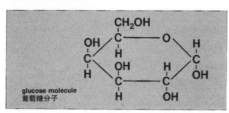

glucose molecule
葡萄糖分子

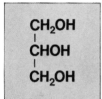

$$CH_2OH$$
$$|$$
$$CHOH$$
$$|$$
$$CH_2OH$$

molecule of glycerol
甘油分子

$$CH_3-(CH_2)_{14}-COOH$$

molecule of palmitic acid
棕櫚酸的分子

fat (*n*) true fat is an ester (p.132) of glycerol (↓) and one or more fatty acids (↓). On hydrolysis (p.136) it is decomposed to glycerol and a mixture of fats. A fat is solid at room temperature, an oil (p.134) is liquid. Other substances which can be dissolved in hot alcohol are also known as fats, but they are not true fats. Butter is a true fat. **fatty** (*adj*).

glycerol (*n*) an alcohol (p.132) with three hydroxyl groups and formula $C_3H_8O_3$, *see diagram*; it is a sweet, sticky, odourless, colourless liquid.

fatty acid an organic acid with a straight hydrocarbon (p.131) chain. The fatty acids present in plants and animals usually have an even number of carbon atoms; they have a carboxyl group of atoms (−COOH) at the end of the chain. Common acids in plants are stearic acid ($C_{17}H_{35}COOH$); palmitic acid ($C_{15}H_{31}COOH$); oleic acid ($C_{17}H_{33}COOH$).

diet (*n*) the different kinds of food and the amount of each kind of food eaten by a person or by a group of people. It is often described by the name of the food which provides most of the energy for nutrition, e.g. a rice diet, a wheat diet.

Calorie[2] (*n*) a kilocalorie, i.e. 1 Calorie (capital C) = 1000 calories, a unit of energy used to measure the energy value of different kinds of foodstuffs

foodstuff (*n*) a chemical substance in food used by animals. Foodstuffs are carbohydrate (p.173), protein (p.172) and fat (↑); they provide the energy for living, and the substances needed for growth and for replacing worn out tissues.

calorific value the number of Calories provided by a known mass of a foodstuff.

energy value another name for calorific value.

vitamin (*n*) an organic (p.131) substance which an animal must obtain in its food in order to be healthy. Vitamins are only needed in small amounts, and part of the need may be synthesized (p.136) by an animal, although this does not often happen. Different animals require different vitamins. Every vitamin is available from another plant or animal and many can be synthesized in factories. **vitaminize** (*v*).

脂肪(名) 真正的脂肪是甘油(↓)和一個或多個脂肪酸(↓)形成的酯(第 132 頁)。脂肪水解(第 136 頁)時分解爲甘油和多種脂肪酸的混合物。脂肪在室溫下是固體,油(第 134 頁)則是液體。能溶解於熱醇中的其他物質也稱爲脂肪,但不是真正的脂肪。奶油是真正的脂肪。(形容詞爲 fatty)

甘油(名) 含三個羥基的醇(第 132 頁),化學式爲 $C_3H_8O_3$,(見圖),是一種味甜、黏性。無臭、無色的液體。

脂肪酸 一種含直鏈烴(第 131 頁)的有機酸。動、植物體內存在的脂肪酸通常含偶數的碳原子;其鏈端有一羧基(−COOH)原子團。植物中常有的酸是硬脂肪酸($C_{17}H_{35}COOH$)、棕櫚酸($C_{15}H_{31}COOH$)、油酸($C_{17}H_{33}COOH$)。

飲食(名) 指一個人或一群人食用的各種食物和或每種食物的份量。人們常用能提供營養中大部分能量的食物的名字來描述飲食,例如米食、麥食。

大卡(名) 1 千卡即 1 大卡(大寫 C = 1000 卡),一種測量各種食料能值的能量單位。

食料(名) 動物飲食中的化學物質。食料含碳水化合物(第 173 頁)、蛋白質(第 172 頁)和脂肪(↑);它們提供生活所需能量、生長所需物質、及代替壞死組織所需的各種物質。

熱值;卡值 已知量食料物質所提供的大卡數。

能值 卡值的別稱。

維他命;維生素(名) 動物爲其健康必須從食物中獲得的有機(第 131 頁)物質。維他命的需要量不大,部分需要還可由動物來合成(第 136 頁),雖然這種情況不多。不同的動物需要不同的維他命。每種維他命都可從植物或動物獲得。許多種維他命可在工廠中合成(動詞爲 vitaminize)。

balanced diet a diet which supplies enough
energy for a person to live, enough protein for
him to grow new tissues with a balance between
carbohydrate (p.173), protein (p.172) and fat
(p.175). In addition a person needs enough
vitamins (p.175), mineral salts and roughage (↓).

malnutrition (*n*) a condition in which the body of
a person does not get a suitable diet. There can
be a lack of protein or a lack of carbohydrate
and fat or a lack of vitamins, or a lack of them
all.

roughage (*n*) a part of food which cannot be
digested, such as cellulose (p.138) in man's diet.
It helps peristalsis (p.169), and is a part of a
balanced diet (p.175).

deficiency disease a disease caused by the lack
of a vitamin or a mineral element such as iron or
calcium, or an amino acid needed in the diet.
The lack must be great to cause the disease.
Deficiency diseases arise from malnutrition
(p.176).

night blindness a deficiency disease (↑) in which
a person sees by day but badly by night or in
a bad light. It is caused by a lack of vitamin A.

xerophthalmia (*n*) a deficiency disease (↑) in
which the cornea (p.204) of the eye becomes dry
and the person finally becomes blind. It is
caused by a lack of vitamin A, greater than the
lack which causes night blindness (↑).

beri-beri (*n*) a deficiency disease (↑) in which
the nerves fail to act, particularly in the legs,
and a person is not able to walk. It is caused by
a lack of vitamin B_1.

pellagra (*n*) a deficiency disease (↑) in which the
skin becomes rough and brown, and a person's
mind becomes ill. It is caused by a lack of
vitamin B_7.

scurvy (*n*) a deficiency disease (↑) in which the
teeth become loose, and a person can readily fall
ill from other diseases. It is caused by a lack
of vitamin C.

rickets (*n*) a deficiency disease (↑) in which the
bones of young children become soft and their
legs are deformed. It is caused by a lack of
vitamin D.

平衡飲食　爲人的生存提供足夠能量的一種飲
食。這種飲食能提供足夠的蛋白質（第 172
頁）以促進新組織生長。同時保持體內碳水
化合物（第 173 頁）、蛋白質和脂肪（第 175
頁）之間的平衡。此外一個人還需要足夠的
維他命（第 175 頁）、礦物鹽和粗糙食物
（↓）。

營養不良（名）　人體沒有獲得適宜飲食時的狀
況。可能是缺乏蛋白質或缺乏碳水化合物和
脂肪，或缺乏維他命，或者全都缺乏。

粗糙食物（名）　不能消化的食物部分，例如人們
飲食中的纖維素（第 138 頁）。它有助於蠕動
（第 169 頁），是平衡飲食（第 175 頁）的一部
分。

營養缺乏症　由於缺乏維他命或礦物元素如鐵或
鈣，或缺乏飲食中必需的氨基酸而引起的一
種疾病。必定是嚴重營養缺乏才引起這種疾
病。營養不良（第 176 頁）引發營養缺乏症。

夜盲症　一種營養缺乏症（↑），病人白天能視物，
但晚間或光線弱時，視力很差。病因是缺乏
維他命 A。

乾眼症（名）　一種營養缺乏症（↑），病人的眼角
膜（第 204 頁）變得乾燥，最後失明。病因是
缺乏維他命 A，缺乏的量比引致夜盲症（↑）的
量還大。

腳氣病（名）　一種營養缺乏症（↑），病人的神
經，特別是腿部神經喪失功能，以致不能行
走，病因是缺乏維他命 B_1。

糙皮病（名）　一種營養缺乏症（↑），病人皮膚變
粗糙，呈褐色，記憶不佳，病因是缺乏維他
命 B_7。

壞血病（名）　一種營養缺乏症（↑），病人牙齒鬆
動，病人很易感染其他疾病。病因是缺乏維
他命 C。

佝僂病（名）　一種營養缺乏症（↑），幼童病人骨
骼變軟，兩腿變形，病因是缺乏維他命 D。

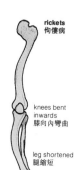

rickets
佝僂病

knees bent
inwards
膝向內彎曲

leg shortened
腿縮短

marasmus (*n*) a deficiency disease (↑) in children under the age of 5 years. The child is hungry, always crying, fails to grow, and its legs and body become thin. It is caused by a diet lacking in carbohydrates (p.173) and fats (p.175), i.e. a lack of foods which provide energy.

kwashiorkor (*n*) a deficiency disease (↑) in children under the age of 4 years. The child does not want to eat, has a swollen body, its hair becomes soft and changes colour, and it fails to grow. The disease is caused by a lack of protein (p.172) in the diet.

消瘦症（名）　五歲以下孩童所患的一種營養缺乏症（↑）。孩子常覺饑餓，常啼哭，發育受障，腿和身體都消瘦。病因是飲食中缺乏碳水化合物（第 173 頁）和脂肪（第 175 頁）。即缺乏提供能量的食物。

加西卡嚴重蛋白質缺乏綜合症（名）　四歲以下孩童所患的一種營養缺乏症（↑）。病孩厭食，身體浮腫，頭髮變軟，髮色改變，生長受障。病因是飲食中缺乏蛋白質（第 172 頁）。

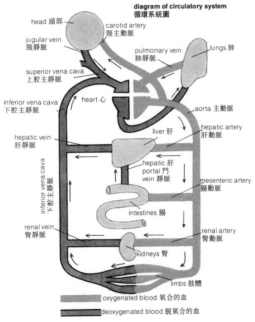

diagram of circulatory system
循環系統圖

circulatory system a system (p.162) of tubes and spaces in an animal, through which a liquid flows taking dissolved substances for the purpose of metabolism (p.172). It usually has an organ which pumps (p.33) the liquid round the system; in vertebrates the organ is the heart (p.181).

循環系統　動物體內管道和空腔組成的系統（第162 頁）。液體流過該系統時吸收溶解的物質進行新陳代謝（第 172 頁）。循環系統通常有一個器官（第 33 頁）泵送液體繞系統循環；脊椎動物體內的這一器官是心臟（第 181頁）。

vessel (*n*) a tube for conducting (p.45) fluids in an organism, e.g. blood-vessels in vertebrates (p.148) conduct blood to all parts of the body.

artery (*n*) a blood vessel conducting (p.45) blood from the heart (p.181) to the tissues of the body. In vertebrates (p.148) an artery has a thick, muscular wall. One artery, the aorta (↓) leaves the heart (p.181) and branches again and again until·smaller arteries reach every part of the body.

arteriole (*n*) a small artery (↑) with walls formed from smooth muscle. The autonomic nervous system (p.200) controls the muscle and thus controls the blood supply to the capillaries (↓).

vein (*n*) (1) a blood vessel conducting (p.45) blood from the tissues to the heart (p.181); its diameter is smaller than that of an artery. In vertebrates (p.148) a vein has a thin wall and contains valves (p.33) which allow blood to flow in one directory only. (2) a vascular bundle (p.160) in a leaf, *see* venation (p.158).

vascular system a system of vessels for the conduction (p.45) of fluids. In vertebrates (p.148) the fluid is usually blood and lymph (p.182), and the vascular system consists of the circulatory system (p.177) and the lymphatic system (p.182). In plants, the vascular system conducts dissolved mineral salts, water and synthesized (p.136) food materials.

aorta (*n*) in mammals (p.150) the main artery, which leaves the heart (p.181) and supplies blood to all parts of the body except the lungs. In human beings, blood passes through the aorta at a rate of 4 litres per minute.

vena cava in vertebrates, except fish, the main vein entering the heart (p.181) and bringing back blood from all parts of the body, except the lungs. The vein is divided in two, one vessel bringing blood from the head and arms, and the other vessel bringing blood from the rest of the body.

portal system a system of veins conducting blood from one capillary network (↓) to another capillary network, e.g. the hepatic portal system conducts blood from capillaries in the intestines (p.164) to capillaries in the liver (p.166).

管；脈管（名） 生物體內輸導（第 45 頁）流體的管道。例如脊椎動物（第 148 頁）的血管輸送血液到身體各部位。

動脈（名） 從心臟（第 181 頁）輸導（第 45 頁）血液到身體各組織的血管。脊椎動物（第 148 頁）的動脈有厚的肌肉管壁。一根動脈（即主動脈（↓））從心臟（第 181 頁）伸出，沿途一再分枝，直到較小的動脈分佈達身體的每個部分。

小動脈（名） 管壁由平滑肌構成的小動脈（↑）。自主神經系統（第 200 頁）控制肌肉，因此也控制向毛細血管（↓）的血液供應。

靜脈；脈（名） （1）從各組織輸導（第 45 頁）血液回到心臟（第 181 頁）的血管。其直徑比脈的小。脊椎動物（第 148 頁）的靜脈管壁薄並有只讓血液向一個方向流動的瓣膜（第 33 頁）；（2）葉片的維管束（第 160 頁），（ 參見 " 葉脈序"（158 頁））。

血管系統；維管系統 輸導（第 45 頁）流體的導管系統。在脊椎動物（第 148 頁），此流體通常是血液和淋巴（第 182 頁）。血管系統由循環系統（第 177 頁）和淋巴系統（第 182 頁）組成。植物的維管系統輸導溶解的礦物鹽、水分和合成（第 136 頁）養料。

主動脈（名） 哺乳動物（第 150 頁）的主要動脈，它自心臟（第 181 頁）發出，將血液供給除肺以外的身體各部分。人體內血液以每分鐘 4 升的速度流過主動脈。

主靜脈 脊椎動物（魚除外）的主要靜脈，它進入心臟（第 181 頁）帶回身體各部分（肺除外）的血。靜脈分兩根，一根血管帶回頭部和臂部的血，另一血管帶回身體其他部分的血。

門靜脈系統 將血從一個毛細血管網（↓）輸導入另一毛細血管網的靜脈系統，例如，肝門靜脈系統將腸（第 164 頁）毛細血管的血輸導入肝臟（第 166 頁）的毛細血管內。

thick muscular wall
厚的肌肉壁

an artery
動脈

artery
動脈

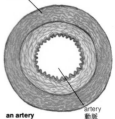

thin muscular wall
薄的肌肉壁

a vein
靜脈

vein
靜脈

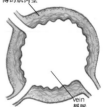

a valve in a vein
靜脈中的瓣膜

a network of capillaries
毛細血管網

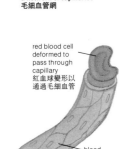

red blood cell
deformed to
pass through
capillary
紅血球變形以
通過毛細血管

blood
capillary
毛細血管

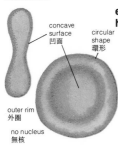

human red blood cells
人類的紅血球

concave
surface
凹面

circular
shape
環形

outer rim
外圈

no nucleus
無核

capillary (n) a blood vessel with a very small diameter (about 10 μm) and very thin walls. The capillary walls are permeable (p.156) to water and ions of inorganic salts, together with dissolved oxygen, glucose, amino acids and carbon dioxide.

capillary network the arrangement of capillaries (↑) in tissues, in which the vessels branch and rejoin and so cover all the tissues.

sinus (n) a hollow space; it differs from a vessel in having a varying diameter. Blood sinuses are present in the circulatory system of some animals, especially invertebrates (p.148).

blood (n) in animals, a liquid contained in vessels or sinuses (↑), passed round a circulatory system (p.177) by a muscular pumping action. It contains dissolved products of digestion and excretions (p.186), and carries oxygen in blood cells (↓). In vertebrates, it consists of plasma (↓), red (↓) and white blood cells (p.180) and platelets (p.180).

plasma (n) a clear, almost colourless liquid left after blood cells have been taken out of blood of vertebrates (p.148).

red blood cell in vertebrates (p.148), a cell in the shape of a flat disc, containing haemoglobin (↓), which gives it a red colour. Red blood cells possess cytoplasm and a membrane, but, in mammals, have no nucleus; they have a life of about four months. They are elastic (p.27) and easily deformed to pass through capillaries; they have no power of motion. In human beings, there are about five million red blood cells in 1 mm³ of blood.

erythrocyte (n) another name for red blood cell.

haemoglobin (n) a red-coloured substance present in the red blood cells (↑) of vertebrates (p.148) and in the blood of some invertebrates, e.g. the earthworm. Each animal species has a different kind of haemoglobin; all combine readily with oxygen to form oxyhaemoglobin. Oxyhaemoglobin readily decomposes to set free oxygen, so haemoglobin acts as a carrier of oxygen from the lungs to the tissues. Haemoglobin is dark red and oxyhaemoglobin is bright red.

毛細血管　直徑極小(約 10 μm)，管壁極薄的血管。水和無機鹽離子，以及溶解的氧、葡萄糖、氨基酸和二氧化碳都可滲透過(第 156 頁)毛細血管壁。

毛細血管網　毛細血管(↑)在組織內的分佈，網內血管先分枝，然後再滙合，直至遍佈所有的組織。

竇　(名)　一個中空空間，與血管不同之處是竇的直徑是變化的。一些動物，特別是無脊椎動物(第 148 頁)的循環系統都有血竇。

血液(名)　動物血管內或竇(↑)內的液體，由於肌肉的泵送作用在循環系統(第 177 頁)中流動。血內含有溶解的消化物和排泄物(第 186 頁)並以血細胞(↓)運送氧。脊椎動物的血液由血漿(↓)、紅血球(↓)、白血球(第 180 頁)和血小板(第 180 頁)組成。

血漿(名)　脊椎動物(第 148 頁)血液中的血細胞被取走後留下的一種透明、幾乎無色的液體。

紅血球　脊椎動物(第 148 頁)體內一種扁平碟狀的細胞，含有血紅蛋白(↓)使之呈紅色。紅血球有細胞質和細胞膜，但哺乳動物的紅血球沒有細胞核；紅血球的壽命約爲四個月。它們有彈性(第 27 頁)，容易變形以通過毛細血管；它們沒有運動能力。人體內 1 mm³ 的血液中含有五百萬個紅血球。

紅血細胞(名)　紅血球的另一名稱。

血紅蛋白(名)　脊椎動物(第 148 頁)的紅血球(↑)中和一些無脊椎動物如蚯蚓的血液中含有的一種紅色物質。每一動物物種都有種類不同的血紅蛋白，且都容易與氧結合形成氧合血紅蛋白。氧合血紅蛋白容易還原釋放出氧。所以血紅蛋白起着將氧從肺輸送到各組織的載體作用。血紅蛋白呈深紅色，氧合血紅蛋白呈鮮紅色。

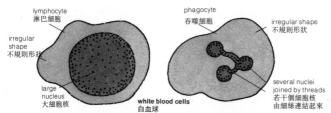

lymphocyte
淋巴細胞

irregular
shape
不規則形狀

large
nucleus
大細胞核

white blood cells
白血球

phagocyte
吞噬細胞

irregular shape
不規則形狀

several nuclei
joined by threads
若干個細胞核
由細絲連結起來

white blood cell in animal blood, a cell without colouring material; in vertebrates (p.148), it may be a phagocyte or a lymphocyte (p.182). In human beings, there are about 7000 white blood cells in 1 mm³ of blood.

leucocyte (*n*) another name for white blood cell.

platelet (*n*) a small, flat part of a cell from bone marrow (p.192), present in the blood of mammals. In human beings, there are about 250 000 platelets in 1 mm³ of blood. Their function is to start the process of blood clotting (↓).

thrombin (*n*) an enzyme formed from a protein in the blood by the action of blood platelets, or injured tissues.

fibrin (*n*) an insoluble protein formed from a soluble protein, **fibrinogen**, by the action of the enzyme, thrombin (↑). Fibrin forms long fibres (p.195) in blood clotting (↓).

clot (*n*) a twisted net of fibrin (↑) fibres, which traps red blood cells to form a solid mass. A clot prevents blood escaping from a wound, and bacteria (p.145) from entering a wound; serum (↓) leaks out of a clot. **clot** (*v*).

serum (*n*) (*sera*) the liquid obtained from clotted (↑) blood; it is blood plasma without fibrin and the other substances needed to clot blood.

tissue fluid the liquid bathing all cells in an animal. It supplies the cells with glucose, amino acids, and fats in solution, i.e. the products of digestion. It takes away from cells carbon dioxide and any other unwanted products. Tissue fluid bathes capillaries and the capillary wall acts as a permeable membrane between blood and tissue fluid, allowing diffusion of dissolved substances.

白血球 動物血液中一種無着色物的細胞；脊椎動物（第 148 頁）體內的這種細胞可以是吞噬細胞或是淋巴細胞（第 182 頁）。人體內每 1 mm³ 的血液中約有 7000 個白血球。

白血細胞（名） 白血球的另一名稱。

血小板（名） 由骨髓（第 192 頁）產生的細胞中小而扁的部分，它存在於哺乳動物的血液中。人體每 1 mm³ 的血液中含有 250,000 個血小板。其功能是起動血凝（↓）過程。

凝血酶（名） 由於血小板或受傷組織的作用，血液中的一種蛋白質所形成的一種酶。

纖維蛋白（名） 由於酶，即凝血酶（↑）的作用，由一種可溶性蛋白質即纖維蛋白原所形成的一種不溶性蛋白質。血液凝固（↓）時纖維蛋白形成長纖維（第 195 頁）。

凝塊；血塊（名） 纖維蛋白（↑）的纖維纏結的網，它將紅細胞網住形成一硬塊。凝塊阻止血液從傷口流出，防止細菌（第 145 頁）侵入傷口；血清（↓）從凝塊滲出。（動詞爲 clot）

血清（名） 從凝固（↑）的血液中獲得的液體；它是一種血漿其中不含纖維蛋白和凝血所需要的其他物質。

組織液 動物體內浸浴全部細胞的液體。它供給細胞葡萄糖、氨基酸、脂肪溶液等各種消化物。它從細胞中帶走二氧化碳和其他不需要的產物。組織液清洗毛細血管，管壁起着血液和組織液間的滲透膜作用，使溶解的物質能滲透。

homeostasis (*n*) the state of equilibrium (p.20) of the concentration (p.90) of dissolved substances in tissue fluid with the concentration of these substances in blood. The composition (p.95) of blood is controlled by various organs of the body so that it is kept constant (p.18) concerning: (a) osmotic pressure (p.139); (b) pH value (p.116); (c) concentration (p.90) of glucose; (d) concentration of amino acids. Homeostasis is the state of maintaining a constant composition of blood. Any change from these constant values causes damage to the body cells. **homeostatic** (*adj*).

heart (*n*) a hollow organ with muscular walls in the circulatory system (p.177) of an animal. Contractions (p.38) of the muscular walls pump blood round the system. In vertebrates (p.148), the heart is divided into auricles (p.182) and ventricles (p.182). Fishes have one auricle and one ventricle; amphibians have two auricles and one ventricle; birds and mammals have two auricles and two ventricles.

體內平衡（名） 組織液中溶解物質的濃度（第 90 頁）和血液中這些物質的濃度的平衡（第 20 頁）狀態。血液的成分（第 95 頁）受身體不同器官控制，所以能保持以下幾方面穩定（第 18 頁）：(a) 滲透壓（第 139 頁）；(b) pH 值（第 116 頁）；(c) 葡萄糖的濃度（第 90 頁）；(d) 氨基酸的濃度。體內平衡是保持血液成分穩定的狀態。這些恆定值的任何改變都會損傷身體細胞。（形容詞為 homeostatic）。

心臟（名） 動物循環系統（第 177 頁）中一個有肌肉壁的中空器官。當其肌肉壁收縮（第 38 頁）時泵送血液進入循環系統。脊椎動物（第 148 頁）的心臟分爲心房（第 182 頁）和心室（第 182 頁）。魚類有一個心房和一個心室；兩棲動物有兩個心房和一個心室；鳥類和哺乳動物有兩個心房和兩個心室。

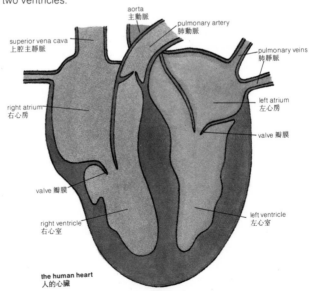

aorta
主動脈

pulmonary artery
肺動脈

superior vena cava
上腔主靜脈

pulmonary veins
肺靜脈

right atrium
右心房

left atrium
左心房

valve 瓣膜

valve 瓣膜

right ventricle
右心室

left ventricle
左心室

the human heart
人的心臟

auricle (*n*) a chamber in the heart which receives blood from veins (p.178) and passes it to a ventricle (↓). It has muscular walls which are thinner than those of a ventricle. In vertebrates (p.148) with two auricles, one receives blood from the body and the other receives blood from the lungs. **auricular** (*adj*).

atrium (*n*) (*atria*) alternative name for auricle.

ventricle (*n*) a chamber in the heart which receives blood from an auricle (↑) and pumps blood, with its thick, strong, muscular walls, round the circulatory system (p.177). In vertebrates (p.148) with two ventricles, one pumps blood to the body and the other pumps blood to the lungs. **ventricular** (*adj*).

systole (*n*) the stage of contraction (p.38) of heart muscle in the action of the heart. Ventricles contract after auricles. **systolic** (*adj*).

diastole (*n*) the stage of relaxation (p.196) of heart muscle in the action of the heart. **diastolic** (*adj*).

lymph (*n*) a colourless liquid consisting of tissue fluid and white blood cells (p.180), mainly lymphocytes (↓). **lymphoid** (*adj*).

lymphatic system an arrangement of very small tubes, called lymph capilaries, take away tissue fluid (p.180). The lymph capilaries join to form larger tubes, called lymphatics; and the lymphatics join to form a lymph vessel which passes lymph into a main vein (p.178) near the heart. The lymph capilaries, lymphatics and lymph vessels form the lymphatic system. Lymph capilaries have walls which are more permeable (p.156) than blood capilaries, so even bacteria pass into the lymphatic system. Lymphatics have valves similar to those in veins.

lymphatic (*n*) a tube conducting lymph (↑); it collects lymph from lymph capilaries.

lymph node a small organ on a lymphatic (↑) consisting of lymphoid tissue. It produces lymphocytes (↓), removes bacteria from lymph, and filters out foreign bodies. Present in mammals and birds.

lymphocyte (*n*) a white blood cell (p.180) with a large nucleus and little cytoplasm; it produces antibodies (p.238).

心耳 (名) 心臟內接收流自靜脈(第178頁)的血接着送入心室(↓)的一個腔室。心房的肌肉壁比心室的薄。脊椎動物(第148頁)的心臟有兩個心房，一個接收從身體來的血，另一個接收從肺部來的血。(形容詞為auricular)

心房 (名) 心耳的另一名稱。

心室 (名) 心臟內接收流自心房(↑)的血，並以其厚而有力的肌肉壁泵送血液流入循環系統(第177頁)的一個腔室。脊椎動物(第148頁)的心臟有兩個心室，一個將血泵送到全身，另一個泵血入肺。(形容詞為ventricular)

收縮期 (名) 心臟動作中的心肌收縮(第38頁)期，心室收縮在心房收縮之後。(形容詞為systolic)

舒張期 (名) 心臟動作中的心肌鬆弛(第196頁)期。(形容詞為diastolic)

淋巴 (名) 由組織和白血球(第180頁)，主要是淋巴細胞(↓)組成的無色液體。(形容詞為lymphoid)

淋巴系統 稱為淋巴毛細管的許多極小管的配置。這些管運送組織液(第180頁)。淋巴毛細管滙合成較大的管，稱爲淋巴管；淋巴管滙合成大淋巴管，它將淋巴送入心臟附近的主靜脈(第178頁)。淋巴毛細管，淋巴管和大淋巴管組成淋巴系統。淋巴毛細管壁的滲透力(第156頁)比毛細血管的大得多，甚至細菌都能進入淋巴系統。淋巴管還有與靜脈瓣膜相似的瓣膜。

淋巴管 (名) 輸送淋巴(↑)的導管；它滙集從淋巴毛細管流來的淋巴。

淋巴結 位於淋巴管(↑)上由淋巴組織構成的小器官。它產生淋巴細胞(↓)，清除淋巴管中的細菌並濾出異物。哺乳動物和鳥類的體內都有淋巴結。

淋巴細胞 (名) 一種細胞核大而細胞質很少的白血球(第180頁)；它產生抗體(第238頁)。

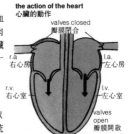

the action of the heart
心臟的動作

valves closed
瓣膜閉合

r.a 右心房 — l.a. 左心房
r.v. 右心室 — l.v. 左心室

valves open
瓣膜開啟

atria contract ventricles relax
心房收縮心室鬆弛

valves open
瓣膜開啟

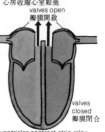

valves closed
瓣膜閉合

ventricles contract atria relax
心室收縮心房鬆弛

lymphatic system
淋巴系統

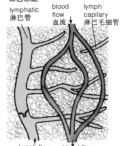

lymphatic 淋巴管
blood flow 血流
lymph capilary 淋巴毛細管

lymph flow 淋巴流
blood flow 血流

cells bathed in tissue fluid
細胞浸浴在組織液中

lymph vessels
大淋巴管

blood capilaries, oxygenated blood
毛細血管，氧合的血

blood capilaries, deoxygenated blood
毛細血管，脫氧合的血

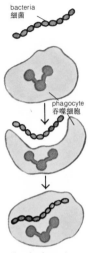

bacteria
細菌

phagocyte
吞噬細胞

action of a phagocyte
吞噬細胞的作用

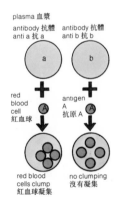

plasma 血漿

antibody 抗體
anti a 抗 a

antibody 抗體
anti b 抗 b

a

b

red
blood
cell
紅血球

antigen
A
抗原 A

A

A

red blood
cells clump
紅血球凝集

no clumping
沒有凝集

agglutination
凝集作用

phagocyte (*n*) a white blood cell (p.180), usually with several nuclei, with the power of amoeboid movement (p.144); it closes pseudopodia (p.144) round bacteria (p.145) and digests them. Phagocytes are important in defending animals against the attack of bacteria. **phagocytic** (*adj*).

phagocytosis (*n*) the action of phagocytes.

spleen (*n*) in vertebrates (p.148), an organ composed of lymphoid (↑) tissue, situated near the stomach. It is connected to the circulatory system (p.177). The spleen produces lymphocytes and takes worn out red blood cells (p.179) from the blood; it also stores red blood cells and supplies them to blood. **splenic** (*adj*).

septicaemia (*n*) a condition in which bacteria gets into the blood, and the blood is poisoned; the bacteria may come from a wound or may be bacteria causing a disease. Septicaemia causes a high temperature and red places on the skin.

blood group a group of people whose blood can be mixed without agglutination (↓). There are four blood groups, A; B; AB; O.

agglutination (*n*) sticking together. Red blood cells show agglutination, i.e. they stick together, when blood of different groups is mixed.

rhesus factor a substance present in the red blood cells of most people; such people are called rhesus positive (rh positive). The rest of the people do not have the substance present; they are rhesus negative. Rhesus negative persons do not possess an antibody (p.238) in their plasma (p.179) against the rhesus factor, but they can be given an antibody by a blood transfusion (↓). A rhesus negative woman bearing a rhesus positive child develops the antibody against the rhesus factor in the child. If the woman bears a second rhesus positive child, the unborn child can be damaged by the antibody in its mother's blood.

transfusion (*n*) the action of transferring blood from one person into another person. The blood given must be compatible (p.184). **transfuse** (*v*).

donor (*n*) a person who gives blood for a transfusion (↑) or who gives tissue or an organ to another person. **donate** (*v*).

吞噬細胞（名） 一種白血球（第 180 頁），通常有數個核，有變形運動（第 144 頁）的能力；能用偽足（第 144 頁）圍住細菌並將之消化。吞噬細胞極為重要，它保衛動物不受細菌侵襲。（形容詞為 phagocytic）

吞噬作用（名） 吞噬細胞的行動。

脾臟（名） 脊椎動物（第 148 頁）體內由淋巴（↑）組織構成的器官，位於胃附近。脾與循環系統（第 177 頁）相連。脾產生淋巴細胞並吞食血液中衰老的紅血球（第 179 頁）；它還貯存並將紅血球供給血液。（形容詞為 splenic）

敗血病（名） 細菌侵入血液令血液中毒的疾病；細菌可能從傷口侵入或者本身可能就是致病的細菌。敗血病引起高燒和皮膚上出現紅瘢。

血型 一些人的血能相混合而不凝集（↓）。血型有四種：A；B；AB；O。

凝集作用（名） 黏在一起。紅血球表現出凝集作用，即將不同血型的血相混，紅血球會黏在一起。

Rh 因子 大多數人紅血球中存在的一種物質；這些人稱為屬猴因子陽性的（Rh 陽性的）。其餘沒有這種物質的人，屬猴因子陰性的（Rh 陰性的）。猴因子陰性的人的血漿（第 179 頁）中沒有抗猴因子的抗體（第 238 頁），但可藉輸血（↓）給予抗體。胎兒猴因子陽性的猴因子陰性婦女能在孩子身上產生抗猴因子的抗體。如該婦女懷第二個猴因子陽性的孩子時，則該未出生的孩子可能受其母體血液中抗體的損害。

輸血（名） 將血液從一個人身上輸入另一人身上的行動。提供的血必須是可相容的（第 184 頁）。（動詞為 transfuse）

供血者；供體（名） 為輸血（↑）提供血液者，或將組織或器官獻給他人者。（動詞為 donate）

recipient (*n*) a person who is given blood in a transfusion (p.183), or is given tissue or an organ in a medical operation. **receive** (*v*).

受血者；受者(名)　輸血(第 183 頁)時接受血液者，或在醫療手術時接受組織或器官者。(動詞爲 receive)

human blood groups 人類血型

blood group of donor 供血者的血型	blood group of recipient 受血者的血型			
	O	A	B	AB universal recipient 萬能受血者
O universal donor 萬能供血者	✓	✓	✓	✓
A	✕	✓	✕	✓
B	✕	✕	✓	✓
AB	✕	✕	✕	✓

✓ compatible transfusions 相容的輸血
✕ incompatible transfusions 不相容的輸血

compatible (*adj*) of blood, being able to be given in a transfusion. Red blood cells can have antigens (p.237) present called A and B. Red blood cells of group A have antigen A; of group B have antigen B; of group AB have antigens A and B; and of group O have no antigens. The plasma can have antibodies (p.238) present, called anti-A (or a) and anti-B (or b). Plasma of group A has antibody b; of group B has antibody a; of group AB has no antibodies; of group O has antibodies a and b. Plasma with antibody a agglutinates (p.183) red blood cells with antigen A; the cells stick together to form a clot (p.180). In a transfusion, donor (p.183) and recipient (↑) should be of the same blood group to prevent agglutination. Other blood groups can be used if the plasma of the recipient does not agglutinate the red blood cells of the donor. In these cases, the donor's blood is compatible with the recipient's. The table shows blood groups compatible for transfusion. **compatibility** (*n*).

incompatible (*adj*) of blood, not being able to be given in transfusion. If incompatible blood is given in a transfusion, the donor's blood forms a clot in the supply tube.

相容的；適合的(形)　指血液能用於輸血。紅血球可含有稱爲 A 和 B 抗原(第 237 頁)。A 型的紅血球含有抗原 A；B 型的含有抗原 B；AB 型的含有抗原 A 和抗原 B；O 型的沒有抗原。血漿中可含有抗體(第 238 頁)，稱爲抗體 A(或 抗體 a)和抗體 B(或 抗體 b)。A 型血漿含有抗體 b；B 型的含有抗體 a；AB 型的沒有抗體；O 型的含有抗體 a 和抗體 b。有抗體 a 的血漿使帶抗原 A 的紅血球起凝集作用(第 183 頁)；細胞黏在一起形成凝塊(第 180 頁)。輸血時供血者(第 183 頁)與受血者(↑)應是同一血型以防止凝集。如果受血者的血漿不使供血者的紅血球凝集。那麼，其他的血型也能使用。在這種情況下供血者的血和受血者的血是相容的。上表表示輸血時相容的血型。(形容詞爲 compatibility)

不相容的(形)　指血液不能用於輸血。如果輸血時提供不相容的血，供血者的血會在供血管中結成凝塊。

antigen A 抗原 A
antibody anti-B
抗體抗 B
blood group A
A 血型

antigen B
抗原 B
antibody anti-A
抗體抗 A
blood group B
B 血型

no antigens
antibodies
anti-A anti-B
無抗原
抗體
抗 A 抗 B
blood group AB
AB 血型

antigens A, B
抗原 A, B
no antibodies
無抗體
blood group O
O 血型

human blood groups
人類的血型

diagram of a kidney
腎的圖示

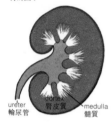

ureter
輸尿管

cortex
腎皮質

medulla
髓質

universal donor a person of blood group O, who can donate (p.183) blood to people with blood of any group.

universal recipient a person of blood group AB who can receive (↑) blood in a transfusion (p.183) from people with blood of any group.

kidney (*n*) in vertebrates (p.148), one of two bean-shaped glands which control the amount of water in the body by taking water out of blood. A kidney also takes urea (p.186) and mineral salts out of the blood. It helps maintain homeostasis (p.181). It has an outer layer, the *cortex*, and an inner layer, the *medulla*.

renal (*adj*) concerned with the kidneys, e.g. the renal artery supplies blood to the kidneys.

Bowman's capsule a cup-shaped structure, in a kidney, about 0.1 mm in diameter in man, with a knot of blood capillaries inside it. Urea (p.186), glucose (p.174), mineral salts, and water filter (p.91) through the walls of the capillaries and are taken away by a uriniferous tubule (↓).

萬能供血者 屬 O 型血的人；此類人能爲任何血型的人提供（第 183 頁）血液。

萬能受血者 屬 AB 型血的人；此類人能接受（↑）任何血型的人所輸（第 183 頁）的血。

腎（名） 脊椎動物（第 148 頁）體內的兩個豆形腺，其功能是排出血液中的水分以調節體內水量，並排出血液中的尿素（第 186 頁）和礦物鹽。它幫助保持體內平衡（第 181 頁）。腎有外層（即" 腎皮質 "）和內層（即" 髓質 "）。

腎的（形） 與腎有關的，例如腎動脈給腎脈供血。

鮑曼氏囊 腎內的一個杯狀結構，人的鮑曼氏囊直徑約爲 0.1 mm，內有一毛細血管結。尿素（第 186 頁）、葡萄糖（第 174 頁）、礦物鹽和水經毛細管壁濾出（第 91 頁），被腎小管（↓）送走。

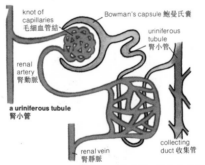

knot of capillaries
毛細血管結

Bowman's capsule 鮑曼氏囊

uriniferous tubule
腎小管

renal artery
腎動脈

a uriniferous tubule
腎小管

renal vein
腎靜脈

collecting duct 收集管

uriniferous tubule a narrow, coiled tube leading from the Bowman's capsule to collecting-ducts in the medulla of the kidney (p.185). The filtrate from the capsule passes along the tubule and glucose, mineral salts and water are absorbed (p.163) in it, and returned to the blood stream; the amounts absorbed keep the composition of the blood constant, see **homeostasis** (p.181).

uriniferous (*adj*) describes a tissue producing urine.

腎小管 從鮑曼氏囊引至腎（第 185 頁）髓質內收集管的一條彎曲細狹管。濾液從鮑曼氏囊流向腎小管，葡萄糖、礦物鹽和水分則在腎小管內被吸收（第 163 頁）後回到血液中，所吸收的量保持血液成分恆定。（參見**體內平衡**（第 181 頁））。

輸尿的（形） 描述輸尿的組織。

deamination (*n*) the taking away of an amino group from an amino acid (p.172), leaving an organic (p.131) acid. This action is done by the liver so that the concentration of amino acids in the blood is kept constant. The action forms ammonia (↓) from the amino group. **deaminate** (*v*).

ammonia (*n*) an inorganic compound with the formula of NH_3. It is poisonous to animals, so it is converted to urea (↓), a harmless compound, by the liver.

urea (*n*) an organic compound, soluble in water, with a formula of $CO(NH_2)_2$, formed from ammonia (↑) in animals; also present in plants.

urine (*n*) a liquid containing dissolved urea (↑) and some inorganic (p.116) salts; it is the liquid leaving a uriniferous tubule (p.185) and the kidney (p.185). **urinate** (*v*).

excrete (*v*) to send out waste products (i.e. those no longer needed) from the body, e.g. urea (↑) is a waste product of metabolism (p.172) as it is of no use to an animal; urea is excreted in urine. **excretion** (*n*), **excreta** (*n*).

去氨基作用(名) 從氨基酸(第 172 頁)中取走氨基，留下有機(第 131 頁)酸。肝的這種作用使血液中氨基酸的濃度保持恆定，並由氨基形成氨(↓)。(動詞爲 deaminate)

氨(名) 一種無機化合物，分子式爲 NH_3。氨對動物有毒性，所以被肝轉變成一種無害的化合物，即尿素(↓)。

尿素(名) 一種有機化合物，溶於水，化學式爲 $CO(NH_2)_2$，係由動物體內的氨(↑)所形成，植物體內也有氨存在。

尿(名) 一種溶解有尿素(↑)和一些無機(第 116 頁)鹽的液體，是從腎小管(第 185 頁)和腎臟(第 185 頁)排出的液體。(動詞爲 urinate)

排泄(動) 從體內送出廢物(即不需要之物)，例如尿素(↑)是新陳代謝(第 172 頁)的廢物，因爲它對動物毫無用處；它隨尿排出體外。(名詞爲 excretion, excreta)

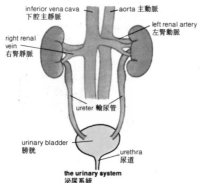

inferior vena cava 下腔主靜脈
aorta 主動脈
left renal artery 左腎動脈
right renal vein 右腎靜脈
ureter 輸尿管
urinary bladder 膀胱
urethra 尿道
the urinary system 泌尿系統

urinary bladder a bag-like structure for storing urine until it is sent out of the body.

bladder (*n*) a shorter, less correct, name for urinary bladder.

ureter (*n*) a tube leading from a kidney (p.185) to the urinary bladder (↑).

膀胱 尿排出體外前貯尿的袋狀結構。

尿泡(名) 膀胱的別稱，此稱不甚準確。

輸尿管(名) 由腎(第 185 頁)導入膀胱(↑)的管道。

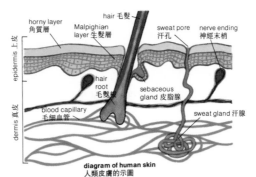

horny layer
角質層
Malpighian
layer 生髮層
hair 毛髮
sweat pore
汗孔
nerve ending
神經末梢
epidermis 上皮
hair
root
毛髮根
sebaceous
gland 皮脂腺
blood capillary
毛細血管
sweat gland 汗腺
dermis 真皮

diagram of human skin
人類皮膚的示圖

skin (*n*) the outer covering of an animal. In
invertebrates (p.148) it consists of epithelium
(p.192) on a membrane. In vertebrates (p.148) it
consists of two layers (↓), epidermis (↓) and
dermis (p.188).

layer (*n*) a flat, thin piece of material with
similar flat pieces above and below it. The layers
are thin in relation to their area; they may act
as a cover to material beneath them. The strata
(p.124) of rocks are in layers.

epidermis (*n*) the outside layer (↑) of cells of a
plant or animal. In plants and invertebrates
(p.148) the epidermis is only one cell thick. In
vertebrates (p.148) the epidermis consists of a
horny layer (↓) and a Malpighian layer (↓).
epidermal (*adj*).

horny layer the outside layer (↑) of the epidermis
(↑) of vertebrates (p.148) except fishes. It
consists of dead cells which are slowly rubbed
off and replaced by cells beneath. Its function is
to prevent the entry of bacteria, and the loss of
water from the body.

stratum corneum another name for horny layer (↑).
cornified layer another name for horny layer (↑).
Malpighian layer the layer (↑) of the epidermis
between the horny layer (↑) and the dermis
(p.188). It consists of actively dividing cells.
(p.138). The outer cells of the Malpighian layer
die and they replace the cells lost from the horny
layer.

皮膚(名) 動物身體的外皮層，無脊椎動物(第
148頁)的皮膚由膜和一層上皮(第192頁)組
成。脊椎動物(第148頁)的皮膚由表皮(↓)
和真皮(第188頁)兩層(↓)組成。

層(名) 一片薄而平坦的物質，其上、下都有相
似的薄平面。相對於其面積，層是薄的；層
的作用是覆蓋其下方的物質。岩層(第124
頁)是分層的。

表皮層(名) 動、植物細胞的外層(↑)。植物和
無脊椎動物(第148頁)的表皮層只有一個細
胞那麼厚。脊椎動物(第148頁)的表皮層由
角質層(↓)和生髮層(↓)組成。(形容詞爲 epidermal)

角質層 除魚類外的脊椎動物(第148頁)的表皮
(↑)外層(↑)。它由死亡的細胞組成，死細胞
逐漸磨擦脱落，由下面的細胞來補充。其功
能是防止細菌侵入和保持身體的水分。

角質層(↑)的拉丁文名稱爲 **stratum corneum**。
角質化層 即角質層(↑)。
生髮層 角質層(↑)和真皮(第188頁)之間的表
皮層(↑)。由積極分裂的細胞(第138頁)組
成。生髮層的外層細胞死亡後補充角質層脱
落的細胞。

dermis (*n*) the inside layer (p.187) of cells in the skin of a vertebrate (p.148); it is much thicker than the epidermis (p.187). In this layer there are blood capillaries (p.179), nerve endings, hair roots, sebaceous glands (↓) and sweat glands (↓). The dermis provides the elastic strength of skin. **dermal** (*adj*).

sebaceous gland a gland (p.162) in the dermis (↑) usually opening on to a hair root. It produces sebum (↓).

sebum (*n*) an oily secretion (p.162) which keeps hair and skin soft and waterproof.

pore (*n*) a very small hole in a surface. The skin of mammals (p.150) has many pores in it. **porous** (*adj*).

sweat (*n*) a dilute solution of common salt, together with small amounts of other mineral salts, which is secreted (p.162) by sweat glands (↓). **sweat** (*v*).

sweat gland in mammals, a tube in the form of a knot, with blood capillaries around it; water and mineral salts are taken from the blood and sweat (↑) is formed. The sweat leaves the gland by a duct, a narrow tube, and passes through a pore (↑) to spread out over the skin. The production of sweat is controlled by the autonomic nervous system (p.200). The function of sweat glands and sweat is to control the temperature of warm-blooded (↓) animals, *see diagram* (p.187).

warm-blooded describes an animal which keeps its body at a constant (p.18) temperature, usually higher than that of the environment (p.226). Some heat is always lost by radiation (p.45); sweating, which cools the body by evaporation, provides the extra control to keep the temperature constant. Birds and mammals are warm-blooded.

homoiothermic (*adj*) warm-blooded (↑).

真皮（名）　脊椎動物（第 148 頁）皮膚的內側細胞層（第 187 頁），比表皮（第 187 頁）厚得多。真皮層內有許多毛細血管（第 179 頁）、神經末稍、毛髮根、皮脂腺（↓）和汗腺（↓）。真皮使皮膚具有彈力。（形容詞爲 dermal）

皮脂腺　真皮（↑）內產生皮脂（↓）的一種腺體（第 162 頁），其出口通常通到毛髮根部。

皮脂（名）　一種油質分泌物（第 162 頁），它保持毛髮和皮膚柔軟，並可防水。

毛孔（名）　表皮上極細小的孔。哺乳動物（第 150 頁）的皮膚上有許多毛孔。（形容詞爲 porous）

汗液（名）　汗腺（↓）所分泌（第 162 頁）的食鹽稀釋液，內摻有少量的其他礦物鹽。（動詞爲 sweat）

汗腺　哺乳動物體內一個周圍有毛細血管的節狀管道，血液中排出的水分和礦物鹽形成汗液（↑）。汗液通過一條窄導管從汗腺流出，並經由毛孔（↑）擴散到皮膚上。汗液的產生受自主神經系統（第 200 頁）控制。汗腺和汗液的功能是控制溫血動物（↓）的體溫（見圖（第 187 頁））。

溫血的　描述身體保持恆溫（第 18 頁）的一種動物，其體溫通常高於周圍環境（第 226 頁）的溫度，總是通過輻射（第 45 頁）散失一些熱量；汗液的蒸發又進一步控制體溫以保持恆溫。鳥類和哺乳動物都屬溫血動物。

恆溫的（形）　見溫血的（↑）。

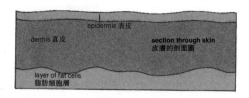

epidermis 表皮
dermis 真皮
section through skin 皮膚的剖面圖
layer of fat cells 脂肪細胞層

thorax
胸廓

the thorax
胸廓

thorax
胸部

gill cover
腮蓋

cold-blooded describes an animal whose body temperature is usually about the same as its environment (p.226). All animals except birds and mammals are cold-blooded. Animals living in water have a temperature very close to that of the water. Animals living on land have a temperature which may be, under some conditions, very different from that of the environment.

poikilothermic (*adj*) cold-blooded (↑).

thorax (*n*) (1) in vertebrates (p.148), except fish, the part of the body containing the lungs (↓) and heart; in mammals (p.150) it is separated from the abdomen (p.162) by a diaphragm (p.190), but not in other vertebrates. (2) in insects, the part of the body between the head and the abdomen; it carries the legs and wings. **thoracic** (*adj*).

lung (*n*) an organ for breathing air in vertebrates (p.148); there are two lungs, one on each side of the heart. In the lung, oxygen is given to blood and carbon dioxide taken from blood, in capillaries.

gill (*n*) an organ of respiration (p.191) in most animals living in water; there are usually two gills, one on each side of the animal. Thin membranes (p.138) separate vessels conducting water and blood capillaries; through these membranes, dissolved oxygen from the water enters the blood, and carbon dioxide leaves the blood and dissolves in the water. The gills are inside the body of most organisms, but larvae (p.151) may have gills outside the body.

branchial (*adj*) describes anything to do with the gills, e.g. the branchial artery.

trachea (*n*) (*tracheae*) (1) in vertebrates (p.148) with lungs (↑), a tube leading from the mouth and nose, passing down the throat to the chest, where it branches into two bronchi (p.190); it conducts air down to the lungs. (2) in insects, a network of tubes reaching all parts of the body. Openings (spiracles) in the skin allow air to pass into the tracheae, which form the respiratory (p.191) system.

pulmonary (*adj*) concerned with the lungs, e.g. the pulmonary artery taking blood to the lungs.

冷血的　描述體溫通常與周圍環境(第 226 頁)大致相同的一種動物,除鳥類和哺乳動物外,一切其他動物都是冷血的。水生動物體溫與水溫非常相近。陸生動物的體溫在某些情況下可能與周圍環境溫度大不相同。

變溫的(形)　冷血的(↑)。

胸廓(名)　(1)脊椎動物(第 148 頁)(魚除外)包含肺(↓)和心臟的身體部分;在哺乳動物(第 150 頁)體內,膈膜(第 190 頁)將胸腔與腹部(第 162 頁)分隔。而在其他的脊椎動物則不分隔;(2)昆蟲的頭部與腹部之間的身體部分;其上長有腿和翼。(形容詞爲 thoracic)

肺(名)　脊椎動物(第 148 頁)體內呼吸空氣的器官;肺分兩葉,心臟兩側各一葉,肺裏的氧氣通過毛細管進入血液,並從血液中排出二氧化碳。

腮(名)　大部分水生動物的呼吸(第 191 頁)器官;一般有兩個腮,身體兩側各一個。腮的薄膜(第 138 頁)將導水管和毛細血管分開,溶解在水中的氧透過這些薄膜進入血液,二氧化碳從血液中出來並溶於水。腮位於大部分生物體的體內,但幼體(第 151 頁)可以有體外的腮。

腮的(形)　描述有關腮的任何東西,例如腮動脈。

氣管(名)　(1)有肺(↑)的脊椎動物(第 148 頁)體內有一根導管上接口和鼻,下經喉頭通入胸腔,再分爲兩條支氣管(第 190 頁);它輸送空氣下達肺部;(2)昆蟲體內一個網狀管通到身體各部。皮膚上的氣孔讓空氣進入氣管形成呼吸(第 191 頁)系統。

肺的(形)　與肺有關的,例如肺動脈將血送入肺部。

bronchus (n) (bronchi) in vertebrates with lungs (p.189), a tube leading from the trachea (p.189) to each lung. A bronchus has plates of cartilage (p.192), as does the trachea, to prevent the tube closing. Both bronchus and trachea have glands secreting mucus (↓) and walls bearing cilia (p.144). The mucus removes dust, and the cilia beat to drive the dust back up to the mouth.

bronchiole (n) inside a lung a bronchus branches again and again, forming small tubes called bronchioles.

alveolus (n) (alveoli) a very small bag-like structure at the end of a bronchiole (↑). It has a network of blood capillaries on its surface, *see diagram*. Air inside an alveolus gives oxygen to blood and receives carbon dioxide from blood; these gases pass through the thin surface of the alveolus and the wall of a capillary. **alveolar** (adj).

pleura (n) (pleurae) a bag-like structure of very thin skin, which is the outer cover of a lung and also the inner coat of the space containing the lung. The skin is a membrane producing a watery liquid, which fills the space between the outer cover and the inner coat and prevents friction (p.22) between the two surfaces. **pleural** (adj).

diaphragm² (n) in mammals (p.150), a cup-shaped muscle between the thorax (p.189) and the abdomen (p.162). When the muscle contracts, the diaphragm becomes flatter, and air is drawn into the lungs.

inspire (v) to take air into lungs (p.189) or water into gills. Air is drawn into vertebrate (p.148) lungs by movement of the ribs. In mammals (p.150), the diaphragm is also used to inspire (inhale) air. **inspiration** (n).

expire (v) to push air out of the lungs (p.189) (exhale) or pass water out of the gills. Air is pushed out of vertebrate lungs by the relaxation (p.196) of the rib muscles and the diaphragm. **expiration** (n).

mucus (n) a sticky liquid secreted by special cells in membranes of vertebrates (p.148). Its function is lubrication (p.22). **mucous** (adj).

支氣管（名） 有肺（第 189 頁）的脊椎動物體內，從氣管（第 189 頁）通到每一葉肺的一根管。支氣管和氣管一樣，都有軟骨（第 192 頁）片防止氣管關閉。支氣管和氣管都有分泌黏液（↓）的腺體而管壁長滿纖毛（第 144 頁）。黏液清除灰塵，纖毛擺動來將灰塵推回口腔。

細支氣管（名） 支氣管在肺內一再分支形成細管。稱爲細支氣管。

肺泡（名） 細支氣管（↑）一端的一個很小的袋狀組織。其表面佈滿毛細血管網（見圖）。肺泡內的空氣給血液供氧，同時接收血液內的二氧化碳；這些氣體都可通過肺泡的薄壁和毛細血管的管壁。（形容詞爲 alveolar）

胸膜（名） 一個極薄的袋狀結構，分兩層，外層包覆着肺，內層包圍肺的空間。薄皮層是一層膜，產生水狀液體充滿外層和內層之間的空間以防止二表面磨擦（第 22 頁）。（形容詞爲 pleural）

橫隔膜（名） 哺乳動物（第 150 頁）體內胸廓（第 189 頁）和腹部（第 162 頁）之間的杯狀肌。肌肉收縮時，隔膜變得較平坦，將空氣吸入肺部。

吸入（動） 將空氣吸入肺部（第 189 頁）或將水吸入腮內。脊椎動物藉肋骨的運動將空氣吸入（第 148 頁）肺內。哺乳動物（第 150 頁）也用隔膜吸入（呼出）空氣。（名詞爲 inspiration）

呼出（動） 將空氣排出（呼出）肺部（第 189 頁）或將水排出腮外。脊椎動物靠鬆弛（第 196 頁）肋肌和隔膜將空氣排出體外（名詞爲 expiration）

黏液（名） 脊椎動物（第 148 頁）膜細胞所分泌出的一種黏性液體。它起潤滑作用（第 22 頁）。（形容詞爲 mucous）

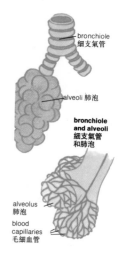

bronchiole
細支氣管

alveoli 肺泡

bronchiole and alveoli 細支氣管和肺泡

alveolus 肺泡

blood capillaries 毛細血管

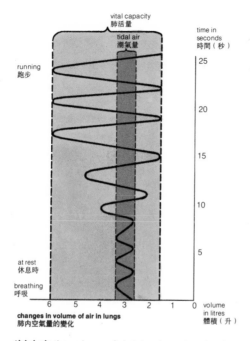

vital capacity
肺活量

tidal air
潮氣量

time in
seconds
時間（秒）

running
跑步

at rest
休息時

breathing
呼吸

changes in volume of air in lungs
肺內空氣量的變化

volume
in litres
體積（升）

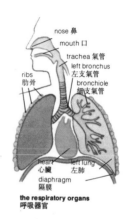

nose 鼻
mouth 口
trachea 氣管
left bronchus
左支氣管
ribs
肋骨
bronchiole
細支氣管
heart
心臟
left lung
左肺
diaphragm
隔膜

the respiratory organs
呼吸器官

tidal air the volume of air taken in and sent out of lungs (p.189) when breathing ordinarily. In man, this is about 500 cm³.

vital capacity the volume of air taken in, after having fully breathed out, until the lungs (p.189) are completely full. In man, this is about 4000 cm³.

respiration (*n*) the whole process from the inspiration (↑) of oxygen by lungs (p.189), gills (p.189) or tracheae (p.189), through the use of oxygen to supply energy in cells of the body, to the expiration (↑) of carbon dioxide by lungs, gills or tracheae. **External respiration** is the process by which oxygen in the air is taken to cells of the body, and carbon dioxide taken from the cells and passed back to the air. **Internal** (or **tissue**) **respiration** is the processes taking place in the cell to use oxygen to provide energy for metabolism (p.172). **respire** (*v*), **respiratory** (*adj*).

潮氣量 正常呼吸時，肺部（第 189 頁）呼出的空氣量。人體的潮氣量約爲 500 cm³。

肺活量 完全呼盡空氣後，肺部（第 189 頁）完全充滿空氣時所吸入的空氣量。人體的肺活量約爲 4000 cm³。

呼吸作用（名） 從肺（第 189 頁）、腮（第 189 頁）或氣管（第 189 頁）吸進（↑）氧，以利用氧給體內細胞提供能量，然後肺、腮或氣管排出（↑）二氧化碳的整個過程。外呼吸是空氣中的氧被吸入體內細胞，二氧化碳被排出（↑）細胞，返回空氣中的過程。內（或組織）呼吸是發生在細胞內部的，利用氧爲新陳代謝（第 172 頁）提供能量的過程。（動詞爲 respire，形容詞爲 respiratory）

epithelium (*n*) (*epithelia*) a tissue which forms the surface or the outside of an organism, or is a covering for the inside of tubes and cavities in an organism. Examples of epithelium are the surface cells of the skin and the inside of blood vessels. Secretory (p.162) cells in glands are mostly epithelial tissue. Epithelium can also bear cilia (p.144), e.g. the inside of the walls of a bronchus are made of ciliated epithelium. **epithelial** (*adj*).

connective tissue a tissue in vertebrates, containing fibres or a matrix (↓); it provides support for other tissues and for organs, e.g. cartilage, bone, fatty tissue.

matrix (*n*) a solid material which is made by cells and secreted (p.162) round them, so that the cells are pushed apart leading to cells scattered here and there in the matrix.

cartilage (*n*) a connective tissue (↑) with cells scattered in an elastic (p.27) matrix containing polysaccharides (p.173) and many protein (p.172) fibres; the cells produce the matrix. Some animals, e.g. sharks, have cartilage instead of bone in their skeleton (↓); young children have cartilage which becomes changed to bone as they grow. **cartilaginous** (*adj*).

bone (*n*) a connective tissue of vertebrates (p.148) only; cells are scattered in a matrix (↑) which consists of fibres (p.195) and calcium salts. The inorganic (p.116) salts provide the hardness and the fibres the strength of bone. The bone cells are joined by small tubes carrying blood vessels and nerves. **bony** (*adj*).

Haversian canals small tubes in bone which carry blood vessels and nerves. Bone cells are arranged in circles round a Haversian canal.

marrow (*n*) soft material forming the inside of a bone. In long bones, *yellow marrow* fills the centre; it consists mainly of fat cells. The ends of the bone are filled with *red marrow*. Red blood cells (p.179) are produced in red marrow. Some white blood cells (p.180) are formed in yellow marrow.

ossification (*n*) the formation of bone, usually by the change of another tissue, such as cartilage, into bone. **ossify** (*v*).

上皮（名） 構成生物體表面或外面部分的一種組織，或組成管和腔內面蒙皮的一種組織。例如皮膚的表層細胞和血管的內壁都是上皮。腺體中的分泌（第 162 頁）細胞大多是上皮組織。上皮也可長纖毛（第 144 頁）。例如氣管內壁是由纖毛上皮構成的。（形容詞爲 epithelial）

結締組織 脊椎動物體內含有纖維或基質（↓）的組織；它支持其他組織或器官。例如軟骨、硬骨和脂肪組織爲結締組織。

基質（名） 由細胞組成的固體物質。因爲它是在細胞周圍分泌（第 162 頁）出來的，所以細胞被推離，以至基質處散佈着細胞。

軟骨（名） 一種結締組織（↑），其中的細胞散佈在含有多醣（第 173 頁）和很多蛋白質（第 172 頁）纖維的彈性（第 27 頁）基質內；這些細胞產生基質。一些動物（如鯊魚）的骨骼（↓）裏有軟骨而無硬骨；幼兒的軟骨在發育過程中長成硬骨。（形容詞爲 cartilaginous）

硬骨（名） 脊椎動物（第 148 頁）才有的一種結締組織；細胞散佈在由纖維（第 195 頁）和鈣鹽組成的基質（↑）內，無機（第 116 頁）鹽使骨頭變硬，纖維使骨頭變得堅固。骨細胞藉載有血管和神經的小血管連結。（形容詞爲 bony）

哈佛氏細管 骨內載有血管和神經的小管。骨細胞成環狀排在哈佛氏小管周圍。

骨髓（名） 構成骨頭內部的軟物質。長骨的中段充滿“黃骨髓”；它主要由脂肪細胞構成。骨的兩端充滿“紅骨髓”。紅血球（第 179 頁）產生於紅骨髓中。一些白血球（第 180 頁）是在黃骨髓中形成的。

骨化作用（名） 通常是由另一組織發生變化而形成骨頭，例如軟骨變成骨頭。（動詞爲 ossify）

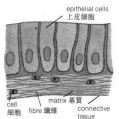

epithelium and connective tissue
上皮和結締組織

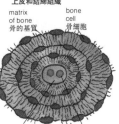

structure of bone
骨的結構

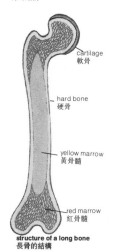

structure of a long bone
長骨的結構

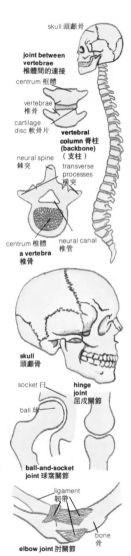

skull 頭顱骨

joint between vertebrae 椎體間的連接

centrum 框體

vertebrae 椎骨

cartilage disc 軟骨片

vertebral column 脊柱 (backbone) （支柱）

neural spine 棘突

transverse processes 橫突

centrum 椎體

neural canal 椎管

a vertebra 椎骨

skull 頭顱骨

socket 臼

hinge joint 屈戍關節

ball 球

ball-and-socket joint 球窩關節

ligament 韌帶

bone 骨

elbow joint 肘關節

skeleton (*n*) a hard structure, either inside or outside an animal, which supports the animal's organs and tissues. Vertebrates (p.148) have an internal skeleton; some invertebrates (p.148), e.g. insects, have an external skeleton, while other invertebrates, e.g. earthworms, have no skeleton. **skeletal** (*adj*).

axial skeleton in vertebrates (p.148), the bones of the head and body.

appendicular skeleton the skeleton of the limbs of an animal.

vertebra (*n*) (*vertebrae*) a bone with a large central mass (the **centrum**), a hollow (the **neural canal**), and finger-like pieces of bone that stand out from the vertebra (**transverse processes** and **neural spine**), *see diagram*. The spinal cord (p.198) passes through the neural canal of a vertebra and is protected by the bone around it. **vertebral** (*adj*).

vertebral column the arrangement in a line of vertebrae (↑) joined by ligaments (↓) and separated by elastic cartilage (↑). It forms the main support of the body. The movement of any two vertebrae in relation to each other is small, but it is enough to allow the whole column to bend. Muscles are fixed to the finger-like pieces of bone which stand out from each vertebra, and control movement.

skull (*n*) the bones which protect the brain, together with those that form the face.

joint (*n*) the structure which joins two bones so that they can move in relation to one another.

ligament (*n*) a strong band of fibre (p.195) which keeps two bones of a joint held together.

ball-and-socket joint a joint in which the round end of one bone (the ball) fits into the hollow (the socket) of another bone; the bone can be turned in any direction by means of the ball turning in the socket, e.g. hip joint.

hinge joint a joint in which the round end of one bone turns on the flat surface of another bone, allowing the joint to bend in one plane only, e.g. knee joint.

gliding joint (*n*) a joint in which the surface of one bone moves over the surface of another bone.

骨骼（名） 動物體內或體外支持着器官和組織的硬結構。脊椎動物（第 148 頁）有體內骨骼；一些無脊椎動物（第 148 頁）如昆蟲有體外骨骼，而另一些無脊椎動物如蚯蚓則無骨骼。（形容詞爲 skeletal）

中軸骨骼 脊椎動物（第 148 頁）體的頭骨和軀幹骨。

附屬骨骼 動物肢體的骨骼。

脊椎骨（名） 一塊有大的中心塊（椎體）、中空（椎管）的骨和從椎骨伸出的指狀骨塊（橫突和棘突），（見圖）。脊髓（第 198 頁）通過脊椎骨的椎管，並受椎管外骨保護。（形容詞爲 vertebral）

脊柱 靠韌帶（↓）連結，由彈性軟骨（↑）分隔並排成一行的許多椎骨（↑），它組成人體主要的支架。雖然任何兩根椎骨相互間的運動很小，但足以使整個脊椎彎曲。肌肉附在從每塊椎骨伸出的指狀骨塊上，並控制椎骨的運動。

頭顱骨（名） 保護腦並形成面部的骨。

關節（名） 連接兩塊骨，使之能相互運動的結構。

韌帶（名） 結實的纖維（第 195 頁）帶，它將關節的兩塊骨保持在一起。

球窩關節；杵臼關節 一種關節，關節內一塊骨的圓端（球端）嵌入另一塊骨的凹處（臼）。藉着球端在臼內轉動，使骨能向任何方向轉動，例如髖關節。

屈戍關節 一種關節，關節內一塊骨的圓端在另一塊骨的扁平面上轉動，使關節僅能向一個面彎曲，例如膝關節。

滑動關節（名） 一種關節，關節內一塊骨的表面移過另一塊骨的表面。

movement (*n*) the action carried out when the bones in a joint are moved, e.g. waving the hands is a movement: it is not a motion, because the person does not change his place in space.

locomotion (*n*) the action or ability of moving from one place to another, i.e. being in motion without any outside help. In many animals, the movements of legs produce locomotion. **locomotor** (*adj*).

synovial capsule a bag-like membrane in a movable joint; it is fixed to the bones on either side of the joint, *see diagram*. The capsule is filled with a sticky liquid (synovial fluid) which lubricates (p.22) the joint.

pelvis (*n*) a bony structure in vertebrates (p.148) which provides strong support for the back legs or back fins. In mammals (p.150) it consists of several bones fixed together in the shape of a bowl. **pelvic** (*adj*).

動作（名） 骨頭在關節內移動的活動，例如揮手是動作而不是運動，因為人並未改變在空間的位置。

移動；行動 從一處移向另一處的行動或能力，即在無任何外力的幫助下移動。很多動物腿的動作產生移動。（形容詞爲 locomotor）

滑液囊 活動關節內的袋狀膜；它固定在關節兩側的骨上，（見圖）。囊內充滿潤滑（第 22 頁）關節的黏性液（滑液）。

骨盆（名） 脊椎動物（第 148 頁）體內的一種硬骨結構，它强有力地支持着後腿或後鰭。哺乳動物（第 150 頁）體內，骨盆由數塊固定在一起形成盆狀的硬骨構成。（形容詞爲 pelvic）

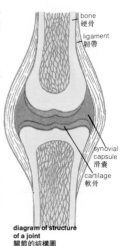

diagram of structure of a joint
關節的結構圖

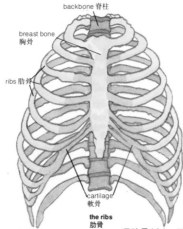

the ribs
肋骨

rib (*n*) a narrow curved bone fixed to the vertebral column (p.193). The ribs form a protective structure round the thorax (p.189).

intercostal muscle a muscle (↓) between two ribs (↑). When the muscle contracts the ribs are raised and air is drawn into the lungs.

硬肋骨（名） 固定在脊柱上（第 193 頁）的狹而彎的硬骨。肋骨組成一個包圍着胸廓（第 189 頁）的保護結構。

肋間肌 兩相鄰肋骨（↑）間的肌肉（↓）。肋間肌收縮時，肋骨上升，肺部吸入空氣。

fibre of voluntary muscle
隨意肌的纖維

muscle (*n*) a tissue consisting·of cells which have the power to contract. The contraction of muscles causes all movement of joints.

voluntary muscle in vertebrates (p.148), a tissue able to contract (p.38) rapidly, consisting of long muscle fibres (↓) with many nuclei (p.139) covered by a membrane. On receiving a stimulus (p.201) from the brain by way of a nerve, voluntary muscle contracts by becoming shorter and thicker. Muscle fibres are held together by fibres (↓) or connective tissue (p.192). Voluntary muscles are concerned with locomotion (↑) and movement by moving the bones of joints.

fibre (*n*) in animals, a thread-like structure of protein; it is very strong. In plants, a long thread-like structure of cellulose, e.g. cotton fibres. **fibrous** (*adj*).

肌肉（名） 由具收縮能力的細胞所組成的組織。關節的全部運動都是肌肉的收縮引起的。

隨意肌 脊椎動物（第 148 頁）體中能迅速收縮（第 38 頁）的組織，係由許多爲膜所包覆的細胞核（第 139 頁）的長肌肉纖維（↓）所組成。一接到大腦通過神經傳來的刺激（第 201 頁），隨意肌立即收縮，變粗變短。纖維（↓）或結締組織（第 192 頁）將許多肌肉纖維聚集在一起。隨意肌通過關節中骨頭的運動參與移動（↑）或動作。

纖維（名） 動物體內的一種蛋白質絲狀結構；它非常結實。植物體的長纖維素絲狀結構，例如棉花纖維。（形容詞爲 fibrous）

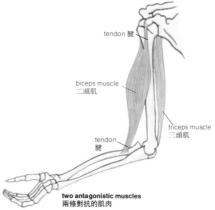

tendon 腱

biceps muscle
二頭肌

triceps muscle
三頭肌

tendon
腱

two antagonistic muscles
兩條對抗的肌肉

antagonism (*n*) of two muscles, producing movement in opposite directions, so that contraction of one muscle stretches the other. The only voluntary action of a muscle is contraction, so contractions of muscles in opposite directions are needed to control the movement at a joint. **antagonistic** (*adj*).

striped muscle another name for voluntary muscle, because it has stripes across it.

頡頏作用（名） 兩塊肌肉作相反方向運動，結果一塊肌肉收縮而拉長另一塊肌肉。肌肉唯一的隨意動作是收縮，所以肌肉必須作相反方向收縮以控制關節的運動。（形容詞爲 antagonistic）

橫紋肌 隨意肌的另一名稱；因有條紋橫過肌肉而得名。

involuntary muscle in vertebrates (p.148), a tissue which contracts (p.38) slowly; it consists of long spindle-shaped cells bound together by connective tissue (p.192), and does not have the stripes of voluntary muscle (p.195). Involuntary muscle is usually found in sheets round hollow organs, e.g. intestines, blood vessels. This kind of muscle is controlled by the autonomic nervous system (p.200) and waves of contraction can pass along a muscle as in peristalsis (p.169).

smooth muscle another name for involuntary muscle.

cardiac muscle the muscle in the walls of the heart. It consists of a network of striped muscle fibres (p.195), but no membrane covers the fibres. The fibres contain separate cells, each with its nucleus (p.139). Cardiac muscle has the characteristics of both voluntary and involuntary muscle; its action is automatic (↓) and regular, faster than involuntary muscle but slower than voluntary muscle.

automatic (*adj*) describes an action or a process which, once started, continues to act by itself without being controlled by outside conditions, e.g. the action of heart muscle, the process of breathing.

relax (*v*) of muscles, to go from a state of acting to a state of rest. A muscle is either contracted or relaxed; in antagonistic (p.195) pairs of muscles, one contracts while the other relaxes. **relaxation** (*n*).

tendon (*n*) a band or string-like piece of connective tissue which attaches a muscle to a bone. A tendon consists of parallel fibres (p.195).

coordination (*n*) of muscles, the state of acting together to produce the same effect, e.g. when walking, the muscles of the legs act together to make each leg move in turn, and the muscles of the body keep it upright; all the muscles used in this process are working in coordination. **coordinate** (*v*).

flex² (*v*) to become bent; when a joint is flexed the angle between the two bones which form the joint becomes smaller. The opposite action is to extend a joint.

不隨意肌　脊椎動物(第 148 頁)體內一種收縮 (第 38 頁)緩慢的組織，它由結締組織(第 192 頁)束在一起的許多長紡錘形細胞所組 成，它沒有隨意肌(第 195 頁)的條紋。不隨 意肌通常呈片狀存在於中空器官例如腸、血 管的周圍。不隨意肌由自主神經系統(第 200 頁)控制，收縮波沿肌肉傳播時就像蠕動(第 169 頁)。

平滑肌　不隨意肌的另一名稱。

心肌　心臟壁的肌肉，它由橫紋肌的纖維(第 195 頁)網組成，但纖維上無膜包覆。纖維的細 胞各自獨立，有自己的細胞核(第 139 頁)。 心肌兼有隨意肌和不隨意肌二者的特徵。它 的動作是自動(↓)而有規律的，比不隨意肌 快，但比隨意肌慢。

自動的(形)　描述一個一經發動，就不受外界條 件控制，本身能繼續進行的動作或過程，例 如心肌的動作和呼吸過程。

鬆弛(動)　指肌肉的鬆弛，從動作狀態變成靜止 狀態。肌肉或收縮或鬆弛；一對頡頏的(第 195 頁)肌肉中，一塊肌肉收縮，另一塊則鬆 弛。(名詞爲 relaxation)

腱(名)　帶狀或線狀的結締組織塊，它使肌肉附 着在骨上。腱由平行的纖維(第 195 頁)組 成。

協調(名)　指肌肉的協調一同動作以產生同一效 果的狀態，例如步行時，腿部肌肉一起動作 使每條腿輪流移動，身體的肌肉則保持身體 挺直；在這過程中全部肌肉都協調地工作。 (動詞爲 coordinate)

屈曲(動)　變成彎曲；當關節屈曲時，構成關節 的兩塊骨之間的角度變小。相反的動作是關 節伸展。

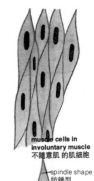

muscle cells in
involuntary muscle
不隨意肌 的肌細胞

spindle shape
紡錘型

oval nucleus
卵形核

membrane 膜

a single muscle cell
單個的肌細胞

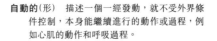

muscle
fibre
肌纖維

nucleus
細胞核

cardiac muscle
心肌

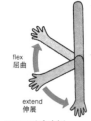

flex
屈曲

extend
伸展

movement of a joint
關節的運動

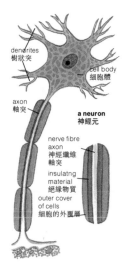

dendrites
樹狀突

cell body
細胞體

axon
軸突

a neuron
神經元

nerve fibre
axon
神經纖維
軸突

insulatng
material
絕緣物質

outer cover
of cells
細胞的外覆層

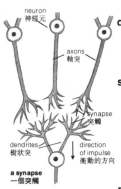

neuron
神經元

axons
軸突

synapse
突觸

dendrites
樹狀突

direction
of impulse
衝動的方向

a synapse
一個突觸

neuron (*n*) a single cell (p.138) in a system of nerves (p.198) in an animal; it conducts impulses (↓) by which all nerves function (p.140). A neuron has a nucleus (p.139) and cytoplasm (p.138) which forms the body of the cell. From the body of the cell arise thread-like parts which can vary in length and in number. One of these thread-like parts is an axon (↓) and the others are dendrites (↓).

process² (*n*) a part of a cell that stretches out, e.g. an axon (↓).

impulse (*n*) a change, partly chemical and partly physical, in a neuron, which is conducted by axons, dendrites and cell bodies; it acts as a message sent through a neuron (↑). An impulse is like a travelling wave and it acts along a length of between 2 and 5 cm of a nerve-fibre (↓); it moves with a speed between 1 and 100 cm/sec, depending on the kind of nerve (p.198) and the species (p.148) of animal. The energy for the impulse is provided by the neuron, so the impulse leaves a neuron with the same energy as it entered the neuron.

axon (*n*) a long, thread-like part of a neuron (↑), *see diagram*. It conducts impulses (↑) away from the cell body of a neuron. A neuron has only one axon.

dendrite (*n*) a thread-like part of a neuron (↑), usually short in length, with branches at its end, *see diagram*. It conducts impulses (↑) towards the cell body of a neuron. A neuron can have one or several dendrites.

synapse (*n*) the meeting place of dendrites (↑) and axons (↑). The end of an axon touches the end of a dendrite, and an impulse (↑) jumps from the axon to the dendrite. An axon can have several synapses, each with dendrites (↑) of different neurons (↑). A dendrite can have synapses with several axons. Through synapses, an impulse can travel along many paths through different neurons. **synaptic** (*adj*).

nerve-fibre an axon or a dendrite with a cover of a fatty material for insulation (p.74). The diameter of a nerve-fibre in vertebrates (p.148) is between 1 and 20 μm. Some very large nerve-fibres in invertebrates are up to 1 mm in diameter.

神經元（名）　動物神經（第 198 頁）系統中的單個細胞（第 138 頁）；它傳導衝動（↓）使全部神經起作用（第 140 頁）。神經元有細胞核（第 139 頁）和組成細胞體的細胞質（第 138 頁）。細胞體中產生長度和數目都可不同的絲狀部分。這些絲狀部分，有一個是軸突（↓），其他部分是樹狀突（↓）。

突起（名）　細胞向外伸出的部分，例如軸突（↓）。

衝動（名）　神經元的變化，部分是化學變化，部分是物理變化，這種變化由軸突、樹狀突和細胞體傳導；衝動的作用相當於通過神經元（↑）傳送信息。衝動像在傳播的波，沿着長度爲 2 至 5 cm 的神經纖維（↓）起作用，衝動以每秒 1 至 100 cm 的速度運動，視神經（第 198 頁）的種類和動物的種（第 148 頁）而定。神經元提供所需的能量，所以衝動進出神經元所用的能量相同。

軸突（名）　神經元（↑）的長絲狀部分，（見圖），它從神經元的細胞體中傳出衝動（↑）。一個神經元只有一條軸。

樹狀突（名）　神經元（↑）的絲狀部分，通常較短，末端分支（見圖），它將衝動（↑）傳到神經元的細胞體。一個神經元可有一個或數個樹狀突。

突觸（名）　樹狀突（↑）和軸突（↑）的會合處。軸的一端接觸樹狀突的一端，衝動（↑）從軸突跳向樹狀突。一個軸突有數個突觸，每個突觸有不同的神經元（↑）的樹狀突（↑）。一個樹狀突可以有帶數個軸突的突觸。衝動可以通過突觸經由不同的神經元，沿許多路線傳導。（形容詞爲 synaptic）

神經纖維　外覆蓋脂肪物質作爲隔離物（第 74 頁）的軸突或樹狀突。脊椎動物（第 148 頁）體的神經纖維，直徑在 1 至 20 μm 之間。無脊椎動物體中有一些非常大的神經纖維，直徑約爲 1mm。

nerve (*n*) a bundle of nerve-fibres (p.197) supported by connective tissue (p.192) with blood vessels (p.178). Each nerve-fibre conducts impulses independently of the others. All nerves are motor (p.201) or sensory (p.200) or both in function (p.140). A nerve can be a millimetre long or as long as the animal. **nervous** (*adj*).

ganglion (*n*) (*ganglia*) a solid mass of cell bodies of neurons (p.197); nerves enter and leave it. **ganglionic** (*adj*).

spinal cord in vertebrates (p.148) a cylindrical mass of nervous tissue containing cell bodies of neurons, nerve-fibres and synapses; it is divided into white matter (↓) and grey matter (↓). The spinal cord runs through and is protected by the vertebrae (p.193); a pair of spinal nerves leave the spinal cord through holes in each vertebra. Very simple coordination (p.196) by reflex arcs (p.202) is effected in the spinal cord.

神經(名) 有血管(第 178 頁)的結締組織(第 192 頁)支持的一束神經纖維(第 197 頁)。每根神經纖維能互不依賴而單獨地傳導衝動。全部神經包括運動神經(第 201 頁),感覺神經(第 200 頁)或兼有兩功能(第 140 頁)的神經。一根神經可以長 1 毫米或與動物身體等長。(形容詞爲 nervous)

神經節(名) 神經元(第 197 頁)的細胞體的硬團塊;神經進出神經節。(形容詞爲 ganglionic)

脊髓 脊椎動物(第 148 頁)體內圓柱體形的神經組織,內含神經元的細胞體、神經纖維和突觸;脊髓分爲白質(↓)和灰質(↓)。脊髓流過脊椎(第 193 頁)並受其保護;一對脊髓神經自脊髓處開始通過洞孔進入每根椎骨。通過反射弧(第 202 頁),可在脊髓中產生極簡單的協調(第 196 頁)。

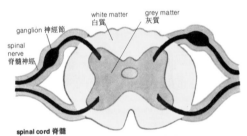

ganglion 神經節
spinal nerve 脊髓神經
white matter 白質
grey matter 灰質

spinal cord 脊髓

cerebrospinal fluid a liquid which fills the hole in the middle of the spinal cord and fills spaces in the brain. It is a solution of glucose and mineral salts, and contains a few blood cells but no protein in solution.

white matter a kind of nervous tissue consisting of nerve fibres (p.197) and connective tissue (p.192); it is the outer part of the spinal cord (↑) and the inner part of the brain (↓).

grey matter a kind of nervous tissue consisting mainly of cell bodies of neurons (p.197) with dendrites and synapses and blood vessels; it is the inner part of the spinal cord (↑) and the outer part of the brain.

腦髓液 充滿脊髓中部孔洞和腦部空間的液體。它是葡萄和礦物鹽的溶液,其中含有一些血細胞但無蛋白質。

白質 由神經纖維(第 197 頁)和結締組織(第 192 頁)組成的一種神經組織;它是脊髓(↑)的外面部分和腦(↓)的裏面部分。

灰質 主要由神經元(第 197 頁)細胞體組成的一種神經組織,有樹狀突、突觸和血管;它是脊髓(↑)的裏面部分和腦的外面部分。

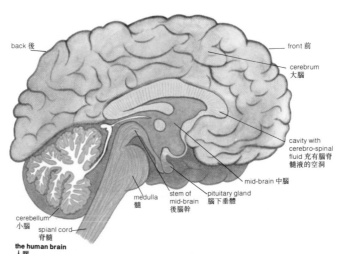

back 後　　　　　　　　　　　　　front 前

cerebrum
大腦

cavity with
cerebro-spinal
fluid 充有腦脊
髓液的空洞

mid-brain 中腦

medulla
髓

stem of
mid-brain
後腦幹

pituitary gland
腦下垂體

cerebellum
小腦

spianl cord
脊髓

the human brain
人腦

brain (n) a large mass of ganglia (↑) protected in
vertebrates (p.148) by the bones of the skull. It
coordinates (p.196) actions of the whole body
particularly those in response to stimuli (p.201).

hind-brain this part of the vertebrate (p.148) brain
has two divisions, the medulla (↓) and the
cèrebellum (↓); it joins the spinal cord (↑).

mid-brain a small part joining the hind-brain to
the fore-brain.

fore-brain the part containing the cerebrum
(p.200); in man it is the largest part of the brain.

medulla (n) part of the brain (↑) joined to the
spinal cord. It is concerned with (a) the
coordination (p.196) of impulses (p.197) from
hearing, tasting and touching; (b) the control of
respiratory (p.191) movements, cardiac muscle
(p.196), and blood vessels.

medulla oblongata fuller, more correct name for
medulla (↑).

cerebellum (n) a part of the brain (↑) growing out
from the stem of the hind-brain. It is concerned
with the coordination (p.196) of all muscular
movement, particularly the muscles fixed to
bones, e.g. coordination as needed in
locomotion (p.194).

腦(名)　脊椎動物(第 148 頁)體內由頭顱骨保護
着的一大團神經節(↑)。它協調(第 196 頁)
全身的動作，特別是對刺激(第 201 頁)作出
反應的動作。

後腦　脊椎動物(第 148 頁)腦的這一部分又分為
兩部分：腦髓(↓)和小腦(↓)；後腦與脊髓
(↑)相連。

中腦　連接後腦與前腦的一個小的部分。

前腦　這部分包括大腦(第 200 頁)；人腦的前腦
佔腦子的大部。

腦髓(名)　腦(↑)與脊髓相連的部分。它與兩種
功能有關：(a) 協調(第 196 頁)聽覺、味覺
和觸覺的衝動(第 197 頁)；(b) 控制呼吸(第
191 頁)運動、心肌(第 196 頁)和血管。

延髓　腦髓(↑)的正式名稱。

小腦(名)　腦(↑)的一部分，從後腦幹長出。它
與協調(第 196 頁)一切肌肉運動，特別是與
骨骼肌的運動有關。例如移動(第 194 頁)時
肌肉的協調。

cerebrum (*n*) in vertebrates (p.148) this consists of a pair of outgrowths from the front of the fore-brain. It is concerned with stimuli (↓) from receptors (↓) and sends impulses (p.197) by motor (↓) nerves to cause action in voluntary muscle (p.195) in answer to stimuli.

cerebral (*adj*) describes any tissue or effect which is to do with the cerebrum (↑).

meninges (*n.pl.*) in vertebrates (p.148), the three membranes (p.138) covering the spinal cord and the brain.

dura mater strong connective tissue (p.192) containing blood vessels; it is the outermost membrane (p.138) of the meninges (↑).

arachnoid (*n*) a membrane (p.138) between the dura mater (↑) and the pia mater (↓). It is separated from the pia mater by a space filled with cerebrospinal fluid (p.198).

pia mater thin membrane (p.138) containing many blood vessels; the innermost membrane of the meninges (↑).

central nervous system the nervous tissue which coordinates (p.196) all the activities of an animal. In vertebrates (p.148) it consists of the brain (p.199) and spinal cord (p.198). In many invertebrates it consist of a few large nerve-fibres (p.197) joined to several ganglia (p.198).

peripheral nervous system the nervous tissue in an animal, other than the central nervous system (↑). It consists of nerves and nerve fibres (p.197) which leave the central nervous system and branch to every part of the body.

autonomic nervous system in vertebrates, (p.148) motor (↓) nerves supplying involuntary muscle (p.196), the heart (p.181) and glands (p.162) together with sensory (↓) nerves from receptors (↓) inside the body. A lot of the co-ordination (p.196) between these nerves takes place in the spinal cord (p.198) and the medulla (p.199). The actions caused by the nerves are mainly automatic and take place independently of outside stimuli (↓).

sensory (*adj*) of nerves, concerned with receptors (↓). A stimulus (↓) causes an impulse (p.197) to travel along a sensory nerve.

大腦(名)　脊椎動物(第148頁)的大腦由前腦前部長出的一對分支組成。它與感受器(↓)傳來的刺激(↓)有關，並通過運動(↓)神經傳送衝動(第197頁)，使隨意肌(第195頁)產生動作，對刺激作出反應。

大腦的(形)　描述與大腦(↑)有關的任何組織或效應。

腦膜(名、複)　脊椎動物(第148頁)體內覆蓋脊髓和腦的三層膜(第138頁)。

硬腦膜　含有血管的堅韌結締組織(第192頁)；是腦膜(↑)最外層的膜(第138頁)。

蛛網膜(名)　位於硬腦膜(↑)和軟腦膜(↓)之間的一層膜(第138頁)。由充滿腦脊液(第198頁)的空間，將它和軟腦膜分開。

軟腦膜　含有很多血管的薄膜(第138頁)；是腦膜(↑)中最裏層的膜。

中樞神經系統　動物體內協調(第196頁)一切活動的神經組織。脊椎動物(第148頁)的這一系統是由腦(第199頁)和脊髓(第198頁)構成。在很多無脊椎動物體內，它由連在一些神經節(第198頁)上的大神經纖維(第197頁)組成。

周圍神經系統　動物體內除中樞神經系統(↑)以外的神經組織。它由神經和神經纖維(第197頁)組成，神經和神經纖維從中樞神經系統出發，分佈到全身各個部分。

自主神經系統　脊椎動物(第148頁)體內，將不隨意肌(第196頁)、心臟(第181頁)和腺體(第162頁)的運動(↓)神經和身體內感受器(↓)相連的感覺(↓)神經。這些神經之間的許多協調(第196頁)均在脊髓(第198頁)和腦髓(第199頁)內進行。神經支配引致的行動主要是自主的，與體外刺激(↓)無關。

感覺的(形)　指神經，與感受器有關(↓)。刺激(↓)使衝動(第197頁)傳到感覺神經。

skull (bone) 頭顱骨
dura mater 硬腦膜
arachnoid 蛛網膜
cerebro-spinal fluid 腦脊液
pia mater 軟腦膜
brain 腦

the meninges
腦脊膜

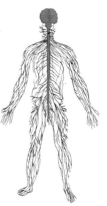

peripheral nervous system
周圍神經系統

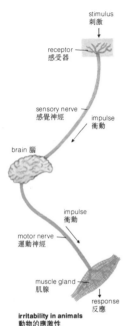

stimulus 刺激
receptor 感受器
sensory nerve 感覺神經
impulse 衝動
brain 腦
impulse 衝動
motor nerve 運動神經
muscle gland 肌腺
response 反應

irritability in animals
動物的應激性

stimulus (*n*) (*stimuli*) any change in the outside conditions of an organism which produces an effect in the organism, but does not provide energy for the effect. In animals, any change in the outside conditions which is the cause of an impulse in its nervous system, e.g. (a) the direction of light is a stimulus for many plants; (b) the smell of food is a stimulus for the salivary glands (p.166); (c) seeing a snake is a stimulus for a bird. **stimulate** (*v*), **stimulation** (*n*).

response (*n*) the change in an organism, or the effect on an organism, produced by a stimulus (↑), e.g. (a) under the stimulus of the direction of light, the response of a plant is to grow towards the light (see *phototropism*, p.161); (b) the stimulus of seeing a snake produces a response in a bird of flight. **respond** (*v*).

irritability (*n*) the ability to make a response (↑) to a stimulus (↑). All organisms have irritability while alive; it is a characteristic of life. **irritate** (*v*).

irritable (*adj*) describes an organism which makes a response (↑) to a stimulus (↑).

motor² (*adj*) of nerves, concerned with producing action in muscles, glands (p.162), cilia (p.144), e.g. (a) a motor nerve stimulates (↑) a muscle to contract and flex (p.196) a joint; (b) a motor nerve stimulates glands in the stomach wall to secrete gastric juice (p.166).

receptor (*n*) an organ containing nervous tissue, which responds (↑) to a particular stimulus (↑). Different kinds of receptors respond to different kinds of stimuli. When a stimulus is strong enough to have an effect on a receptor, the receptor sends an impulse (p.197) along the sensory (↑) nerve connected to the receptor, e.g. the eye is a receptor for the stimulus of light and it sends impulses which travel along an optic nerve to the brain. **receptive** (*adj*), **receptivity** (*n*), **receive** (*v*).

sense-organ another name for receptor.

sense (*n*) the ability to receive stimuli (↑) from the environment (p.226). The senses are: seeing; hearing; smelling; tasting; touching; feeling pain, heat, cold. **sense** (*v*), **sensory** (*adj*).

刺激(名) 指生物體外界條件的任何改變在生物體內產生的一種效應，但並不爲此效應提供能量。外界條件的任何變化都是動物神經系統產生衝動的原因。例如 (a) 光的方向對很多植物都是一種刺激；(b) 食物的氣味對唾液腺(第 166 頁)是一種刺激；(c) 對鳥而言，見到一條蛇是一種刺激。(動詞爲 stimulate，名詞爲 stimulation)

反應(名) 生物體受刺激(↑)所發生的變化或受到的影響。例如：(a) 植物對光方向刺激的反應是向光生長(參見第 161 頁"向光性")；(b) 鳥看到蛇的刺激反應是飛逃。(動詞爲 respond)

應激性(名) 對刺激(↑)作出反應(↑)的能力。一切活着生物體都有應激性；這是生命的特徵。(動詞爲 irritate)

應激性的(形) 描述生物體對刺激(↑)作出反應(↑)。

運動的(形) 指神經而言，與肌肉、腺體(第 162 頁)、及纖毛(第 144 頁)的活動有關。例如(a) 運動神經刺激(↑)肌肉收縮和關節彎曲(第 196 頁)；(b) 運動神經刺激胃壁腺體分泌胃液(第 166 頁)。

感受器(名) 含神經組織並能對特定刺激(↑)作出反應(↑)的一種器官。不同的感受器對不同類的刺激作出反應。當刺激足夠強能使感受器起作用時，感受器沿着與之相連的感覺(↑)神經傳出衝動(第 197 頁)。例如眼是光刺激的感受器，它將衝動沿視神經傳到腦部。(形容詞爲 receptive，名詞爲 receptivity，動詞爲 receive)

感覺器官 感受器的別稱。

感覺；官能(名) 接受環境(第 226 頁)刺激(↑)的能力。感覺分視覺、聽覺、嗅覺、味覺、觸覺以及對痛、熱、冷的感覺。(動詞爲 sense，形容詞爲 sensory)

tactile (*adj*) concerned with touch, e.g. a tactile corpuscle is an end-organ (↓) of touch.

end bulb (*n*) a small end organ (↓) which is a receptor (p.201) in the skin.

end-organ a small organ containing one or a few cells; it is connected to the central nervous system (p.200) by a nerve-fibre (p.197). It may be a receptor (p.201) or it may change an impulse (p.197) into a stimulus (p.201) for a muscle or gland.

taste-bud the receptor (p.201) of taste, present in four groups on the surface of the tongue. Each group responds to one of the four tastes: bitter, sweet, sour, salty. When food is tasted, the flavour is sensed by the receptors for smell in the nose.

觸覺的(形) 有關接觸的,例如觸覺小體是觸覺的終器官(↓)。

終球(名) 微小的終器官(↓),爲皮膚內的感受器(第201頁)。

終器官 含一個或數個細胞的小器官;它以神經纖維(第197頁)與中樞神經系統(第200頁)相連。它可以是一感受器(第201頁),即可以將一個衝動(第197頁)變成對肌肉或腺體的刺激(第201頁)。

味蕾 味覺的感受器(第201頁),在舌面共分四組,每一組分別對苦、甜、酸、鹹四味中的一種味道起反應。品嘗食物時,鼻內的嗅覺感受器感覺到香味。

end-organ 終器官
receptor 接受器

direction of nervous impulse 神經衝動的方向

sensory nerve 感覺神經

association neuron 聯絡神經元

ganglion 神經節

mixed nerve 混合神經

motor nerve 運動神經

spinal card 脊髓

motor neuron cell body 運動神經元細胞體

muscle 肌肉

simple reflex arc 簡單的反射弧

reflex (*n*) a very simple kind of behaviour seen in all animals with a nervous system. A particular stimulus (p.201) always causes the same response (p.201) with no delay, e.g. an object aimed at the eye causes the reflex action of the eyelid closing.

reflex arc the path followed by an impulse (p.197) from receptor to end-organ in a reflex, *see diagram*. The path is: receptor, sensory (p.200) nerve, synapse, association neuron (↓), synapse, motor (p.201) nerve, end-organ.

association neuron a neuron in the spinal cord (p.198) which, by synapses, joins a sensory (p.200) nerve to a motor (p.201) nerve. It also has synapses to pass an impulse (p.197) to other parts of the spinal cord and to the brain (p.199).

反射作用(名) 在有神經系統的所有動物中都可見的一種非常簡單的行爲。一種特定的刺激(第201頁)常立即引起同樣的反應(第201頁),例如一個物體對準眼睛,引起眼瞼閉合的反射動作。

反射弧 反射時衝動(第197頁)從感受器到終器官所經的路線,(見圖)。這條路線是:感受器-感覺(第200頁)神經-突觸-聯絡神經元(↓)-突觸-運動(第201頁)神經-終器官。

聯絡神經元 脊髓(第198頁)中的神經元,它靠突觸將感覺(第200頁)神經和運動(第201頁)神經聯接起來。它有突觸用來將衝動(第197頁)傳送到脊髓的其他部分和腦部(第199頁)。

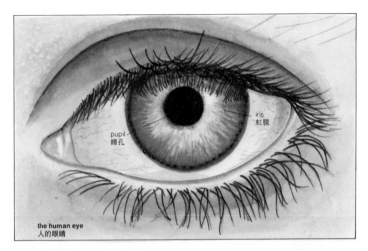

iris
虹膜

pupil
瞳孔

the human eye
人的眼睛

eye (*n*) in animals, the receptor for light. The structure of an eye varies greatly from a simple organ – that responds (p.201) only to light, such as seen in some protozoa (p.143) – to the human eye.

eyeball (*n*) in vertebrates, the ball-shaped eye containing nervous tissue stimulated by light.

iris (*n*) the coloured circular muscle in the front of the eye; it contains a central opening, the pupil (↓). It controls the amount of light entering the eye.

pupil (*n*) the opening in the iris (↑) through which light enters the eye. The size of the pupil is controlled by the iris. The amount of light is a stimulus (p.201) for a reflex (↑) controlling the response (p.201) of the iris.

lachrymal gland in vertebrates (p.148), other than fish, a gland which secretes (p.162) a slightly antiseptic (p.240) liquid (tears) which keep the cornea (p.204) wet.

sclerotic coat in vertebrates (p.148), the strong outer layer of the wall of the eyeball, consisting of fibrous or cartilaginous (p.192) connective tissue (p.192). It gives shape to the eyeball and protects the inner parts.

眼睛(名)　動物對光的感受器。動物眼的結構有很大的不同，一些原生動物(第 143 頁)的眼睛是只對光有反應(第 201 頁)的簡單器官，人的眼睛則較複雜。

眼球(名)　脊椎動物的球狀眼，內含有能感受光刺激的神經組織。

虹膜(名)　眼睛前部的有色環形肌；其中央有一個孔，即瞳孔(↓)。瞳孔控制進入眼內的光量。

瞳孔(名)　虹膜(↑)中的小孔，光線通過它進入眼內，瞳孔的大小由虹膜控制。光量對控制虹膜反應(第 201 頁)的反射作用(↑)的刺激(第 201 頁)。

淚腺　脊椎動物(第 148 頁)(魚類除外)體內的一種腺，能分泌(第 162 頁)有輕微殺菌作用(第 240 頁)的液體(眼淚)以保持角膜(第 204 頁)濕潤。

鞏膜　脊椎動物(第 148 頁)眼球壁的堅韌外層，由纖維狀的或軟骨(第 192 頁)結締組織(第 192 頁)構成。鞏膜使眼球成形並保護眼球內部。

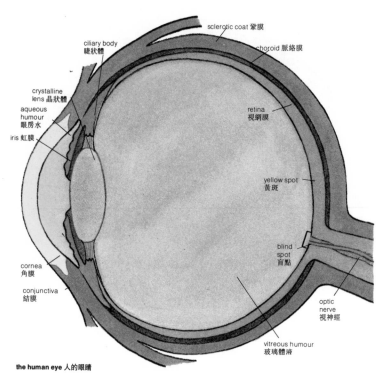

sclerotic coat 鞏膜

choroid 脈絡膜

ciliary body
睫狀體

crystalline
lens 晶狀體

aqueous
humour
眼房水

iris 虹膜

retina
視網膜

yellow spot
黃斑

blind
spot
盲點

cornea
角膜

conjunctiva
結膜

optic
nerve
視神經

vitreous humour
玻璃體液

the human eye 人的眼睛

cornea (*n*) in vertebrates (p.148), the transparent (p.52) covering over the iris (p.203) and the crystalline lens (p.59); it is part of the sclerotic coat. The cornea in land vertebrates produces most of the refraction (p.57) needed to focus (p.58) an image (p.56) on the retina (↓). The crystalline lens completes the refraction. **corneal** (*adj*).

conjunctiva (*n*) in vertebrates (p.148), a layer of epidermis (p.187) which covers the cornea (↑) and the part of the sclerotic coat (p.203) seen as the white part of the eye. The conjunctiva secretes (p.162) mucus (p.190); it is transparent (p.52) and prevents bacteria from entering the eyeball.

角膜（名） 脊椎動物（第 148 頁）眼虹膜（第 203 頁）和晶狀體（第 59 頁）外面的透明（第 52 頁）覆蓋層；它屬鞏膜的一部分。對陸上脊椎動物而言，聚焦（第 58 頁）在視網膜上（↓）成像（第 56 頁）所需之大部分折射（第 57 頁）係由角膜產生。晶狀體則完成折射。（形容詞爲 corneal）

結膜（名） 脊椎動物（第 148 頁）眼上蓋住角膜（↑）和部分鞏膜（第 203 頁）的一層表皮（第 187 頁），即在眼上所見的白色部分。結膜分泌（第 162 頁）黏液（第 190 頁），它是透明（第 52 頁）的並能防止細菌侵入眼球。

choroid coat the middle layer of the wall of the eyeball in some vertebrates (p.148). It is dark coloured and absorbs (p.163) light. The choroid coat carries many blood vessels to supply the retina (↓) and other parts of the eye.

retina (*n*) in vertebrates (p.148), the inner layer of the wall of the eyeball. It contains nervous tissue which is stimulated (p.201) by light, to send impulses by the optic (↓) nerve to the brain.

optic (*adj*) concerned with the eye, e.g. the optic nerve, which goes from the eye to the brain.

blind spot the place where the optic nerve enters the eye of vertebrates (p.148). There is no retina at this point, so light does not stimulate nervous impulses at the blind spot.

yellow spot an area of the retina where sight is clearest, present in man and some apes. When a person looks at an object, the image (p.56) is focused (p.58) on the yellow spot.

ciliary body a ring of muscle and supporting tissue at the edge of the choroid coat (↑). It contains the ciliary muscle, which alters the shape of the crystalline lens (p.59) and gives the lens its power of accommodation (p.59). The iris is fixed to the ciliary body. The ciliary body secretes (p.162) aqueous humour (↓).

aqueous humour a watery liquid which fills the space between the cornea (↑) and the crystalline lens (p.59). It helps to keep the shape of the eyeball.

vitreous humour a jelly-like material which fills the eyeball from the crystalline lens (p.59) to the retina. It keeps the shape of the eyeball and also helps the lens by further refraction (p.57) of light.

脈絡膜　一些脊椎動物(第 148 頁)眼球壁的中層，色暗，能吸收(第 163 頁)光。上面有許多血管供血給視網膜(↓)和眼睛的其他部分。

視網膜(名)　脊椎動物(第 148 頁)眼球壁的內層，含有感受光刺激(第 201 頁)的神經組織。通過視(↓)神經將衝動傳到腦部。

眼的；視覺的(形)　與眼睛有關的，例如從眼睛伸向腦的視神經。

盲點　脊椎動物(第 148 頁)視神經入眼之處，此點無視網膜，所以在盲點上光不能刺激神經衝動。

黃斑　視網膜上視力最清晰的區域。人和一些猿的眼都有黃斑。人凝視物體時，物像(第 56 頁)聚焦(第 58 頁)在黃斑上。

睫狀體　一種環形肌，它支持着脈絡膜(↑)末端的組織，有能改變晶狀體(第 59 頁)的形狀並賦晶狀體調節(第 59 頁)能力的睫狀體肌。虹膜固定在睫狀體上，睫狀體分泌(第 162 頁)眼房水(↓)。

眼房水；水狀液　充滿角膜(↑)和晶狀體(第 59 頁)之間的水狀液體。它幫助保持眼球的形狀。

玻璃體液　一種充滿晶狀體(第 59 頁)直至視網膜整個眼球的膠狀物質。它保持眼球的形狀，也幫助晶狀體進一步折射(第 57 頁)光線。

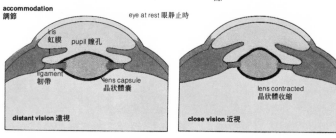

accommodation
調節

eye at rest 眼靜止時

iris
虹膜　　pupil 瞳孔

ligament
韌帶　　lens capsule
晶狀體囊

lens contracted
晶狀體收縮

distant vision 遠視

close vision 近視

outer ear in mammals (p.150), a short tube leading to the ear-drum from the pinna (↓). In birds, there is no pinna. In amphibians (p.149) and most reptiles (p.150) there is no outer ear; the ear-drum (↓), if present, is in the skin. Fish do not possess an ear.

pinna (*n*) in mammals, structure of skin and cartilage fixed to the head; the only part of the ear that can be seen. It helps to collect sound vibrations (p.64).

ear-drum a thin membrane (p.138) stretched over and closing the tube of the outer ear. It vibrates (p.64) to sound, and passes the vibrations to the ossicles (↓).

middle ear in vertebrates (p.148), other than fish, a narrow air-filled space containing the ossicles (↓).

ossicle (*n*) a small bone in the middle ear. There are three ossicles in mammals and they pass vibrations (p.64) from the ear-drum (↑) to the oval window (↓). In birds, reptiles (p.150) and many amphibians (p.149) there is only one ossicle.

Eustachian tube a tube leading from the middle ear (↑) to the throat in land vertebrates (p.148). Its function is to make the air pressures equal on each side of the ear-drum (↑).

外耳 哺乳動物(第 150 頁)體上一根從耳廓(↓)通至鼓膜的短管。鳥類無耳廓,兩棲動物(第 149 頁)和大部分爬蟲動物(第 150 頁)無外耳,即使有鼓膜(↓),也是在皮膚裏,魚類無耳朵。

耳廓(名) 哺乳動物體上由皮膚和固定在頭部的軟骨組成的一種結構,是耳朵中唯一可見的部分。耳廓幫助收集聲音的振動(第 64 頁)。

鼓膜 延伸並封閉外耳道的一層薄膜(第 138 頁),它隨着聲音而振動(第 64 頁),並將振動傳給聽小骨(↓)。

中耳 除魚類以外的脊椎動物(第 148 頁)體內充滿空氣的狹小空間,此空間內有聽小骨(↓)。

聽小骨(名) 中耳內的小骨頭。哺乳動物的耳有三塊聽小骨。聽小骨將振動(第 64 頁)從鼓膜(↑)傳至卵圓窗(↓)。鳥類、爬蟲動物(第 150 頁)和許多兩棲動物(第 149 頁)只有一塊聽小骨。

歐氏管;耳咽管 陸上脊椎動物(第 148 頁)體內從中耳(↑)通向咽喉的管道,其機能是使鼓膜(↑)兩側面的氣壓平衡。

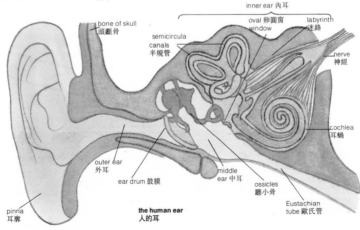

inner ear 內耳

oval 卵圓窗 window

labyrinth 迷路

bone of skull 頭顱骨

semicircula canals 半規管

nerve 神經

cochlea 耳蝸

outer ear 外耳

ear drum 鼓膜

middle ear 中耳

ossicles 聽小骨

Eustachian tube 歐氏管

pinna 耳廓

the human ear 人的耳

inner ear in land vertebrates (p.148), the receptor (p.201) for sound, and the sense organ of balance (p.18); it contains the labyrinth (↓).

oval window a thin membrane (p.138) between the middle ear and the inner ear (↑). The ossicles pass vibrations (p.64) to the oval window. The oval window passes vibrations to the labyrinth (↓).

labyrinth (n) a system (p.162) of tubes and hollows in the inner ear (↑). The bony labyrinth is in the side of the skull (p.193). The membranous (p.138) labyrinth consists of a membrane filled with liquid and fits inside the bony labyrinth.

cochlea (n) a tube in the shape of a spiral (p.219), part of the labyrinth (↑). The liquid in the labyrinth conducts sound vibrations (p.64) to the cochlea, and it responds to the pitch (p.65) of the sound. The cochlea senses both loudness and pitch, and sends impulses to the brain by a nerve.

organ of Corti (n) the part of the ear that hears sounds; it is in the cochlea. Sound vibrations enter the pinna (↑) and are passed by the eardrum and ossicles to the oval window (↑) which then passes sounds to the fluid in the cochlea. Each part of the organ of Corti responds to a different pitch (p.65). Hair cells on the organ send impulses through nerve fibres to the auditory nerve, connected to the brain.

cochlea
耳蝸

內耳　陸上脊椎動物（第 148 頁）的聲音感受器（第 201 頁）和平衡（第 18 頁）感覺器官；內有迷路（↓）。

卵圓窗　中耳和內耳（↑）之間的薄膜（第 138 頁）。聽小骨將振動（第 64 頁）傳至卵圓窗；卵圓窗將振動傳至迷路（↓）。

迷路（名）　內耳（↑）裏的管道和空腔系統（第 162 頁）。骨質的迷路位於頭蓋骨（第 193 頁）的一側。膜（第 138 頁）迷路由充滿液體的膜構成，貼在骨質迷路內。

耳蝸（名）　迷路（↓）中的螺旋狀（第 219 頁）管，迷路中的液體將聲音的振動（第 64 頁）傳入耳蝸，令它對聲音的音調（第 65 頁）作出反應。耳蝸能感覺響度和音調並由神經將衝動傳入腦部。

柯蒂氏器官（名）　耳中能聽到聲音的部分；位於耳蝸內。聲音振動進入耳廓（↑），通過鼓膜和聽小骨傳入卵圓窗（↑），卵圓窗再將聲音傳入耳蝸中的液體。柯蒂氏器官的各個部分對不同的音調（第 65 頁）作出反應。器官上的毛細胞使衝動經神經纖維傳入與腦連通的聽神經。

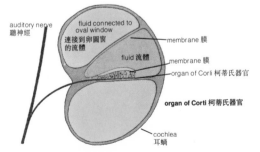

auditory nerve
聽神經

fluid connected to oval window
連接到卵圓窗的流體

membrane 膜

fluid 流體

membrane 膜

organ of Corli 柯蒂氏器官

organ of Corti 柯蒂氏器官

cochlea
耳蝸

semicircular canals in vertebrates (p.148), three tubes, semicircular in shape; placed at right angles to each other and fixed to the labyrinth (↑). They are concerned with the sense of balance, and detect any turning movement of the animal.

半規管　脊椎動物（第 148 頁）耳內的三條半圓形管；彼此位置互成直角，並固定於迷路（↑）。這些半規管與平衡感覺有關，能覺察出動物身體的任何轉動動作。

endocrine gland in vertebrates (p.148), a gland (p.162) which secretes (p.162) a hormone (↓). The gland has no duct (p.162), the hormone diffuses (p.27) into the blood stream from the gland.

ductless gland another name for endocrine gland (↑).

hormone (*n*) an organic (p.131) substance produced in very small amounts by endocrine glands (↑) in one part of an animal and carried by the blood stream to another part where it has an important effect. Plants also produce hormones in very small amounts and hormone movement in the plant is controlled by the plant cells.

adrenalin (*n*) a hormone (↑) secreted by the *adrenal glands* of vertebrates (p.148). Some invertebrates (p.148) also secrete adrenalin. Adrenalin has the following effects: (1) increases the rate of heart beat; (2) widens the blood vessels of muscles, brain and heart; (3) narrows the blood vessels of skin and viscera (p.162); (4) makes the pupil (p.203) of the eye larger; (5) makes hair stand up; (6) increases the amount of sweat (p.188); (7) speeds the change of glycogen (p.174) to glucose (p.174). The stimulus (p.201) for the nervous system is fear, anger or pain. The autonomic nervous system (p.200) then stimulates the adrenal gland, and the body is made ready to escape from danger.

thyroid gland in vertebrates (p.148) an endocrine gland (↑) secreting (p.162) a hormone, **thyroxin**. Thyroxin is a substance containing iodine; it controls the rate of metabolism (p.172) and so controls growth and the production of heat in the body. Lack of iodine in the diet causes the thyroid gland to become very large, a condition known as **goitre**. The thyroid gland is controlled by a hormone from the pituitary gland (↓).

pituitary gland in vertebrates (p.148) an endocrine gland (↑) beneath the floor of the brain. It secretes (p.162) a number of hormones (↑) which control the action of other endocrine glands. The pituitary gland is the most important endocrine gland; it is under the direct control of the central nervous system (p.200).

内分泌腺　脊椎動物(第 148 頁)體內一種分泌(第 162 頁)激素(↓)的腺體(第 162 頁)。它無導管(第 162 頁)，激素從腺體滲進(第 27 頁)血液中。

無管腺　內分泌腺(↑)的另一名稱。

激素；荷爾蒙(名)　動物體內一個部分的內分泌腺(↑)所產生的微量有機(第 131 頁)物質被血流輸送到它起重要作用的另一個部分。植物也能產生微量激素。激素在植物中的活動由植物細胞控制。

腎上腺素(名)　脊椎動物(第 148 頁)的腎上腺分泌的一種激素(↑)。有些無脊椎動物(第 148 頁)也能分泌腎上腺素。腎上腺素有下列作用：(1)增加心搏率；(2)擴張肌肉、腦和心臟的血管；(3)收縮皮膚和內臟(第 162 頁)的血管；(4)使眼睛的瞳孔(第 203 頁)放大；(5)使毛髮豎立；(6)增加汗(第 188 頁)量；(7)加速糖原(第 174 頁)轉變成葡萄糖(第 174 頁)。對神經系統的刺激(第 201 頁)是恐懼、憤怒和疼痛。自主神經系統(第 200 頁)隨後刺激(第 201 頁)腎上腺，令身體準備逃避危險。

甲狀腺　脊椎動物(第 148 頁)體內分泌(第 162 頁)激素即甲狀腺素的一種內分泌腺(↑)。甲狀腺是一種含碘物質；它控制新陳代謝(第 172 頁)率，從而也控制身體的生長和體內熱量的產生。食物中缺碘會引起甲狀腺肥大，稱爲甲狀腺腫。甲狀腺受垂體腺(↓)產生的激素控制。

垂體腺；腦下垂體　脊椎動物(第 148 頁)體內大腦皮層下的一種內分泌腺(↑)。它分泌(第 162 頁)多種激素(↑)控制其他內分泌腺的活動。垂體腺是最重要的內分泌腺；它直接受中樞神經系統(第 200 頁)控制。

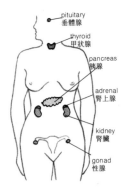

some endocrine glands in a human
人體中的一些內分泌腺

insulin (*n*) a hormone (↑) secreted (p.162) by special cells in the pancreas (p.166) of vertebrates (p.148). Secretion is stimulated (p.201) by a high concentration (p.90) of glucose (p.174) in the blood. Insulin changes glucose to glycogen (p.174) and controls the concentration of glucose in the blood. Lack of insulin causes the disease **diabetes**, in which glucose is present in urine; the kidneys excrete glucose to lower the concentration of glucose in blood. A complete lack of insulin causes death. Insulin and adrenalin (↑) are antagonistic (p.195) in their actions.

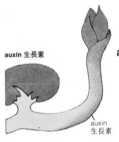

auxin 生長素

auxin
生長素

auxin (*n*) one of a group of plant hormones produced by actively dividing cells at the ends of stems and roots. An auxin increases the growth in length of a plant cell, thus causing the curving of a stem, or root, in phototropism (p.161) and geotropism (p.161). Auxins also control fruit growth, leaf fall, growth of buds (p.213) and other kinds of plant growth.

reproduction (*n*) of organisms (p.147), the process of producing young organisms which have the same kind of characteristics (p.147) as the parent organism. Reproduction can be either sexual or asexual (↓). **reproduce** (*v*), **reproductive** (*adj*).

sexual reproduction reproduction (↑) in which two organisms (p.147) take part, one called *male* and the other called *female*. Male and female organisms have different reproductive structures (p.144); a male gamete (p.210) is given by a male to a female to unite with a female gamete to produce a new organism.

asexual reproduction reproduction (↑) in which only one organism takes part; it is reproduction without gametes (p.210). In plants it is carried out by forming spores (p.146) or by vegetative reproduction (p.213). In animals it is carried out by binary fission (p.141) or by budding (p.213).

fertile (*adj*) of gametes (p.210) and organisms (p.147), capable of reproducing a young organism. **fertility** (*n*).

sterile (*adj*) (1) of organisms (p.147), not capable of sexual reproduction (↑). (2) free from bacteria, viruses, fungi and protozoa (p.143).

胰島素（名） 脊椎動物（第 148 頁）胰臟（第 166 頁）內特殊細胞所分泌（第 162 頁）的一種激素（↑）。血內高濃度（第 90 頁）的葡萄糖（第 174 頁）刺激（第 201 頁）分泌出胰島素。胰島素將葡萄糖轉變成爲醣原（第 174 頁）以控制血中葡萄糖的濃度。缺乏胰島素會引致**糖尿病**。病人的尿中含有葡萄糖；腎排泄葡萄糖以降低血中葡萄糖的濃度。完全缺乏胰島素會導至死亡。胰島素和腎上腺素（↑）的作用是對抗的（第 195 頁）。

植物生長素（名） 由莖端和根端積極分裂的細胞產生的數種植物激素之一。植物生長素能加快植物細胞的生長，因此使莖或根在向光性（第 161 頁）和向地性（第 161 頁）作用下變彎曲。激素還控制果實的生長、葉子的脫落和發芽（第 213 頁）以及植物生長的其他方面。

生殖；繁殖（名） 指生物體（第 147 頁）的生殖，產生與親體相同特徵（第 147 頁）的生物幼體的過程。生殖可以是有性生殖或無性殖（↓）。（動詞爲 reproduce，形容詞爲 reproductive）

有性生殖 兩個生物體（第 147 頁）參與生殖（↑）方式。一個稱爲雄性，另一個稱爲雌性。二者有不同的生殖結構（第 144 頁）；雄性生物體將雄配子（第 210 頁）給予雌性生物體，使與雌配子結合而產生新的生物體。

無性生殖 只有一個生物體參加的生殖（↑）過程；是無需配子（第 210 頁）的生殖。植物生殖是通過形成孢子（第 146 頁）或營養生殖（第 213 頁）來完成的。動物的生殖是通過二分裂（第 141 頁）或出芽生殖（第 213 頁）來完成的。

能生育的（形） 指能生殖幼體的配子（第 210 頁）和生物體（第 147 頁）。（名詞爲 fertility）

不育的；無菌的（形） (1)指不能進行有性生殖（↑）的生物體（第 147 頁），(2)沒有細菌、真菌和原生動物（第 143 頁）的。

gamete (*n*) a reproductive (p.209) cell with a haploid (p.220) number of chromosomes (p.142) in its nucleus (p.139). In many organisms (p.147) the male and female gametes are different. The female gamete usually has a large cytoplasm (p.138) and is not capable of locomotion (p.194). The male gamete usually has very little cytoplasm and is capable of locomotion by means of a flagellum (p.144).

sex-cell another name for gamete.

gonad (*n*) in animals, the organ which produces gametes (↑). In some animals, gonads also produce hormones.

fusion (*n*) the process of two nuclei (p.139) from gametes (↑) joining and becoming one nucleus.

fertilization (*n*) the union of two gametes (↑), which takes place in two stages (p.152); (1) the fusion (↑) of the nuclei of two gametes; (2) the start of the development of a new organism.

external fertilization is the union of gametes outside the body of the parents. **fertilize** (*v*).

internal fertilization is the union of gametes inside the body of the female parent.

zygote (*n*) a fertilized female gamete (↑) before it starts to grow by binary fission (p.141). It is the first cell of the new individual in sexual reproduction.

offspring (*n*) (*offspring*) the young animals reproduced (p.209) by their parents.

generation (*n*) a single stage (p.152) in a family; the parents form one generation and their offspring (↑) the next generation.

配子(名) 細胞核(第 139 頁)中帶有單倍數(第 220 頁)染色體(第 142 頁)的生殖(第 209 頁)細胞。很多生物體(第 147 頁)中雄配子與雌配子是不同的。雌配子通常有很多細胞質(第 138 頁),而且不能移動(第 194 頁)。雄配子的細胞質通常很少而且能依靠鞭毛(第 144 頁)移動。

性細胞 配子的另一名稱。

性腺(名) 動物產生配子(↑)的器官。一些動物的性腺也能產生激素。

融合(名) 兩個配子(↑)的細胞核(第 139 頁)合併成一個核的過程。

受精(名) 兩個配子(↑)的結合分兩個階段(第 152 頁)進行:(1)兩個配子的細胞核融合(↑);(2)開始發育成爲新的生物體。

體外受精 配子在親體外的結合。(動詞爲 fertilize)

體內受精 配子在母體內的結合。

受精卵;合子(名) 通過二分裂(第 141 頁)開始發育之前的雌配子(↑)。它是有性生殖中新個體的第一個細胞。

後代(名) 親代生殖(第 209 頁)的幼小動物。

一代;世代(名) 家族的一個階段(第 152 頁);親代 組成一代,他們的後代(↑)組成下一代。

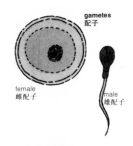

gametes
配子

female
雌配子

male
雄配子

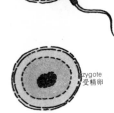

fertilization
受精

zygote
受精卵

sexual fertilization
有性受精

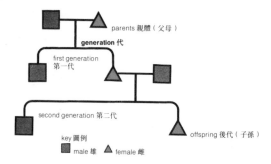

parents 親體 (父母)

generation 代

first generation
第一代

second generation 第二代

offspring 後代 (子孫)

key 圖例

■ male 雄　▲ female 雌

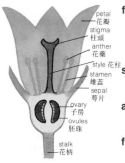

diagram of a flower
花的圖示

petal
花瓣
stigma
柱頭
anther
花藥
style 花柱
stamen
雄蕊
sepal
萼片
ovary
子房
ovules
胚珠
stalk
花柄

flower (*n*) a reproductive structure in monocotyledon (p.155) and dicotyledon (p.155) plants. The *essential* flower parts are concerned with reproduction, while the *accessory* flower parts are not directly concerned with reproduction.

stamen (*n*) the male essential flower part. It consists of a thin stalk (p.157), called a filament, bearing an anther (↓). **staminate** (*adj*).

anther (*n*) a structure at the end of the stamen (↑) stalk consisting of two parts, each containing pollen (↓) sacs. An anther produces pollen (↓).

filament² (*n*) a thin stalk (p.157) which supports an anther (↑) in a position to effect cross-pollination (↓).

pollen (*n*) a small grain which produces two male gametes. Some pollen grains are very light and easily blown by the wind; other pollen grains are heavier and sticky, and are carried by insects.

pollination (*n*) the process by which pollen grains are carried from anther (↑) to stigma (↓). This is done either by wind or by insects (p.151). The pollen is carried from the anther (↑) of one plant to the stigma (↓) of another plant in **cross-pollination**. In **self-pollination**, pollen is carried from anther to stigma of the same flower or to a stigma of another flower on the same plant.

carpel (*n*) a female essential part of a flower (↑). It consists of a stigma (↓), a style (↓) and an ovary (↓) containing ovules (↓). A flower may contain more than one carpel.

stigma (*n*) the surface of a carpel (↑) which is usually sticky; pollen grains stick to it and germinate (p.156).

style (*n*) a short stalk bearing a stigma (↑). After a pollen (↑) grain has germinated (p.156), a pollen tube grows down the style to an ovule (↓), and male gametes pass down the tube.

ovary¹ (*n*) a hollow space at the bottom of a carpel (↑) with a thick wall around it. The ovary contains one or more ovules (↓). It develops into a fruit (p.155).

ovule (*n*) in seed plants, a structure that contains a female gamete (p.210) which, after fertilization by a male gamete, forms an embryo. Each ovule is fixed to the ovary (↑) wall by a stalk.

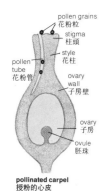

pollinated carpel
授粉的心皮

pollen grains
花粉粒
stigma
柱頭
style
花柱
pollen tube
花粉管
ovary wall
子房壁
ovary
子房
ovule
胚珠

花（名） 單子葉植物（第 155 頁）和雙子葉植物（第 155 頁）的生殖構造。花的主要部分與生殖有關。而花的附屬部分與生殖無直接關係。

雄蕊（名） 雄花的基本部分。它由一細長的花柄（第 157 頁）構成，稱爲花絲，其上有花藥（↓）。（形容詞爲 staminate）

花藥（名） 雄蕊（↑）花柄端的構造，由兩部分組成，每部分都含有花粉（↓）囊。花藥產生花粉（↓）。

花絲（名） 支撐着花藥（↑）在適當位置以便進行異花授粉（↓）的細長花柄（第 157 頁）。

花粉（名） 產生兩個雄配子的小顆粒，有些花粉顆粒非常輕，易被風吹播，有些則較重和有黏性，靠昆蟲傳播。

授粉作用（名） 花粉顆粒由花藥（↑）傳播到柱頭（↓）的過程。或由風、或由昆蟲（第 151 頁）傳播。**異花授粉**係由一株植物的花藥（↑）傳播到另一株植物的柱頭（↓）。**自花授粉**則是花粉由花藥傳到同一朵花的柱頭上或傳到同一株植物的另一朵花的柱頭上。

心皮（名） 花（↑）的雌性主要部分。它由柱頭（↓）、花柱（↓）和含有胚珠（↓）的子房（↓）組成。一朵花可以有一個以上的心皮。

柱頭（名） 心皮（↑）的表層，通常是黏性的；花粉顆粒黏到柱頭上，開始萌發（第 156 頁）。

花柱（名） 上有柱頭（↑）的短柄。花粉（↑）顆粒萌發（第 156 頁）後，花粉管沿花柱向下長，伸入子房（↓），同時雄配子下入花管中。

子房（名） 心皮（↑）底部一有厚壁圍着的中空空間。子房內有一個或多個胚珠（↓）。它發育成果實（第 155 頁）。

胚珠（名） 種子植物中一種內有雌配子（第 210 頁）的結構。它在雄配子授精後形成胚。每個胚珠靠花柄固定在子房（↑）上。

gynoecium (*n*) the whole female reproductive (p.209) organ of a flower, consisting of one or more carpels (p.211).

pistil (*n*) another name for either carpel (p.211) or gynoecium (↑). **pistillate** (*adj*).

androecium (*n*) the whole male reproductive (p.209) part of a flower, i.e. all the stamens (p.211).

receptacle (*n*) the top of a flower-stalk; its shape can vary from convex (p.59) to concave (p.59) in different flowers. The receptacle bears the perianth (↓), carpels (p.211) and stamens (p.211).

perianth (*n*) the accessory parts of a flower. It usually consists of an outer whorl (↓) of sepals (↓) and an inner whorl of petals (↓). The carpels (p.211) and stamens (p.211) lie inside the perianth.

whorl (*n*) a circle of like parts growing at the same level on the stem of a plant, e.g. a whorl of leaves on a stem, a whorl of petals on a receptacle (↑).

corolla (*n*) all the petals (↓) of a flower.

雌蕊群(名) 花的整個雌性生殖(第 209 頁)器官，由一個或多個心皮(第 211 頁)組成。·

雌蕊(名) 心皮(第 211 頁)或雌蕊群(↑)的別稱。(形容詞爲 pistillate)

雄蕊群(名) 花的整個雄性生殖(第 209 頁)部分，即全部雄蕊(第 211 頁)。

花托(名) 花柄的頂端；不同的花有不同的花柄形狀，從凸形(第 59 頁)到凹形(第 59 頁)。花托上長有花被(↓)、心皮(第 211 頁)和雄蕊(第 211 頁)。

花被(名) 花的附屬部分。通常由萼片(↓)的外輪(↓)和花瓣(↓)的內輪構成。花被內側長有心皮(第 211 頁)和雄蕊(第 211 頁)。

輪(名) 植物莖同一水平處所長的一圈圈相似的部分，例如莖上的葉片輪、花托(↑)上的花瓣輪。

花冠(名) 一朵花的全部花瓣(↓)。

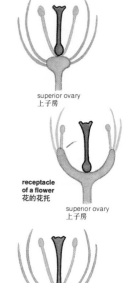

superior ovary
上子房

receptacle of a flower
花的花托

superior ovary
上子房

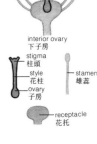

interior ovary
下子房

stigma
柱頭

style
花柱

ovary
子房

stamen
雄蕊

receptacle
花托

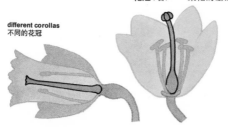

different corollas
不同的花冠

petal (*n*) a leaf-like structure, the inner part of the perianth (↑), usually coloured to attract (p.17) insects (p.151); there are many different shapes of petals to suit the insects pollinating the flower. The petals are the flower parts most easily seen. Dicotyledons (p.155) generally have flower parts in groups of five, while monocotyledons (p.155) generally have flower parts in groups of three.

calyx (*n*) all the sepals (↓) of a flower.

sepal (*n*) in dicotyledons (p.155), a green, leaf-like structure, the outermost part of the perianth (↑). In dicotyledons (p.155) sepals are generally grouped in fives.

花瓣(名) 花被(↑)內部的一種葉狀結構，通常有顏色以吸引(第 17 頁)昆蟲(第 151 頁)；花瓣的形狀極不相同，以適應給花傳粉的昆蟲。花瓣是花最顯眼的部分。雙子葉植物(第 155 頁)的花部通常是一簇五瓣，而單子葉植物(第 155 頁)的花部是一簇三瓣。

花萼(名) 花的全部萼片(↓)。

萼片(名) 雙子葉植物(第 155 頁)的綠色葉狀結構，即花被(↑)的最外部分。雙子葉植物(第 155 頁)的萼片通常是五片爲一簇。

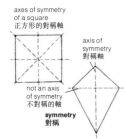

axes of symmetry
of a square
正方形的對稱軸

axis of
symmetry
對稱軸

not an axis
of symmetry
不對稱的軸

symmetry
對稱

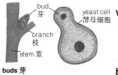

bud
芽

yeast cell
酵母細胞

branch
枝

stem 莖

buds 芽

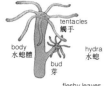

tentacles
觸手

body
水螅體

hydra
水螅

bud
芽

rhizome 根莖

fleshy leaves
肉質葉

flowering shoot
開花的苗

ground level
地面

bud
芽

scale
leaf
鱗狀葉

adventitious
roots 不定根

stem 莖

bulb
鱗莖

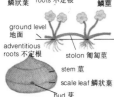

ground level
地面

adventitious
roots 不定根

stolon 匍匐莖

stem 莖

scale leaf 鱗狀葉

bud 芽

corm 球莖

tepal (*n*) in monocotyledons (p.155), a structure of a perianth (↑) in which there is no difference between sepals (↑) and petals (↑).

nectary (*n*) a gland (p.162) which secretes (p.162) a sweet liquid, called **nectar**, which attracts (p.17) insects (p.151) to the flower (p.211).

symmetry (*n*) the state of having a regular shape such that a line can divide the structure into two equal parts which are similar, that is, the parts are balanced about the line. Such a line is called an *axis of symmetry*. For example, a square has symmetry as a line can be drawn through opposite corners, or through the middle of opposite sides, and in each case, the square is divided into equal parts which are balanced about the line. **symmetrical** (*adj*).

vegetative reproduction a kind of asexual reproduction (p.209) in plants. It is the growth of a new plant from a part of an old plant but not from spores (p.146), e.g. as in rhizomes (↓), bulbs (↓), corms (↓).

bud (*n*) (1) a small, pointed structure on a stem. A bud grows into either a leaf or a flower. (2) a bud-like growth from the wall of a cell which becomes large, leaves the parent cell and becomes a daughter cell, e.g. buds on yeast cells. (3) a bud-like growth from the body of certain simple animals. The bud grows into a young animal and leaves the parent, e.g. a bud on hydra, *see diagram*.

bulb (*n*) a modified (p.158), very small, under-ground stem covered in succulent (p.155) fleshy leaves which store food. Buds grow on the stem between the leaves. The new plant forms new bulbs in vegetative reproduction (↑), e.g. onion.

rhizome (*n*) an underground stem with buds (↑) in the axils (p.158) of scale-like leaves. The rhizome grows year after year, and plants grow from the buds. A way of vegetative reproduction (↑).

corm (*n*) a short, thick, round, underground stem, which stores food. The corm has buds in the axils (p.158) of scale-like leaves. A new plant grows from the bud using the stored food, and produces new corms before it dies, e.g. gladioli. A way of vegetative reproduction (↑).

花被片（名） 單子葉植物（第 155 頁）中，萼片（↑）和花瓣（↑）無區別的花被（↑）結構。

蜜腺（名） 分泌（第 162 頁）甜液（又稱**花蜜**）的腺（第 162 頁），它吸引（第 17 頁）昆蟲（第 151 頁）到花（第 211 頁）上。

對稱性（名） 形狀規則的狀態，即可用一條直線將結構分成相似的兩等分，亦即直線兩邊的兩個部分是均衡的。此條線稱爲"對稱軸"。例如正方形有對稱性，因爲畫一條直線通過正方形的兩對角或兩對邊的中點，可將之分爲平衡的兩個相等部分。（形容詞爲 symmetrical）

營養生殖 植物的一種無性生殖（第 209 頁）方式。新株是由老株的一部分，而不是由孢子（第 146 頁）長出。例如在根莖（↓）、鱗莖（↓）或球莖（↓）上長出。

芽（名） (1) 莖上小而尖的結構。芽長成葉或花；(2) 從細胞壁上長出的芽體，它不斷長大，離開親細胞，成爲子細胞，例如酵母細胞上的芽體；(3) 從某些初級動物體上長出的芽體。芽體發育成幼體後離開親體。例如水螅的芽（見圖）。

鱗莖（名） 一種變性的（第 158 頁），非常小的地下莖，上覆有多汁的（第 155 頁）肉質葉，內貯有養料。芽長在葉之間的莖上。新株以營養生殖（↑）長成新鱗莖。例如洋葱。

根莖（名） 鱗狀葉的葉腋（第 158 頁）間長有芽體（↑）的地下莖。根莖年復一年地長大，新株從芽體發育而來。這是一種營養生殖（↑）方式。

球莖（名） 一種貯藏養料的粗短圓形地下莖。球莖鱗狀葉葉腋（第 158 頁）下長有芽體。芽利用貯藏的養料發育成新株，新株在新球莖長出死去。例如菖蒲取營養生殖（↑）方式。

stolon (*n*) a stem that grows along the ground. At a node (p.157) roots grow into the earth and a new plant grows from the node, e.g. strawberry. A way of vegetative reproduction (p.213).

runner (*n*) a stolon (↑) that roots only at its end.

tuber (*n*) the swollen end of an underground stem; it has buds in the axils (p.158) of scale-like leaves. The buds grow into new plants which produce new tubers. The tuber stores food for the new plant, e.g. potato. A way of vegetative reproduction (p.213).

vegetative propagation the use by man of parts of a plant which will grow by vegetative reproduction, e.g. cuttings (↓), grafts (↓).

cutting (*n*) a piece of a stem (p.156) put in earth; roots grow from a node (p.157) and a new plant grows from the roots by vegetative propagation (↑).

graft (*v*) (1) to produce union between the tissues of two different plants, e.g. a piece of stem or a bud is taken from one tree, and put into a cut made in the phloem (p.160) of another tree. A way of vegetative propagation (↑). (2) to produce union between the tissues of two different persons, e.g. skin graft. **graft** (*n*).

transplant (*v*) (1) to take a plant, usually a seedling (p.156), out of the earth and put it in another place. (2) to take an organ out of one animal's body and put it in the body of another animal, e.g. to transplant a kidney.

testis (*n*) (*testes*) in animals, the organ which produces sperms (↓). In vertebrates (p.148), it also produces hormones (p.208); it is a male gonad (p.210).

testicle (*n*) a testis in mammals. It is contained in a bag-like structure.

seminiferous tubule in the testes of vertebrates (p.148), a coiled tube, in man about 50 cm long and 0.2 mm in diameter, which produces sperms (↓). There are several hundred seminiferous tubules in a testis.

sperm (*n*) in animals, a male gamete (p.210); it has a nucleus (p.139), very little cytoplasm (p.138) and, in most animals, a flagellum (p.144) which allows it to swim in a liquid towards a female gamete.

spermatozoon (*n*) (*spermatozoa*) a sperm.

匍匐莖（名）　貼地面生長的莖。根的莖節（第 157 頁）長入地下而新株從莖節上長出。例如草莓。這是營養生殖（第 213 頁）方式。

長匍匐莖（名）　末端生根的匍匐莖（↑）。

塊莖（名）　地下莖肥大的一端。其鱗狀葉的葉腋（第 158 頁）處長芽，芽長成新株，新株生出新的塊莖。塊莖爲新株貯藏養料。例如馬鈴薯。這是營養生殖（第 213 頁）的一種方式。

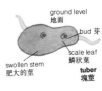

ground level 地面
bud 芽
scale leaf 鱗狀葉
swollen stem 肥大的莖
tuber 塊莖

營養繁殖　人類利用植株的某些部分，通過營養生殖的方法使之繁生，例如插條（↓）和嫁接（↓）。

插條（名）　插入土中的一段莖（第 156 頁）；根從莖節（第 157 頁）上長出。藉營養繁殖（↑）的方式從根上長出新株。

嫁接；移植（動）　(1) 將兩種不同的植株的組織連接起來。例如從樹上摘下一段莖或一個樹芽，放入另一株樹韌皮部（第 160 頁）的切口內。這是一種營養繁殖（↑）的方式；(2) 將兩個人身上的組織連接起來，例如植皮。（名詞為 graft）。

移植（動）　(1) 從地裏拔出一株植物，通常是幼苗（第 156 頁）栽在另一處；(2) 從一動物體內切除一個器官植入另一動物體內。例如腎移植。

精巢；睪丸（名）　動物體內產生精子（↓）的器官。脊椎動物（第 148 頁）的睪丸也產生激素（第 208 頁）；它是雄性腺（第 210 頁）。

睪丸（名）　哺乳動物的睪丸，它納於一袋狀結構內。

精小管　脊椎動物（第 148 頁）的睪丸內產生精子（↓）的螺旋管。人的精小管長約 50 cm，直徑約 0.2 mm，睪丸中有數百根精小管。

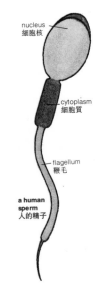

nucleus 細胞核
cytoplasm 細胞質
flagellum 鞭毛
a human sperm 人的精子

精子（名）　動物的雄配子（第 210 頁）；它有細胞核（第 139 頁）和極小的細胞質（第 138 頁）。大部分動物的精子都有鞭毛（第 144 頁），使精子能在精液中游向雌配子。

精子亦稱 **spermatozoon**（spermatozoa）。

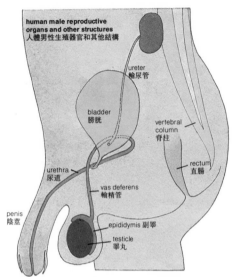

human male reproductive
organs and other structures
人體男性生殖器官和其他結構

ureter
輸尿管

bladder
膀胱

vertebral
column
脊柱

urethra
尿道

rectum
直腸

vas deferens
輸精管

penis
陰莖

epididymis 副睪

testicle
睪丸

epididymis (*n*) in reptiles (p.150), birds, and
mammals (p.150), a long tube which receives
sperms from a testis (↑), stores them, and then
passes them to the vas deferens (↓).

vas deferens one of two tubes, one on each side
of the body, passing sperm from a testis (↑),
in fishes and amphibians (p.149) to a cloaca
(p.165) or from an epididymis to a cloaca (p.165)
in reptiles (p.150) and birds, and the urethra in
mammals (p.150).

urethra (*n*) in mammals (p.150), a tube from the
urinary bladder to outside the body; in males, it
passes through the penis (↓). In males, the
urethra is joined by the vas deferens (↑); it is the
path for urine and for sperms. In females, it is a
tube from the bladder alone.

penis (*n*) in mammals (p.150), an organ with many
blood vessels, which is used for internal fertilization (p.210) by putting sperms into the vagina
(p.217) of females. It contains erectile (↓) tissue.

erectile (*adj*) able to be stiff and upright when
supplied with blood.

副睪(名)　爬蟲動物(第150頁)、鳥類和哺乳動
物(第150頁)體內,接受並貯存來自睪丸
(↑)的精子,然後送入輸精管(↓)的一條長
管。

輸精管　輸送來自睪丸(↑)的精子的兩條管之
一,身體每側各一條。在魚類和兩棲動物
(第149頁)體內精子是送入泄殖腔(第165
頁),在爬蟲動物(第150頁)和鳥類體內精
子是從副睪送入泄殖腔(第165頁);在哺乳
動物(第150頁)體內是送入尿道。

尿道(名)　哺乳動物(第150頁)體內從膀胱通到
體外的管道;雄性的尿道通過陰莖(↓),並
與輸精管(↑)相聯;尿道是尿和精子的通
道。雌性的尿道只是從膀胱通外的管道。

陰莖(名)　哺乳動物(第150頁)體內一個有很多
血管的器官。它將精子送入雌性的陰道(第
217頁)。以便體內受精(第210頁)。它含有
勃起(↓)組織。

勃起的(形)　充血後能挺直。

reproductive system the organs and structures concerned with the production of gametes.

ovary[2] (*n*) in animals, the organ which produces ova (↓). In vertebrates (p.148), it also produces hormones (p.208); it is a female gonad (p.210).

ovum (*n*) (*ova*) in animals, a female gamete (p.210), an unfertilized egg-cell. An ovum contains a nucleus (p.139), a lot of cytoplasm (p.138), yolk (↓) grains, and is covered by a thick membrane (p.138). In reptiles (p.150) and birds, the ovum is covered by a shell.

fallopian tube in female mammals (p.150), a tube leading from an ovary (↑) to the uterus (↓). It has a funnel-shaped opening near the ovary, and is lined with cilia (p.144) which help to conduct ova from the ovary to the uterus. Sperms (p.214) fertilize (p.210) ova in the fallopian tube.

uterus (*n*) (*uteri*) in female mammals (p.150), a hollow organ in which the embryo (↓) develops until it is born. In humans and monkeys, there is only one uterus, but in other mammals there are two, one for each fallopian tube. The walls of the uterus are formed from involuntary muscle, and the inside of the wall is covered with glandular (p.162) tissue. **uterine** (*adj*).

cervix (*n*) a short, narrow tube leading from the uterus (↑) to the vagina (↓). It secretes (p.162) mucus (p.190) to the vagina. **cervical** (*adj*).

生殖系統　與產生配子有關的器官和結構。

卵巢（名）　動物體內產卵（↓）的器官。脊椎動物（第 148 頁）的卵巢也能產生激素（第 208 頁）；它是雌性腺（第 210 頁）。

卵（名）　動物體內的雌配子（第 210 頁），即未受精的卵細胞。卵有細胞核（第 139 頁）、大量細胞質（第 138 頁）、卵黃（↓）顆粒，而且外包一層厚膜（第 138 頁）。爬蟲動物（第 150 頁）和鳥類的卵有一層外殼。

輸卵管　雌性哺乳動物（第 150 頁）體內從卵巢（↑）通向子宮（↓）的管道。在輸卵管旁有一漏斗形開口，管內鞭毛（第 144 頁）幫助將卵從卵巢送入子宮。精子（第 214 頁）在輸卵管內使卵受精（第 210 頁）。

子宮（名）　雌性哺乳動物（第 150 頁）體內一空腔器官，胚胎（↓）出生前一直在子宮內發育。人和猴只有一個子宮；但其他哺乳動物有兩個。每個輸卵管各有一個。子宮壁由不隨意肌構成，壁的裏面覆蓋着腺（第 162 頁）組織。（形容詞爲 uterine）

子宮頸（名）　由子宮（↑）通到陰道（↓）的一條短窄管道。它向陰道分泌（第 162 頁）黏液（第 190 頁）。（形容詞爲 cervical）

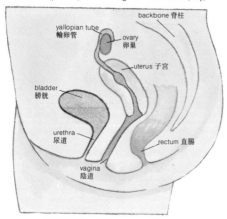

backbone 脊柱

fallopian tube 輸卵管

ovary 卵巢

uterus 子宮

bladder 膀胱

urethra 尿道

rectum 直腸

vagina 陰道

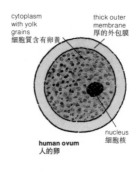

cytoplasm with yolk grains 細胞質含有卵黃

thick outer membrane 厚的外包膜

nucleus 細胞核

human ovum 人的卵

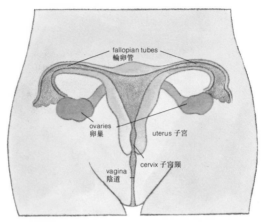

fallopian tubes
輪卵管

ovaries
卵巢

uterus 子宮

cervix 子宮頸

vagina
陰道

vagina (*n*) in female mammals, a tube leading from the cervix (↑) to outside the body. It receives sperm (p.214) for fertilization (p.210), and is the passage for the birth of an offspring (p.210). **vaginal** (*adj*).

egg (*n*) the ovum together with a yolk (↓) covered by a shell in birds, a tough membrane (p.138) in reptiles (p.150), or a jelly-like material in amphibians (p.149). The ovum is an *egg-cell*.

yolk (*n*) a store of food for an embryo (↓) in the eggs (↑) of most animals. It consists of protein and fat. **yolky** (*adj*).

incubate (*v*) to keep eggs under suitable conditions until the offspring (p.210) break out of the shell or membrane.

hatch (*v*) (1) of offspring, to come out of an egg, (2) of eggs, to bring forth young after incubation (↑).

embryo² (*n*) an animal growing in an egg or in its mother's body. Hatching (↑) ends the embryonic stage in animals other than mammals (p.150). **embryonic** (*adj*).

foetus (*n*) in mammals, the stage when an embryo (↑) starts to have the appearance of a fully developed offspring. In man, an embryo changes to a foetus after about two months. **foetal** (*adj*).

**foetus in uterus
子宮內的胎兒**

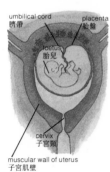

umbilical cord
臍帶

placenta
胎盤

foetus
胎兒

cervix
子宮頸

muscular wall of uterus
子宮肌壁

陰道（名） 雌性哺乳動物體內一根從子宮頸（↑）通向體外的管道。它接收精子（第 214 頁）授精（第 210 頁），並且是後代（第 210 頁）出生的通道。（形容詞爲 vaginal）

卵（名） 鳥類的卵有卵黃（↓），外有蛋殼，爬蟲動物（第 150 頁）的卵有堅韌的膜（第 138 頁），兩棲動物（第 149 頁）的卵有膠狀的物質。卵是一個"卵細胞"。

卵黃（名） 貯藏在大部分動物的卵（↑）內，供胚胎（↓）發育的養料。卵由蛋白質和脂肪組成。（形容詞爲 yolky）

孵育（動） 使卵保持在合適的條件下，直至後代（第 210 頁）破殼或破膜而出。

孵出（動） （1）指後代從卵裏出來；（2）指卵在孵化（↑）後生出幼體。

胚胎（名） 正在卵內或在母體內生長的動物。除哺乳動物（第 150 頁）外，其他動物於孵出時即結束胚胎階段。

胎兒（名） 哺乳動物發育的一個階段，即胚胎（↑）開始具有發育完全的後代的雛形。人的胚胎約在兩個月後才發育成胎兒。（形容詞爲 foetal）

placenta (*n*) an organ of nutrition (p.171) for a foetus (p.217). It consists of foetal (p.217) tissues and mother's tissues with the two tissues interdigitated (↓). The circulatory systems of foetus and mother are quite separate. Oxygen, glucose, amino acids and fats diffuse (p.27) from the mother's capillaries (p.179) into the foetal capillaries, and carbon dioxide and urea diffuse in the opposite direction. **placental** (*adj*).

interdigitate (*v*) of two tissues, to have finger-like structures of one tissue go into the hollows of the other tissue. **interdigital** (*adj*).

crypts (*n.pl.*) small hollows in the thickened wall of a uterus into which the placenta grows.

umbilical cord a tube from the placenta (↑) to the abdomen (p.162) of the foetus. It contains the foetal artery and vein. It breaks, or is broken, at birth.

胎盤(名) 胎兒(第 217 頁)的營養(第 171 頁)器官。它由胎兒的(第 217 頁)組織和母體的組織構成,二者呈指狀咬合(↓)。胎兒的循環系統與母體的循環系統都是相當獨立的。母體的氧、葡萄糖、氨基酸和脂肪從毛細管(第 179 頁)滲(第 27 頁)入胎兒的毛細管中,胎兒的二氧化碳和尿則反向滲透。(形容詞爲 placental)

指狀咬合(動) 指兩個組織的指狀結構彼此嵌入對方的凹處。(形容詞爲 interdigital)

小囊;隱窩(名、複) 增厚的子宮壁的小凹處,胎盤在內生長。

臍帶 從胎盤(↑)通到胎兒腹部(第 162 頁)的管道。臍帶內有胎兒的動脈和靜脈。胎兒出生時臍帶自斷或被切斷。

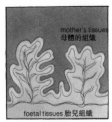

interdigitation in a placenta
胎盤内的指狀咬合

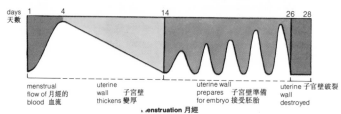

menstrual flow of 月經的 blood 血流 · uterine wall 子宮壁 thickens 變厚 · uterine wall prepares 子宮壁準備 for embryo 接受胚胎 · uterine wall 子宮壁破裂 destroyed · menstruation 月經

menstruation (*n*) in monkeys, apes, and humans, a periodic (p.64) sending out of blood and mucus (p.190) from the uterus; in a woman, this happens every 28 days. Menstruation results from the destruction of the wall of the uterus. After menstruation, the uterine wall thickens and prepares to receive an embryo. If fertilization has not taken place, the uterine wall is destroyed, and finally sent out as blood and mucus through the vagina (p.217). Menstruation stops temporarily during pregnancy. *See diagram.* **menstruate** (*v*), **menstrual** (*adj*).

lactation (*n*) the production of milk in mammary glands (↓). **lactate** (*v*).

mammary glands in female mammals (p.150) glands on the abdomen (p.162) of most mammals, but on the thorax (p.189) of apes and humans.

月經(名) 猿猴類和人類的子宮周期性(第 64 頁)排出的血液和黏液(第 190 頁);婦女每 28 天行經一次。月經是因子宮壁破裂而引起的。月經後子宮壁增厚,準備接收胚胎。如果沒有受精,子宮壁就破裂而最後通過陰道(第 217 頁)排出血液和黏液。懷孕期間月經暫停。(見圖)。(動詞爲 menstruate,形容詞爲 menstrual)

泌乳(名) 乳腺(↓)產生乳汁。(動詞爲 lactate)

乳腺 大部分哺乳動物(第 150 頁)的雌性動物腹部(第 162 頁)的腺體,但猿猴和人的腺體在胸部(第 189 頁)。

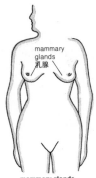

mammary glands 乳腺

mammary glands
乳腺

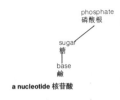

phosphate
磷酸根

sugar
糖

base
鹼

a nucleotide 核苷酸

chain of nucleotides
核苷酸鏈

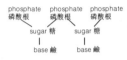

phosphate phosphate phosphate
磷酸根　磷酸根　磷酸根

sugar 糖　sugar 糖

base 鹼　base 鹼

**a nucleic acid
核酸**

**model of a double helix
雙螺旋的模型**

each strand of DNA is a
sugar-phosphate chain
DNA 的每一股都是一
條糖-磷酸根鏈

strand
鏈股

strand
鏈股

bonds
between
bases
鹼基間
的鍵

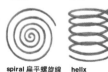

spiral 扁平螺旋線　**helix
直交螺旋線**

nucleic acid a long chain molecule made up of a large number of nucleotides (↓). All organisms (p.147) have nucleic acids present in their cells. There are two kinds of nucleic acid, DNA (↓) and RNA (↓).

nucleotide (n) one part, or unit, of a nucleic acid. It consists of a sugar (p.173), either ribose, or deoxyribose, together with a phosphate radical (p.116) and a base (p.114) containing nitrogen. Each unit combines with two other units and a chain is formed.

DNA this represents deoxyribonucleic acid, a nucleic acid (↑). The sugar (p.173) is deoxyribose. The base (p.114) in each nucleotide (↑) is one of: thymine, cytosine, adenine and guanine. The long chain of nucleotides forms a **strand**, and two strands are coiled round each other to form a double helix (↓), *see diagram*. The strands are joined by bonds (p.109) between pairs of different bases, one on each strand. DNA is found only in the nuclei (p.139) of cells; DNA and protein form chromosomes. From a strand of DNA, a strand of RNA is produced by the action of an enzyme (p.167).

RNA this represents ribonucleic acid, a nucleic acid (↑). The sugar (p.173) is ribose. The base (p.114) in each nucleotide (↑) is one of: uracil, cytosine, adenine and guanine. The chain of nucleotides forms a single strand. RNA is found in the nuclei (p.139) and the cytoplasm (p.138) of cells. Strands of RNA are produced in the nucleus from DNA (↑), passed to the cytoplasm, and then a ribosome (p.141) is joined to the RNA. The ribosome moves along the strand of RNA and produces a polypeptide (p.172); the structure of the polypeptide is controlled by the RNA.

spiral (n) a line which starts at a point and then curves so that it gets farther and farther away from the starting point; it is a flat curve. Any structure of a similar shape. **spiral** (adj).

helix (n) a line which curves in a circle while moving away at right angles to the start of the curve. *See diagram*.

helical (adj) with the shape of a helix or spiral.

核酸　由很多核苷酸(↓)分子組成的長鏈分子。一切生物體(第 147 頁)的細胞中都含有核酸。核酸有兩種類型：DNA(↓)和 RNA(↓)。

核苷酸(名)　核酸的組成部分或一個單元。它由糖(第 173 頁)(即核糖或去氧核糖)與磷酸根(第 116 頁)和含氮鹼(第 114 頁)組成。每一個單元均與兩個其他單元結合形成一條鏈。

DNA　這符號表示去氧核糖核酸，它是一種核酸(↑)。其中的糖(第 173 頁)是去氧核糖。每個核苷酸(↑)中的鹼基(第 114 頁)包括胸腺嘧啶、胞嘧啶、腺嘌呤和鳥嘌呤中的一種。核苷酸長鏈形成一股。兩股相互盤旋而成一雙螺旋(↓)。(見圖)。鏈股通過不同鹼基對間的鍵(第 109 頁)結合在一起，每個鹼基分別位於一條鏈股上。DNA 只存在於細胞核(第 139 頁)中；DNA 與蛋白質一起構成染色體。在酶(第 167 頁)的作用下由一股 DNA 形成一股 RNA。

RNA　這符號表示核糖核酸，它是一種核酸(↑)。其中的糖(第 173 頁)是核糖。每個核苷酸(↑)中的鹼基(第 114 頁)包括尿嘧啶、胞嘧啶、腺嘌呤和鳥嘌呤中的一種。核苷酸鏈形成單股。RNA 存在於細胞核(第 139 頁)和細胞質(第 138 頁)中。RNA 鏈股是在 DNA(↑)為模板在細胞核中產生的，然後進入細胞質，核糖體(第 141 頁)隨之與 RNA 連接。核糖體沿 RNA 鏈股移動並生成多肽(第 172 頁)；多肽的結構受 RNA 控制。

螺旋線(名)　線從一點開始，繞該點旋轉而越轉離起點越遠的線條。它是一條平緩的曲線；或指任何形狀與此相似的結構。(形容詞為 spiral)

直交螺旋線(名)　一根線雖然彎成圈，但它移動時與彎曲的起點成直角。(見圖)。

螺旋狀的(形)　具有直交螺旋線或扁平螺旋線形狀的。

gene (*n*) a short length of a chromosome (p.142) which controls a characteristic (p.147) of an organism (p.147). The gene can be passed on from parent to offspring (p.210), e.g. a gene for eye-colour.

genetic (*adj*) concerned with genes (↑).

diploid (*adj*) of a nucleus, possessing chromosomes (p.142) in pairs. All cells in an organism, except gametes (p.210), have a diploid nucleus.

haploid (*adj*) of a nucleus, possessing unpaired chromosomes (p.142), i.e. half the number of chromosomes. All gametes (p.210) have a haploid nucleus. When two haploid nuclei undergo fusion (p.210), a diploid (↑) nucleus is formed, with all chromosomes paired. In man, there are 23 pairs of chromosomes in all cells except the cells of gametes, which have half the number of a diploid nucleus, i.e. 23 unpaired chromosomes.

allele (*n*) one of a pair of genes (↑) that control the same characteristic (p.147), but have a different effect, e.g. there are two genes for eye-colour, one gene producing brown eyes and the other gene producing blue eyes. These two genes are alleles of each other. Each chromosome of a pair bears a particular gene in a fixed place along its length, so that the two genes controlling a particular characteristic are in corresponding places. The two genes can be identical (p.94) or can be alleles. There can be more than two alleles for a gene, e.g. there are three alleles determining the blood groups of the ABO system (p.162) but any one pair of chromosomes has only two of these alleles.

homozygous (*adj*) of persons, possessing two identical (p.94) genes (↑) for a characteristic (p.147). See heterozygous (↓).

heterozygous (*adj*) of persons, possessing two alleles (↑) of a gene, i.e. two different genes for the same characteristic.

dominant (*adj*) of an allele (↑), determines the effect of a gene, e.g. the allele for brown eyes is dominant, so a person who is heterozygous (↑) for the gene will have brown eyes, and the allele for blue eyes will have no effect. **dominant** (*n*).

基因（名） 控制生物體（第 147 頁）特徵（第 147 頁）的短段染色體（第 142 頁）。基因可由親體遺傳給後代（第 210 頁），例如眼睛顏色的遺傳基因。

遺傳的（形） 有關基因（↑）的。

二倍體的（形） 指有成對染色體（第 142 頁）的細胞核。生物體中除配子（第 210 頁）之外的所有細胞都有二倍體核。

單倍體的（形） 指具有不成對染色體（第 142 頁）的細胞核，即含半數染色體。所有的配子（第 210 頁）都有單倍體核。兩個單倍體核融合（第 210 頁）形成二倍體（↑）核，它有成對的染色體。人體內所有的細胞都有 23 對染色體，但配子的細胞除外，因爲它們只有二倍體核的半數染色體，即 23 個不成對的染色體。

對偶基因；等位基因（名） 控制同一特徵（第 147 頁）但效應不同的一對基因（↑）中的一個。例如控制眼睛的顏色有兩個基因，其一個產生褐色眼睛，另一個產生藍色眼睛，這兩個基因互爲對偶基因。一對染色體中的每一個染色體在沿其長度的固定位置上有特定的基因，因此控制特定特徵的兩個基因處於相應的位置。這兩個基因可以是相同的（第 94 頁），或者是對偶的基因。一個基因可以有兩個以上的對偶基因，例如有三個決定 ABO 血型系（第 162 頁）的對偶基因，但任何一對染色體都只有兩個這類對偶基因。

純合的；同型的（形） 指人而言，對一種特徵（第 147 頁）具有兩個相同的（第 94 頁）基因（↑）的。（參見 " 異型的 "（↓））

雜合的；異型的（形） 指人而言，具有兩個對偶基因（↑）的，即同一特徵有兩個不同的基因。

顯性的（形） 指對偶基因（↑），它決定基因的效應，例如褐眼睛的對偶基因是顯性的，所以具有異型（↑）基因的人有褐眼睛，而對偶基因對藍眼睛無效。（名詞爲 dominant）

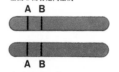

gene 基因

chromosome pair homozygous for gene A and gene B
染色對一對基因 A 和基因 B 而言是同型的

A B

chromosome pair— homozygous for gene A, heterozygous for gene B
染色體一對基因 A 是同型的，對基因 B 則是異型的

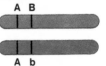

A B

A b

B and b are alleles of the gene
B 和 b 都是基因的對偶基因

some possible combinations of 3 alleles

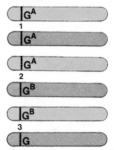

some possible combinations of 3 alleles
三種對偶基因的一些可能的組合

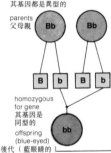

parents both bwown-eyed
both heterozygous for gene
父母都具褐眼睛者
其基因都是異型的

parents
父母親

Bb   Bb

B b B b

homozygous
for gene
其基因是
同型的

offspring
(blue-eyed)
後代（藍眼睛的）

phenotypes 表現型

 brown-eyed
褐眼睛的

 blue-eyed
藍眼睛的

genotypes 基因型

B   allele for borwn eyes
        褐眼睛的對偶基因
        -dominant
        -顯性的

b   allele for blue eyes
        褐眼睛的對偶基因
        -recessive
        -隱性的

BB   homozygous
        for borwn eyes
        對藍眼睛為同型的

bb   homozygous
        for brown eyes
        對褐眼睛為異型的

Bb   homozygous
        for blue eyes
        對褐眼睛為同型的

Inheritance of genes
基因遺傳
to show two brown-eyed
parents may produce a
blue-eye child
表明兩個褐眼睛的父母
可產生一個藍眼睛的孩子

recessive (adj) of an allele (↑), has no genetic (↑) effect in a heterozygous (↑) person, e.g. the allele for blue eyes is recessive, so a person who is heterozygous for the gene will have brown eyes produced by the allele for brown eyes. A person with two identical genes for blue eyes will have blue eyes, as no dominant gene is present.

phenotype (n) a person belonging to a type which has a particular characteristic (p.147), e.g. a brown-eyed person for the characteristic of eye-colour. Such a person can be homozygous (↑) or heterozygous (↑) for a dominant (↑) gene, but must be homozygous for a recessive (↑) gene.

genotype (n) a person as determined by his genetic make-up, e.g. homozygous (↑) for dominant (↑) or recessive (↑) genes, or heterozygous (↑) for the gene.

X-chromosome a sex chromosome (p.142). In humans, a pair of X-chromosomes produce a female. The X-chromosome has many genes for which there are no paired genes on the Y-chromosome (↓).

Y-chromosome a sex-chromosome (p.142). It is smaller in length than an X-chromosome (↑). A nucleus (p.139) can have only one Y-chromosome, the other sex chromosome being an X-chromosome. In humans, an XY chromosome pair produces a male.

inherit (v) of offspring (p.210), to receive characteritistics (p.147) from parents. All characteristics of an organism are determined by the chromosomes (p.142) in the nucleus (p.139) of every cell. In a pair of chromosomes inherited by sexual reproduction, one comes from the father and one from the mother because each gamete (p.210) has a haploid (↑) nucleus, and the offspring has cells with diploid (↑) nuclei. The offspring thus inherits genes from both parents and its characteristics will be determined by the characteristics of both parents. **heritable** (adj), **inheritance** (n).

pass on (v) of parents, to give characteristics (p.147) to offspring (p.210).

隱性的(形)　指對偶基因(↑)而言，基因異型的(↑)人身上無遺傳(↑)效應，例如藍眼睛的對偶基因是隱性的，所以基因異型的人具有由褐眼睛對偶基因產生的褐眼睛。有藍眼睛的兩個相同基因的人有藍眼睛，因為沒有顯性基因存在。

表現型(名)　屬於有特定特徵(第 147 頁)類型的人，例如褐眼睛的人，其眼睛的顏色有特徵。此種人顯性(↑)基因可能是同型的(↑)或異型的(↑)，但隱性(↑)基因必定是同型的。

基因型(名)　由基因的組成決定類型的人，例如對顯性(↑)基因和隱性(↑)基因為同型的(↑)或對此基因為異型的。

X 染色體　為性染色體(第 142 頁)。人類的一對 X 染色體產生的後代為女性。X 染色體有許多在 Y 染色體(↓)上不能配對的基因。

Y 染色體　為性染色體(第 142 頁)。其長度小於 X 染色體(↑)。一個核(第 139 頁)只可以有一個 Y 染色體，另一個性染色體則是 X 染色體。人類的一對 XY 染色體產生的後代為男性。

遺傳(動)　指後代(第 210 頁)獲得父母所遺傳給的特徵(第 147 頁)。生物體全部特徵都是由每個細胞核(第 139 頁)的染色體(第 142 頁)決定的。一對通過性生殖遺傳下來的染色體中，一個來自父本，一個來自母本，因為每個配子(第 210 頁)都有單倍體(↑)核，所以後代就含有二倍體(↑)核的細胞。後代因此通過遺傳得到兩個親本的基因，它的特徵將由兩個親本的特徵來決定。(形容詞為 heritable，名詞為 inheritance)

傳遞(動)　指親本將特徵(第 147 頁)傳給後代(第 210 頁)。

Mendel's laws if in two animals, or plants, one is homozygous (p.220) for a dominant (p.220) gene (p.220) for the characteristic (p.147) and the other is homozygous for a recessive (p.221) gene for the characteristic, then the offspring (p.210) reproduced by sexual reproduction (p.209) will all have the characteristic from the dominant gene. If these offspring, which are heterozygous (p.220) for the gene, reproduce a second generation (p.210), then the offspring of the second generation will have 75% with the characteristic for the dominant gene and 25% with the characteristic for the recessive gene, i.e. a 3:1 ratio. *See diagram.*

heredity (*n*) the passing on (p.221) of characteristics (p.147) from one generation (p.210) to the next generation. **hereditary** (*adj*).

pedigree (*n*) a diagram showing the inheritance (p.221) of particular characteristics (p.147) from one generation (p.210) to later generations.

孟德爾定律　如果兩個動物或植物中，一個是同型的(第 220 頁)顯性(第 220 頁)基因(第 220 頁)決定特徵(第 147 頁)，另一個是同型的隱性狀基因(第 221 頁)決定特徵，那麼有性生殖(第 209 頁)所產生的後代(第 210 頁)都將有顯性基因的特徵。如果這些在基因方面是異型的(第 220 頁)後代再生殖第二代(第 210 頁)，那麼子二代的後代將有 75% 具有顯性基因的特徵，25% 具有隱性基因的特徵。即 3:1 的比率。(見圖)。

遺傳性(名)　特徵(第 147 頁)從一代(第 210 頁)傳給(第 211 頁)下一代。(形容詞為 herediary)

血統；譜系(名)　顯示一種特殊特徵(第 147 頁)從一代(第 210 頁)遺傳給(第 221 頁)下幾代的圖解。

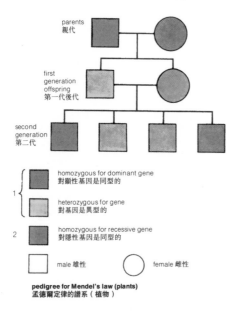

parents
親代

first
generation
offspring
第一代後代

second
generation
第二代

1 {
homozygous for dominant gene
對顯性基因是同型的

heterozygous for gene
對基因是異型的

2
homozygous for recessive gene
對隱性基因是同型的

male 雄性　　　female 雌性

pedigree for Mendel's law (plants)
孟德爾定律的譜系（植物）

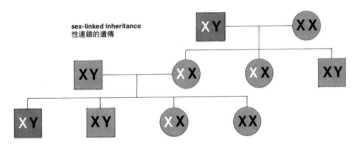

sex-linked inheritance
性連鎖的遺傳

normal female
正常女性

male with haemophilia
患友血病的男性

normal male
正常男性

recessive gene for
haemophilia
血友病的隱性基因

X dominant normal gene
顯性的正常基因

Y no gene on Y chromosome
在 Y 染色體上無基因

sex-linked of a gene, carried on the X-chromosome (p.221) but not on the Y-chromosome (p.221). A man has only one X-chromosome, so that men have the characteristics from recessive (p.221) genes (p.220) in sex linkage more often than women, who have two X-chromosomes. A son cannot inherit (p.221) a sex-linked gene from his father, but his daughter can inherit the gene on the X-chromosome. The daughter in turn can carry or pass on (p.221) the gene to her son, so a recessive gene can appear in male offspring (p.210) every other generation. Colour-blindness is a sex-linked gene. **sex linkage** (adj).

haemophilia (n) a disease of humans in which blood does not clot properly. The sex linkage (↑) of the gene makes it common in men, but uncommon in women. Unless a woman has two X-chromosomes with the recessive (p.221) gene she will not suffer from haemophilia; with only one recessive gene she is a carrier.

birth rate the number of children born in a year per thousand people, e.g. a birth rate of 38 per 1000 means 38 000 children born yearly in a population (p.224) of 1 000 000.

death rate the number of people dying in a year per thousand people, e.g. a death rate of 22 per 1000 means 22 000 people die in a population (p.224) of 1 000 000.

mortality rate the number of deaths in a certain period per 1000 people in an age group; usually the age group is for a five-year period, e.g. the mortality rate for the age group 30–34 years.

性連鎖的；伴性的　指基因而言，載於 X 染色體（第 221 頁）而不載於 Y 染色體（第 221 頁）的。男性只有一個 X 染色體，所以男性從性連鎖隱性（第 221 頁）基因獲得的特徵比女性爲多，女性則有兩個 X 染色體。兒子不能經遺傳（第 221 頁）獲得父親的性連鎖基因，但女兒能經遺傳獲得 X 染色體上的基因。女兒又能將這基因傳給（第 221 頁）她的兒子，所以隱性基因能出現在隔代的男性後代（第 210 頁）中。色盲就是一種性連鎖基因。（形容詞爲 sex-linkage）

血友病（名）　人體血液不能正常凝固的疾病。由於基因的性連鎖（↑）使男性中常見此病，而女性中不常見。除非該女性的細胞中載有兩個帶隱性（第 221 頁）基因的 X 染色體，才會患血友病。如果該女性只載一個隱性基因，則爲攜帶基因者。

出生率　每千人每年所出生的嬰兒數。例如，千分之三十八的出生率表示每百萬人口（第 224 頁）中每年出生三萬八千個嬰兒。

死亡速率　每千人中每年所死亡的人數。例如，千分之二十二的死亡速率表示每百萬人口（第 224 頁）中每年死亡二萬二千人。

死亡率　一個年齡組中在某一段時間內每千人中的死亡人數。一個年齡組通常是以五年來劃分。例如，30–34 歲這一年齡組的死亡率。

population (*n*) the number of organisms (p.147) in a given place or area, e.g. (a) the number of lions in Africa; (b) the number of date palms in Egypt; (c) the mosquito population of a lake. Bacteria and protozoa are not usually considered as a population. If no particular organism is named, the population is the number of persons living in a named place. **populate** (*v*).

survival (*n*) the act of living through unfavourable conditions until the conditions change. **survive** (*v*).

evolution (*n*) the changes over many generations (p.210) by which different kinds of organisms (p.147) have arisen from very early forms, e.g. about 400 million years ago a reptile, called a pterodactyl, flew in the air. About 200 million years ago, many changes in previous generations resulted in an early kind of bird, called an Archaeopteryx. Many more changes resulted in the present kind of birds, which appeared in the last 10 million years or so. These changes have been studied from fossils (↓) and show the evolution of birds from reptiles.

種群量；人口（名） 在一特定地方或區域的生物（第 147 頁）數目。例如 (a) 非洲獅的數目；(b) 埃及椰棗樹的數目；(c) 湖裏蚊子的種群量。細菌和原生動物一般不看作一個種群。如果無指明何種生物時，人口就是指生活在特定地方的人數。（動詞爲 populate）

幸存；殘存（名） 在不利的環境中生存，直到環境改變仍生存下來。（動詞爲 survive）

進化；演化（名） 經過許多代（第 210 頁）產生的變化，通過這些變化，從非常原始的形式產生了各種不同種類的生物（第 147 頁）。例如大約四億年前曾經存在一種在空中飛行稱爲飛龍目的爬蟲動物。大約二億年前經過無數代的變化產生了一種早期的鳥，稱爲始祖鳥。又經過無數的變化才產生了現今的鳥類，約出現在距今一千萬年前。人們已通過化石（↓）對這些變化進行了研究，並證明了鳥是從爬蟲動物進化來的。

evolution
進化

early forms
早期形式

Archaeopteryx
始祖鳥

pterodactyl
飛龍

modern birds
現代鳥

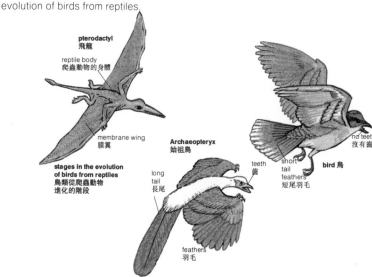

pterodactyl
飛龍

reptile body
爬蟲動物的身體

membrane wing
膜翼

stages in the evolution of birds from reptiles
鳥類從爬蟲動物進化的階段

Archaeopteryx
始祖鳥

long tail
長尾

feathers
羽毛

teeth
齒

short tail feathers
短尾羽毛

no teeth
沒有齒

bird 鳥

fossil (*n*) the remains of an organism (p.147), or the shape of an organism, preserved in rocks. The hard parts of the organism are generally changed into rock by mineral salts dissolved in the water passing through the earth. The organisms that lived many millions of years ago can be studied from their fossils.

化石（名）　保存在岩石中的生物體（第 147 頁）的遺骸或生物體的形狀。由於礦物鹽溶解於流經地下的水中，常使生物的堅硬部分變爲岩石。通過化石可對數百萬年前的生物體進行研究。

fossil
化石

mutation (*n*) a sudden change in the genes (p.220) of an organism (p.147) caused by a change in the DNA (p.219) of the chromosomes (p.142). Mutation does not happen often, but the rate is increased by radiation (p.83), neutrons (p.106) and some chemical substances. Changes of single genes can take place, or a whole chromosome can be altered. Changes in the genes in body cells have an effect only on the person; but changes in the genes in gametes (p.210) have an effect on all the offspring (p.210). All mutations happen by chance. Most mutations have a bad effect. **mutate** (*v*).

突變（名）　生物體（第 147 頁）中基因（第 220 頁）突然變化是染色體（第 142 頁）DNA（第 219 頁）變化引起的。突變並非經常發生，但輻射線（第 83 頁）、中子（第 106 頁）及一些化學物質可增加突變的速率。單個基因可以發生變化，整個染色體也可改變。身體細胞中基因的變化僅對這個人有影響；但配子（第 210 頁）中基因的變化則對所有的後代（第 210 頁）都有影響。一切突變都是偶然發生的。大部分突變都有不良影響。(動詞爲 mutate)

mutant (*n*) (1) a gene (p.220) which has been altered by mutation (↑). (2) an organism (p.147) altered by such a gene. (3) a characteristic (p.147) produced by such a gene.

突變體；突變型（名）　(1)因突變(↑)而改變的基因（第 220 頁）；(2)由於這樣的基因而改變的生物體（第 147 頁）；(3)這樣的基因產生的特徵（第 147 頁）。

natural selection the tendency (p.15) to survive (↑) of those organisms (p.147) which are best suited to their conditions of living. These organisms live longer and reproduce (p.209) offspring (p.210) with inherited (p.221) characteristics (p.147) which are useful for survival (↑). The organisms and offspring which survive are said to be selected by a natural process. Natural selection thus controls the direction of change of inherited characteristics as unsuitable mutants (↑) die.

天擇；自然選擇　最能適應生存環境的生物體（第 147 頁）求生存(↑)的傾向（第 15 頁）。它們生存較長，不斷繁殖（第 209 頁）後代（第 210 頁）。這些後代具有有利於生存(↑)的遺傳（第 221 頁）特徵（第 147 頁）。這些幸存的生物體和後代被認爲是自然過程選擇的結果。因此自然選擇控制着遺傳特徵改變的方向，不適應環境的突變體(↑)不能生存下來。

Darwinism (*n*) the idea that evolution (↑) took place by natural selection (↑).

達爾文學說（名）　自然選擇(↑)產生進化(↑)的學說。

biosphere (*n*) the part of the Earth and the atmosphere (p.51) in which all organisms (p.147) live.

community (*n*) a group of organisms (p.147) living in a certain habitat (↓), having an effect on each other, and reaching a state of equilibrium (p.20) through a food web (p.171).

biome (*n*) a major community (↑) covering a large area of the Earth, e.g. tundra (very cold areas with no trees, and low plant growth); tropical rain forest; savanna (grass lands); desert.

environment (*n*) all the conditions which act on an organism (p.147) while it lives. Besides the physical conditions, there are also the effects of other organisms. The major environments are the sea, the fresh waters, and the land. These are divided into smaller environments by the conditions of climate (p.51). **environmental** (*adj*).

habitat (*n*) the place, or kind of place, where a particular organism (p.147) can survive (p.224), e.g. a sea-shore. Various factors (↓) determine whether a place can be a habitat for an organism, e.g. the habitat of the mangrove tree is a tropical area with brackish (↓) water over a bottom of mud.

brackish (*adj*) describes water less salty than sea water.

生物圈(名)　所有生物體(第 147 頁)都生存於其中的地球和大氣(第 51 頁)部分。

群落(名)　生活在一定棲息地(↓)，相互間有影響，並通過食物網(第 171 頁)達到平衡(第 20 頁)狀態的一群生物(第 147 頁)。

生物群落區(名)　分佈在地球上一大區域內的主要群落，例如凍土帶(即不長樹，只長低矮植物的嚴寒地區)；熱帶雨林；熱帶草原(草地)；沙漠。

環境(名)　生物體(第 147 頁)生存時，對它起作用的全部條件。除了自然界的條件外，還有其他生物體的影響。主要的環境是海洋、淡水和陸地。又可根據氣候(第 51 頁)將條件分爲更小的環境。(形容詞爲 environmental)

棲息地；生境(名)　某種生物體(第 147 頁)能生存(第 244 頁)下來的地方，例如海岸。一個地方是否能成爲一種生物的棲息地決定於各種不同的因素(↓)。例如紅樹的生境是濕土上面有稍帶鹹味(↓)水的熱帶地區。

略有鹹味的(形)　描述不如海水鹹味的水。

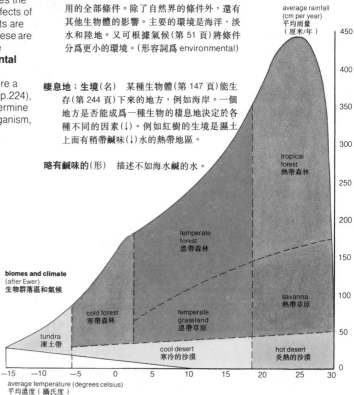

average rainfall
(cm per year)
平均雨量
（厘米/年）

biomes and climate
(after Ewer)
生物群落區和氣候

average temperature (degrees celsius)
平均溫度（攝氏度）

biotope (*n*) an area in which animals and plants having similar habits and needs live together and form a community (↑). Also a very small habitat (↑), e.g. a pond, a tree.

biocoenosis (*n*) a community (↑) living in a biotope (↑).

ecosystem (*n*) a complete ecological (↓) unit which can be studied. It is a community (↑) which acts on the environment (↑) and the environment acts on the community. In a biotope (↑), the bio-coenosis (↑) acts on the biotope, and the biotope acts on the biocoenosis. An ecosystem consists of (a) *producers*, mainly green plants; (b) *consumers*, animals feeding on plants and other animals; (c) *decomposers*, mainly bacteria (p.145) and fungi (p.145) decomposing dead organisms.

territory (*n*) the area in which an animal, or a society (p.153) of animals, lives, e.g. ants from a particular ant hill cover an area for feeding and fight other ants trying to enter their territory. A single bird can have a territory for feeding or mating. Sometimes a territory is held for only part of a life-cycle (p.151). **territorial** (*adj*).

factor (*n*) a factor helps to determine an environment (↑), e.g. climate (p.51), the kind of soil, salinity (p.228) of water, the presence of organisms (p.147) are some of the factors determining an environment. See conditions (p.228).

adaptation (*n*) a change in characteristics (p.147) of an organism (p.147) which increases its chances of survival (p.224). Such changes can come through evolution (p.224), or can be physiological (p.140) or sensory (p.200), e.g. fish living in deep water have eyes stimulated best by blue light, the only light at that depth; this is evolutionary adaptation. **adapt** (*v*).

mimicry (*n*) of an animal, being the same in appearance as another animal, or a plant, e.g. (a) a stick insect looks like a part of a plant; (b) an animal having coloured marks on it so it looks like a poisonous animal. **mimic** (*v*).

ecology (*n*) a study of the relations between animals, plants and the physical conditions of the environment; in particular the study of ecosystems (↑). **ecological** (*adj*).

mimicry
擬態

mimic
模仿

群落生境；生活小區（名）　有相似習性和共同需要的動、植物生活在一起，組成群落（↑）的區域。此區域也可以是一個很小的棲息地（↑），例如一個池塘、一棵樹。

生物群落（名）　生活在一個群落生境（↑）中的群落（↑）。

生態系統；生態區（名）　一個可供研究的完整生態（↓）單元。它是一個對環境（↑）起作用，環境也對它起作用的群落（↑）。在群落生境中（↑），生物群落（↑）對群落生境起作用，而群落生境對生物群落也起作用。生態系統的組成是：(a) 生產者，主要是綠色植物；(b) 消費者，是以植物和其他動物爲食的動物；(c) 分解者，主要是能分解生物屍體的細菌（第145頁）和真菌（第145頁）。

領域；勢力圈；活動區（名）　一隻動物或一群群居（第153頁）動物生活的區域，例如某一個蟻山的螞蟻佔據一片覓食區域，並和試圖侵入其領域的其他螞蟻爭鬥。一隻獨居鳥可以有一覓食和交配的勢力圈。有時只在生活周期（第151頁）的一段時間內保留其領域。（形容詞爲 territorial）

因素（名）　因素有助於決定環境（↑），例如氣候（第51頁）、土壤的性質、水的含鹽量（第228頁），生物體（第147頁）的存在都是決定環境的因素。（參見"環境"（第228頁））。

適應（名）　生物體（第147頁）特徵（第147頁）的改變增加了幸存（第224頁）的機會。此種改變可能是進化（第224頁）的結果，或者是生理的（第140頁）進化，或者是感覺的（第200頁）的進化。例如生活在深水中的魚，其眼睛能感受深海中唯一的光（藍光）的刺激；這就是進化的適應性。（動詞爲 adapt）

擬態（名）　指動物在外表上模仿另一種動物或植物。例如 (a) 桿狀昆蟲看上去像植物的一部分；(b) 某種動物身上帶有顏色的斑點，看似有毒的動物。（動詞爲 mimic）

生態學（名）　研究動、植物和環境自然條件之間的關係，特別是研究生態系統（↑）的學科。（形容詞爲 ecological）

conditions (*n.pl.*) the actual measurable physical factors (p.227) in an environment (p.226), e.g. temperature is a physical factor of climate (p.51) but the actual range (p.108) of temperature in a particular area is a condition.

salinity (*n*) a measure of the amount of common salt in water. The average salinity of sea water is about 2.8 g sodium chloride per 100 g water. **saline** (*adj*).

aquatic (*adj*) living in water; concerned with water.

freshwater (*adj*) living in water which contains no salt; concerned with water which contains no salt, e.g. rivers and lakes are freshwater environments.

sea-water (*n*) water in the seas usually containing 2.8% sodium chloride, 0.4% magnesium chloride, 0.2% magnesium sulphate, 0.1% calcium sulphate, 0.1% potassium chloride.

marine (*adj*) living in sea-water; concerned with the sea, e.g. a marine plant; marine ecology.

estuarine (*adj*) living in an estuary (↓); concerned with estuaries.

條件(名、複) 指一個環境(第 226 頁)中可實際測量的自然因素(第 227 頁)。例如溫度是氣候(第 51 頁)的自然因素，而某特定地區的實際溫度範圍(第 108 頁)就是一種條件。

含鹽量(名) 水中食鹽含量的量度。海水平均含鹽量爲每 100 克水約含氯化鈉 2.8 克。(形容詞爲 saline)

水生的(形) 生活在水中的；與水有關的。

淡水的(形) 生活在不含鹽的水中的；與不含鹽的水有關的，例如江河和湖泊都是淡水環境。

海水(名) 海水一般含氯化鈉 2.8%，氯化鎂 0.4%，硫酸鎂 0.2%，硫酸鈣 0.1%，氯化鉀 0.1%。

海水生的；海水的(形) 生活在海水中的；與海有關的。例如海洋植物，海水生態學。

河口的；港灣的(形) 生活在河口(↓)的；與河口有關的。

aquatic organisms
水生生物

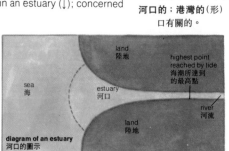
diagram of an estuary
河口的圖示

(labels: land 陸地, sea 海, estuary 河口, highest point reached by tide 海潮所達到的最高點, river 河流, land 陸地)

estuary (*n*) the mouth of a river into which the tides (p.127) of the sea enter. The salinity (↑) of water in an estuary varies, being greatest at high tide and least at low tide. The salinity also varies from the sea to the greatest distance up the river reached by a tide.

littoral (*adj*) living on or near a sea-shore; living in a lake near the shore on the lake bottom, when the lake bottom is at a depth of less than 10 m. The littoral zone of a sea is the part of the sea between high and low tide marks.

河口；港灣(名) 海潮(第 127 頁)湧進的河口。河口水的含鹽量(↑)是變化的，高潮時含鹽量最大，低潮時含鹽量最少。從海到海潮能湧入河流的最遠距離之間的含鹽量也是變化的。

沿岸的；濱海的(形) 生活在海岸或近海岸的；在湖底的深度不到 10 米時，生活在靠岸的湖底的。海的沿岸地帶是高潮的測標和低潮的測標之間的海域。

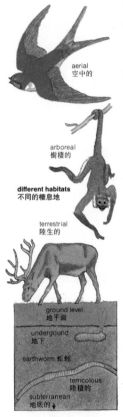

aerial
空中的

arboreal
樹棲的

different habitats
不同的棲息地

terrestrial
陸生的

ground level
地平面

underground
地下

earthworm 蚯蚓

terricolous
陸棲的

subterranean
地底的↓

sessile leaves
無柄的葉

shellfish
貝類

sessile and sedentary
organisms 無柄的和
定棲的生物

amphibious (*adj*) living both on land and in water, e.g. frogs are amphibious animals.

amphibiotic (*adj*) living in water for the first stages (p.152) of a life cycle (p.151) and on land, when fully grown, e.g. mosquitoes are amphibiotic insects.

ground (*n*) the surface of the Earth; the earth on which animals move and plants grow; the support for a structure, e.g. (a) ground speed is the speed of an aeroplane in relation to the Earth; (b) ground level is the height above sea-level of the Earth's surface at a particular point; (c) ground substance is the supporting material between cells, e.g. a matrix (p.192) of fibres (p.195). **ground** (*adj*).

terrestrial (*adj*) living on the ground (↑), concerned with the Earth, e.g. (a) terrestrial magnetism is the magnetism (p.69) of the Earth; (b) a terrestrial animal is one that walks on the ground, e.g. a deer lives in a terrestrial habitat.

subterranean (*adj*) deep down in the Earth, e.g. subterranean rock strata of metamorphic rock (p.121).

underground (*adj*) in the earth, but not deep, e.g. a rhizome (p.213) is an underground stem.

terricolous (*adj*) of animals, living in the soil (p.230), e.g. an earthworm is a terricolous organism.

arboreal (*adj*) of animals, living in trees, e.g. monkeys are arboreal animals. Also describes a habitat (p.226), e.g. monkeys live in an arboreal habitat.

aerial (*adj*) concerned with the air, living in the air, e.g. (a) the aerial roots of a plant grow above ground (↑); (b) birds have an aerial territory (p.227) for feeding.

sedentary (*adj*) of animals, living fixed to a support, not free to move from place to place, e.g. some shellfish are fixed to rocks on the sea-shore.

mobile (*adj*) of animals, free to move from place to place, capable of locomotion (p.194).

sessile (*adj*) (1) of plants and structures, fixed to a stem or support without a stalk (p.157), e.g. a sessile leaf has no petiole (p.158). (2) of animals, sedentary (↑).

motile (*adj*) of protozoa, able to move.

兩棲的(形) 既能生活在陸地上又能生活在水中的。例如青蛙是兩棲動物。

水生陸棲的(形) 生活周期(第 151 頁)的第一階段(第 152 頁)生活在水中，發育成熟後則生活在陸地上。例如蚊子就是水生陸棲的昆蟲。

地面；地；基礎(名) 地球表面；動物活動和植物生長的土地；某種結構的支持物。例如(a) 地面速度是飛機相對於地球的速度；(b) 地平面是地球表面的某一點高於海平面的地方；(c) 基礎物質是細胞間的支持物質，例如纖維(第 195 頁)的基質(第 192 頁)。(形容詞爲 ground)

陸生的；地球上的(形) 生活在地面上的(↑)，關於地球的；例如(a) 地磁是地球的磁力(第 69 頁)；(b) 陸生動物是在地上行走的動物；例如鹿生活在陸上的棲息地。

地底的(形) 深入地下的；例如變質岩(第 121 頁)的地下岩層。

地下的(形) 在地下，但不深，例如根莖(第 213 頁)是地下莖。

陸棲的(形) 指生活在土壤(第 230 頁)中的動物。例如蚯蚓是陸棲生物。

樹棲的；樹的(形) 指生活在樹上的動物。例如猴子是樹棲動物。也描述棲息地(第 226 頁)，例如猴子生活在樹上的棲息地。

氣生的；空中的(形) 關於空氣，生活在空中的，例如：(a) 植物的氣生根生長在離開地面(↑)的地方；(b) 鳥類有一個覓食的空中領域(第 227 頁)。

定棲的(形) 指固定在一個支持物上生活，不能自由地從一處移向另一處的動物，例如一些貝類固定在海岸的岩石上。

活動的(形) 指自由地從一處移向另一處，能行動(第 194 頁)的動物。

座生的；無柄的(形) (1)指植物和結構，即固定在莖上或無柄(第 157 頁)的支持物上的，例如無柄的葉沒有葉柄(第 158 頁)；(2)指定棲的(↑)的動物。

能動的(形) 指能移動的原生動物。

diurnal (adj) (1) active only in the day-time, during the hours of light, e.g. a diurnal animal hunts its food during daylight; (2) happening every day; some flowers have a diurnal rhythm, that is, they have changes that happen every twenty-four hours.

nocturnal (adj) active only at night, e.g. a nocturnal animal hunts its food at night.

soil (n) when earth is considered as an environment (p.226) providing nutrients (p.171) for plants and animals. Soil consists of grains which have water and air between them. The grain structure of soil is important for its properties. The properties of soil determine which kinds of plants can grow in it. Soil is formed by the weathering (p.120) of rocks (p.119) producing particles (p.26) which are mixed with organic (p.131) material.

soil profile the layers (p.187) of different kinds of soil that can be seen if a hole is dug down to the rock beneath.

topsoil (n) the top layer (p.187) of soil. It contains the nutrients (p.171) needed by plants. If the topsoil is taken away by erosion (p.120), plants can no longer grow in the soil left behind.

subsoil (n) the layers of soil beneath the topsoil (↑). Although a subsoil does not provide sufficient nutrients (p.171) for plants, it can be useful by not allowing water to drain (↓) away.

clay (n) soil (↑) with particles (p.26) less than 0.01 mm in diameter, mainly aluminium silicate. Water passes through it very slowly.

sand (n) soil (↑) with particles (p.26) between 0.1 mm and 2 mm in diameter, mainly silicon dioxide. Water passes through it very quickly. **sandy** (adj).

loam (n) a kind of soil (↑) consisting of sand and clay, the best soil for the growth of plants.

humus (n) organic (p.131) material produced by the decay (p.146) of plant and animal tissues; it makes soil dark in colour. It is the most important constituent (p.96) of soil (↑) for plant growth, and also helps soil to hold water.

drain (v) of water, to pass through soil, or to run off a surface.

晝出夜息的；每日的（形）（1）僅在日間有陽光時活動的，例如晝行動物在白天獵食；(2) 每天發生的，一些花卉有晝夜節律，即每隔 24 小時發生一些變化。

晝伏夜出的（形） 僅在夜間活動的，例如夜行動物在夜間獵食。

土壤（名） 當土地被認爲是爲動、植物提供營養素（第 171 頁）的環境（第 226 頁）時便稱之爲土壤。土壤由顆粒組成，顆粒之間有水和空氣。土壤顆粒的結構對它的性質很重要。土壤的性質決定哪種植物可在此土壤生長。土壤是岩石（第 119 頁）風化（第 120 頁）成粒子（第 26 頁），再與有機（第 131 頁）物質混合後形成的。

土壤剖面 挖一個洞深入到岩石下面所看到的不同種類土壤的分佈層（第 187 頁）。

表土（名） 土壤的最上層（第 187 頁）土。它含有植物所需的各種營養素（第 171 頁）。如果因沖蝕（第 120 頁）而沖掉了表土，植物不能在遺留下的土壤生長。

底土（名） 表土（↑）之下的一層土。雖然下層土壤不能爲植物提供充分的營養素（第 171 頁），但它對保持水分不被排（↓）走是有用的。

黏土（名） 粒子（第 26 頁）直徑小於 0.01 毫米的土壤（↑），主要成分是矽酸鋁。水不易滲過黏土。

沙土（名） 粒子（第 26 頁）直徑在 0.1 毫米至 0.2 毫米之間的土壤（↑），主要成分是二氧化矽，水容易滲過沙土。（形容詞爲 sandy）

壤土；沃土（名） 一種由沙土與黏土組成的土壤（↑），最宜於植物生長。

腐殖土（名） 動、植物組織腐爛（第 146 頁）所產生的有機（第 131 頁）物質，它使土壤呈黑色，是植物生長最重要的土壤（↑）成分（第 96 頁），同時它有助於保持水分。

排水（動） 指水通過土壤流走或從地面流走。

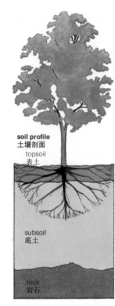

soil profile
土壤剖面
topsoil
表土

subsoil
底土

rock
岩石

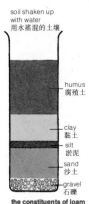

soil shaken up with water
用水搖混的土壤

humus
腐殖土

clay
黏土

silt
淤泥

sand
沙土

gravel
石礫

the constituents of loam
壤土的成分

manure (*n*) the faeces (p.169) and excreta (p.186) of animals when put on soil (↑). It helps to form humus (↑) and provides nutrients (p.171) for plants.

fertilizer (*n*) any substance which is added to the soil (↑) to provide nutrients (p.171) for plants. Manure (↑) and compost (↓) are natural fertilizers. Certain chemical substances are artificial fertilizers, e.g. ammonium sulphate is an artificial fertilizer which is changed to nitrates (p.118) for plants.

compost (*n*) decayed (p.146) plant material put on soil (↑) to provide nutrients (p.171) for plants.

agriculture (*n*) the processes of growing plants and keeping animals to obtain food and other products. To help in these processes, soil conservation (p.44) is needed so that nutrients (p.171) taken out of the soil by plants are put back by the farmer. Erosion (p.120) must also be prevented. **agricultural** (*adj*).

crop (*n*) any kind of plant, grown by agricultural (↑) methods, for food or for other products, e.g. rubber trees grown for rubber. Also the product produced, e.g. rubber is a crop.

pest (*n*) any organism, usually an insect, causing damage to crops (↑).

weed (*n*) a plant which is not wanted, growing in a crop (↑).

escape (*n*) a plant that was grown as a crop (↑) and is found growing wild. The seeds are dispersed (p.155) by wind or animals. The escape is not wanted and is like a weed.

edaphic factors the factors (p.227) in an environment (p.226) which are determined by the characteristics (p.147) of the soil (↑). They arise from the physical and chemical properties of the soil.

biotic factors the factors (p.227) in an environment (p.226) which arise from the activities of organisms (p.147).

climatic factors the factors (p.227) in an environment (p.226) which arise from the climate. The edaphic (↑), biotic (↑) and climatic factors form and determine the environment in a particular area.

pest 害蟲

solitary locust 獨居蝗蟲

swarming locust 婚飛蝗蟲

糞肥（名） 施於土壤（↑）上的動物糞便（第 169 頁）和排泄物（第 186 頁）。它有助於形成腐殖土（↑）和爲植物提供養料（第 171 頁）。

肥料（名） 加入土壤（↑）爲植物提供養料（第 171 頁）的任何物質。糞肥（↑）和堆肥（↓）是天然肥料。某些化學物質是人造肥料。例如磷酸銨是一種人造肥料，它能轉化成植物所需的硝酸鹽（第 118 頁）。

堆肥（名） 腐爛的（第 146 頁）植物物質堆在土壤（↑）上爲植物提供的養料（第 171 頁）。

農業（名） 栽培植物和飼養牲畜以獲得食物和其他產品的操作過程。爲有利於這些過程的進行，必須保持（第 44 頁）土壤，才能補充植物從土壤中攝取的養料（第 171 頁）。還必須防止土壤受侵蝕（第 120 頁）。（形容詞爲 agricultural）

農作物（名） 以農業（↑）方法栽培，用作食物和其他產品的任何植物。例如種植橡膠樹是爲了獲得橡膠。也指以農業方法生產的產品，例如橡膠是農作物。

害蟲（名） 任何損害農作物（↑）的生物體，一般指昆蟲。

雜草（名） 一種生長在農作物（↑）間的不需要的植物。

野化植物（名） 原先作爲農作物（↑）栽培後來變成野生的一種植物。其種子通過風和動物傳播（第 155 頁）。野化植物像雜草一樣是不需要的植物。

土壤因素 土壤（↑）特性（第 147 頁）所決定的環境（第 226 頁）因素（第 227 頁）。這些因素是由土壤的物理性質和化學性質產生的。

生物因素 由生物體（第 147 頁）的活動引起的環境（第 226 頁）因素（第 227 頁）。

氣候因素 由氣候引起的環境（第 226 頁）因素（第 227 頁）。土壤因素（↑）、生物因素（↑）和氣候因素形成並決定特定地區的環境。

cycle (*n*) changes and events which have no beginning and no end, but follow in turn, so that they are repeated continuously. All processes in nature (p.147) must follow a cycle, otherwise a natural process would come to an end.

carbon cycle all organic (p.131) material contains the element (p.103), carbon, and all organisms (p.147) also consist of carbon. A balance (p.18) is needed between carbon as carbon dioxide in the atmosphere (p.51) and carbon in organisms. Respiration (p.191) and decay (p.146) produce carbon dioxide. Green plants use carbon dioxide from the atmosphere to synthesize (p.136) carbohydrates (p.173) used in internal respiration to provide energy for life. The carbon cycle is shown in the diagram below.

循環（名） 指無始無終，一個接一個不斷重複的變化和活動。自然界（第 147 頁）中一切過程必須循環進行，否則自然過程就會終止。

碳循環 一切有機（第 131 頁）物質都含有碳元素（第 103 頁），所有的生物體（第 147 頁）都含有碳。大氣（第 51 頁）中二氧化碳的碳和生物體中的碳之間需要平衡（第 18 頁）。呼吸（第 191 頁）和腐爛（第 146 頁）產生二氧化碳。綠色植物則利用大氣中的二氧化碳合成（第 136 頁）碳水化合物（第 173 頁），在內呼吸中提供生命所需的能量。碳循環如下圖所示。

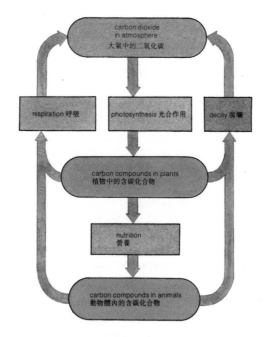

the carbon cycle
碳循環

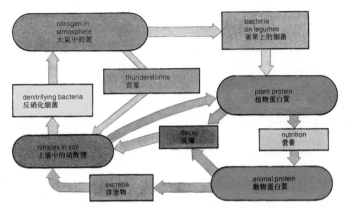

nitrogen in atmosphere 大氣中的氮

bacteria on legumes 莢果上的細菌

thunderstorms 雷暴

denitrifying bacteria 反硝化細菌

plant protein 植物蛋白質

decay 腐爛

nutrition 營養

nitrates in soil 土壤中的硝酸鹽

animal protein 動物蛋白質

excretia 排泄物

main cycle 主循環

other cycles 其他循環

the nitrogen cycle 氮循環

nitrogen cycle the element (p.103), nitrogen, is important as it is a constituent (p.96) of all proteins (p.172). Nitrates (p.118) in the soil (p.230) are used by plants to make proteins. Plant proteins are used by animals to produce the proteins they need for living. The nitrogen in protein is returned to the soil in animal excretion (p.186) as urine (p.186) or by the decay (p.146) of animal and plant tissues. Bacterial (p.145) caused decay produces nitrates from ammonia (p.186) set free from urea (p.186) and protein. The full cycle is shown in the diagram.

wilt (v) of plants, to lose turgor (p.139) so that the plant no longer stands upright. Wilting for a long time results in death.

succession (n) of plants, a change, with time, of the species in a community (p.226). It begins with the start of plant life, often single-cell plants, and grows towards a climax (↓).

sere (n) a plant succession (↑) in a particular kind of environment, e.g. a plant succession in water is a **hydrosere**; a plant succession in a dry environment, such as a desert, is a **xerosere**; a sere on uncovered rock is a **lithosere**.

climax (n) a plant community (p.226) at the end of a succession (↑); it is in equilibrium (p.20) with the environment (p.226) and there is no further change unless there is a change in the climate.

氮循環　氮元素（第 103 頁）十分重要，它是一切蛋白質（第 172 頁）的一種成分（第 96 頁）。植物利用土壤（第 230 頁）中的硝酸鹽（第 118 頁）製造蛋白質，而植物蛋白質被動物用來製造它們生存所需的蛋白質。蛋白質中的氮在動物排泄物（第 186 頁）成爲尿（第 186 頁）或由於動植物組織腐爛（第 146 頁）而返回土壤中。引起腐爛的細菌（第 145 頁）從尿（第 186 頁）和蛋白質釋放出的氨（第 186 頁）產生硝酸鹽。整個循環見圖示。

萎蔫（動）　指植物失去膨壓（第 139 頁）以致不能挺直。長時間枯萎會導致植物死亡。

演替（名）　指植物群落（第 226 頁）中種群隨着時間推移而變化。此變化從植物生命之始起，通常是由單細胞植物發展到演替頂極（↓）。

演替系列（名）　植物在一種特定的環境中的演替（↑），例如植物在水中演替是**水生演替系列**；植物在乾旱環境如沙漠的演替是**旱生演替系列**；在光禿岩石上的演替系列是**石生演替系列**。

演替頂極；顛峰（名）　指演替（↑）結束時的植物群落（第 226 頁）；它與環境（第 226 頁）處於平衡（第 20 頁），如果氣候不變化，就不會再進一步變化。

commensalism (*n*) the state in which animals of two different species (p.148) share the same living place and also the same food, e.g. a species of marine worm makes a hole in the sand; a small shrimp, also lives in the hole; both use the same food, but they have no effect on each other. The two animals live in a state of commensalism. **commensal** (*adj*).

symbiosis (*n*) the state in which organisms (p.147) of two different species (p.148), even a plant and an animal, live together and are useful to each other, e.g. (a) bacteria living in the roots of leguminous (p.155) plants provide nitrates for the plant, while the plant provides carbohydrates (p.173) and other food materials for the bacteria; (b) cattle carry ticks (arthropod parasites (↓)); some species of birds live on the cattle and eat the ticks. The cattle are useful to the birds as they provide food; the birds are useful to the cattle as they free them from ticks. The cattle and the birds live in a state of symbiosis. **symbiotic** (*adj*).

symbiont (*n*) one of the two organisms (p.147) living in symbiosis (↑).

mutualism (*n*) another name for symbiosis (↑).

parasite (*n*) an organism (p.147) living on, or in, another organism and obtaining all its food from its host (↓). Parasites may annoy or harm their host. Examples of parasites are: amoeba (p.143) living in the human gut (p.162); fleas living on human beings, dogs and cats. **parasitic** (*adj*).

host (*n*) a living organism (p.147) on which, or in which, a parasite (↑) lives.

saprophyte (*n*) an organism (p.147) which feeds on dead or decaying organisms. Saprophytes are mainly bacteria and fungi; they are very important as they complete the carbon and nitrogen cycles (p.233). **saprophytic** (*adj*).

epiphyte (*n*) a plant growing on another plant and using it only for support, i.e. it is not a parasite. Examples of epiphytes are ferns growing in the axils (p.158) of branches of trees. **epiphytic** (*adj*).

predator (*n*) an animal which obtains food by hunting other animals. **predatory** (*adj*).

片利共棲（名） 兩不同種（第 148 頁）的動物共生於一處，分享同一種食物的狀態。例如一種海生蠕蟲在沙上打洞，小蝦也住在洞內，二者吃同樣的食物，但互不影響。這兩種動物生活在片利共棲狀態。（形容詞爲 commensal）

共生（名） 兩不同種（第 148 頁）的生物體（第 147 頁），甚至是植物和動物，同生一起而能互惠。例如：(a) 生在豆科（第 155 頁）植物根部的細菌供給植物硝酸鹽，而植物則供給細菌碳水化合物（第 173 頁）和其他食料；(b) 在牛身上長的扁蝨（節肢寄生蟲）（↓），有幾種鳥類生活在牛隻身上以扁蝨爲食。這對鳥類有利，因爲牛爲其提供食物，而鳥類亦對牛有利，它們使牛免受蝨子咬。牛隻和鳥類處於共生狀態。（形容詞爲 symbiotic）

共生生物（名） 處於共生狀態（↑）的兩種生物體（第 147 頁）之一。

互利共生（名） 共生（↑）的別稱。

寄生蟲；寄生生物（名） 寄生在另一生物體內或體上並從此寄主（↓）獲取全部食物的生物（第 147 頁）。寄生蟲可打擾或傷害寄主。例如生活在人腸道（第 162 頁）內的變形蟲（第 143 頁），生活在人、狗和貓身上的蚤都是寄生蟲。（形容詞爲 parasitic）

寄主（名） 寄生蟲（↑）生活在其身上或體內的活生物體（第 147 頁）。

腐生植物；腐生生物（名） 以屍體或腐爛生物體爲食的生物體（第 147 頁），此種生物主要是細菌和真菌；它們對完成碳循環和氮循環（第 233 頁）極爲重要。（形容詞爲 saprophytic）

附生植物（名） 生長在另一植物上，僅用它作爲支持物而不是寄生的植物。例如長在樹枝腋（第 158 頁）中的蕨是附生植物。（形容詞爲 epiphytic）

捕食者；食肉動物（名） 靠捕食其他動物爲食的動物。（形容詞爲 predatory）

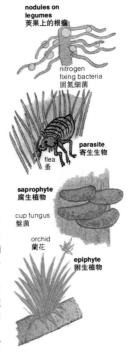

nodules on legumes
英果上的根瘤

nitrogen fixing bacteria
固氮細菌

parasite
寄生生物

flea
蚤

saprophyte
腐生植物

cup fungus
盤菌

orchid
蘭花

epiphyte
附生植物

prey (*n*) the animals attacked and eaten by a predator (↑).

hibernation (*n*) in many mammals (p.150), most reptiles (p.150) and amphibians (p.149) a state of physiological resting, during winter. The rate of metabolism (p.172) is very low, and the temperature of mammals drops to that of the atmosphere.

herbivore (*n*) an animal obtaining all its food from plants. **herbivorous** (*adj*).

carnivore (*n*) an animal obtaining all its food as meat from other animals. **carnivorous** (*adj*).

omnivore (*n*) an animal obtaining its food from both plants and other animals. **omnivorous** (*adj*).

plankton (*n*) very small plants and animals living in water; they float (p.36) in the water. Most of them live near the surface of the water where the plants get sufficient light for photosynthesis (p.159). Plankton are important in aquatic (p.228) food webs (p.171) as they provide food for fishes and for animals as large as whales.

phytoplankton (*n*) plants of plankton (↑), the main producers (p.236) in an aquatic (p.228) food web. They are not capable of locomotion (p.194), so they float (or drift) in the waters of seas and lakes. They are a source of food for fishes and zooplankton (↓).

zooplankton (*n*) animals of plankton (↑). They feed on phytoplankton (↑) and most are capable of locomotion by flagellae (p.144). They provide food for fishes.

nekton (*n*) the name given to all animals that swim in seas or fresh water. Fishes, crabs, octopus, are part of nekton.

benthon (*n*) animals and plants which live on the bottom of a sea or lake. They may be sessile (p.229) on the bottom, or they may move about on the bottom. The bottom stretches from high tide (p.127) mark to the greatest depths of water.

pelagic (*adj*) living in the open waters of a sea or lake as opposed to living on the bottom of the sea or lake. Plankton (↑) and nekton (↑) are pelagic organisms.

被捕食動物（名） 指被食肉動物（↑）獵食的動物。

冬眠（名） 許多哺乳動物（第 150 頁），大部分是爬蟲動物（第 150 頁）和兩棲動物（第 149 頁），在冬季時生理的休息狀態。此時新陳代謝（第 172 頁）率低，該哺乳動物的體溫下降到和大氣相同的溫度。

食草動物（名） 從植物中獲取全部食物的動物。（形容詞爲 herbivorous）

食肉動物（名） 獵取其他動物的肉爲全部食物的動物。（形容詞爲 carnivorous）

雜食動物（名） 既從植物也從其他動物獲取食物的動物。（形容詞爲 omnivorous）

浮游生物（名） 生活在水中並在水中浮游（第 36 頁）的極小的動、植物。大部分浮游生物生活在接近水面處，植物在此處可得到充分的陽光進行光合作用（第 159 頁）。浮游生物在水生（第 228 頁）食物網（第 171 頁）中是很重要的，因爲它們是魚類乃至大如鯨魚的動物的食物。

浮游植物（名） 浮游生物（↑）中的植物爲水生（第 228 頁）食物網中的主要生產者（第 236 頁）。它們不能行動（第 194 頁），只能在海域和湖泊中漂浮（漂流）；它們是魚類和浮游動物（↓）的食物來源。

浮游動物（名） 浮游生物（↑）中的動物。它們以浮游植物（↑）爲食，大部分能靠鞭毛（第 144 頁）行動。它們是魚類的食物。

自泳生物（名） 一切在海水或淡水中游泳的動物的總稱。魚類、蟹類、章魚都是自游動物。

水底生物（名） 生活在海底或湖底的動、植物。它們可以定棲（第 229 頁）在水底或可以在水底附近移動。海底從高潮（第 127 頁）的測標處一直延伸到水的最深處。

海洋的（形） 生活在海、湖開闊水域的而非生活在海底或湖底的。浮游生物（↑）和自泳生物（↑）都是海洋生物。

producers (*n.pl.*) green plants that use inorganic (p.116) materials to produce carbohydrates (p.173), proteins (p.172) and fats (p.175). Energy is obtained from sunlight through photosynthesis (p.159).

consumers (*n.pl.*) animals and plants without chlorophyll (p.159) that obtain their food from other organisms. Primary consumers are herbivores (p.235), secondary consumers are carnivores (p.235). Higher order consumers are large carnivores feeding on smaller carnivores.

decomposers (*n.pl.*) bacteria (p.145), fungi (p.145), and some protozoa (p.143) which cause the decay (p.146) of dead organisms (p.147) into inorganic (p.116) materials so the carbon, nitrogen, and other cycles can be completed.

trophic level the level of producer (↑) or consumer (↑) in a food chain (p.171). The first level contains producers; all other levels are consumers. Herbivores are primary consumers, smaller and larger carnivores are secondary and tertiary consumers. The energy available from carbohydrate in grass for farm animals is shown in the diagram above.

pyramid of numbers the number of organisms at each trophic level in a food web (p.171). At each level, energy is lost from organisms through respiration, heat radiation (p.45) and other metabolic processes (p.172). The energy left is passed on to the next trophic level. At each ascending level, less energy is available; the organisms are also larger, so for both reasons there are fewer organisms at a higher trophic level than at a lower level. *See diagram.*

pathogen (*n*) a bacterium (p.145), virus (p.145), fungus (p.145) or protozoan (p.143) causing disease. Such an organism (p.147) is called an agent of disease; it is a parasite (p.234).

infect (*v*) of pathogens, to enter an organism (p.147) and cause a condition of disease, e.g. a particular virus infects a man and he suffers from yellow fever. **infection** (*n*), **infectious** (*adj*).

infest (*v*) of parasites, to live in large numbers on an animal or plant. Parasites can also infest a room or clothes. **infestation** (*n*), **infested** (*adj*).

生產者（名、複） 用無機（第 116 頁）物質生產碳水化合物（第 173 頁）、蛋白質（第 172 頁）和脂肪（第 175 頁）的綠色植物。此種植物通過光合作用（第 159 頁）從陽光中獲取能量。

消費者（名、複） 從其他生物體中獲取食物的動物和不含葉綠素（第 159 頁）的植物。初級消費者是食草動物（第 235 頁），次級是食肉動物（第 235 頁）。比較高級的消費者是以較小的食肉動物爲食的大食肉動物。

分解者（名、複） 使生物（第 147 頁）屍體腐爛（第 146 頁）而變成無機（第 116 頁）物質，以完成碳、氮和其他循環的細菌（第 145 頁）、真菌（第 145 頁）和某些原生動物（第 143 頁）。

營養級 食物鏈（第 171 頁）中生產者（↑）或消費者（↑）的等級。第一級爲生產者，所有其餘各級均爲消費者。食草動物爲初級消費者，較小的和較大的食肉動物爲次級和三級消費者。農場牲畜從草中的碳水化合物得到能量，如上圖所示。

數塔；數量金字塔 食物網（第 171 頁）中各營養級內的生物數目。各級的生物由於呼吸、熱輻射（第 45 頁）和其他新陳代謝過程（第 172 頁）而失去能量。剩下的能量傳給下一營養級。營養級愈高，可獲得的能量就愈少，而生物則愈大，由於此兩原因，較高的營養級中的生物比較低營養級中的生物少。（見圖）。

病原體（名） 致病的細菌（第 145 頁）、病毒（第 145 頁）、真菌（第 145 頁）或原生動物（第 143 頁）。此種生物體（第 147 頁）稱爲疾病的媒介物，它是一種寄生生物（第 234 頁）。

傳染；感染（動） 指病原體侵入生物體（第 147 頁）內引起病狀。例如某種病毒傳染了人，使人患黃熱病。（名詞爲 infection，形容詞爲 infectious）

侵染（動） 指寄生蟲大量寄生在動、植物上，寄生蟲也可寄生在室內或衣物上。（名詞為 infestation，形容詞為 infested）

1 × 10⁷ kilocalories per year
每年 1 × 10⁷ 千卡

available for 可供
vertebrate 脊椎食
herbivores 草動物

other animals 其他動物
(snails, arthropods)
（螺、節肢動物）

bacteria 細菌
fungi 真菌
protozoa 原生動物

energy from 4000 m² grass
從 4000 m² 草取得的能量

energy available form grass
從草可獲得的能量

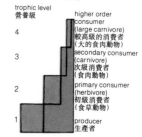

trophic level
營養級

4 — higher order consumer (large carnivore) 較高級的消費者（大的食肉動物）

3 — secondary consumer (carnivore) 次級消費者（食肉動物）

2 — primary consumer (herbivore) 初級消費者（食草動物）

1 — producer 生產者

energy lsot through respiration, heat radiation and other metabolic processes
經呼吸、熱輻射和其他代謝過程失去能量

energy available as food
以食物取得的能量

pyramid of available energy at the trophic levels of a food web
食物網各營養級上可用能量的金字塔

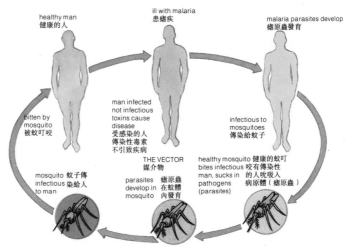

a vector transmits disease
媒介物傳染疾病

healthy man
健康的人

ill with malaria
患瘧疾

malaria parasites develop
瘧原蟲發育

bitten by mosquito
被蚊叮咬

man infected
not infectious
toxins cause
disease
受感染的人
傳染性毒素
不引致疾病

infectious to mosquitoes
傳染給蚊子

THE VECTOR
媒介物

mosquito 蚊子傳
infectious 染給人
to man

parasites 瘧原蟲
develop in 在蚊體
mosquito 內發育

healthy mosquito 健康的蚊叮
bites infectious 咬有傳染性
man, sucks in 的人吮吸入
pathogens 病原體（瘧原蟲）
(parasites)

vector[2] (n) an animal or a physical way of carrying a pathogen (↑) from an infected (↑) person to a new host (p.234), e.g. (a) a mosquito is a vector for a protozoan which is the causative agent (↓) of malaria; (b) contaminated (p.240) water is the physical vector for cholera, the causative agent of which is a bacterium (p.145).

causative agent the particular pathogen which causes a disease, e.g. *Plasmodium* is a protozoan which causes malaria; it is the causative agent of the disease.

toxin (n) any poison produced by a plant or animal, particularly by a bacterium. **toxic** (adj).

venom (n) any poison produced by an animal, and used when attacking other animals to poison them. **venomous** (adj).

antigen (n) a foreign substance, usually a protein or carbohydrate, which enters a vertebrate (p.148) body and stimulates a chemical reaction by the body. When a pathogen (↑) or its toxin (↑) enters the tissues, the animal produces antibodies (p.238). Some pathogens have the same antigen, while a particular pathogen may bear several antigens. Blood is an example of a material bearing an antigen (see **compatible**, p.184).

疾病媒介物（名） 將病原體（↑）從傳染（↑）病患者身上傳到新寄主（第234頁）身上之動物途徑或自然途徑。例如 (a) 蚊子是瘧疾病原體（↓）原生動物的傳病媒介物；(b) 污染的（第240頁）水是霍亂的自然傳染媒介物，霍亂的病原體是一種細菌（第145頁）。

病原體 引起某種疾病的特定致病菌，例如瘧原蟲是引起瘧疾的原生動物，是疾病的病原體。

毒素（名） 動、植物特別是細菌所產生的任何毒質。（形容詞爲 toxic）

毒液（名） 動物所產生用於攻擊其他動物的任何毒質。（形容詞爲 venomous）

抗原（名） 進入脊椎動物（第148頁）體內並刺激機體產生化學反應的一種外來物質，通常爲蛋白質或碳水化合物。當病原體（↑）或它的毒素（↑）侵入組織時，該動物體內產生抗體（第238頁）。某些病原體有同樣的抗原，而特殊的病原體可載有幾種抗原。血液是一種載有抗原物質的例子。參見**相容的**（第184頁）。

antibody (*n*) a chemical substance produced by lymphocytes, and other structures in the body of a vertebrate (p.148), when an antigen (p.237) enters the animal's tissues. The antibody combines chemically with the antigen and makes it harmless. An antibody generally combines with only one particular antigen. Cell walls of bacteria (p.145) bear antigens, and antibodies are formed to make the bacteria harmless; some antibodies make bacteria more easily attacked by phagocytes (p.183). Antibodies are carried round the animal's body by blood and lymph (p.182).

antitoxin (*n*) an antibody (↑) which combines with a toxin (p.237) to make it harmless.

vaccine (*n*) a liquid containing weakened or dead pathogens (p.236). When put in the body of a vertebrate (p.148), it causes antibodies (↑) to be produced, i.e. the same antibodies as combine with the antigens (p.237) of the pathogen.

antiserum (*n*) a liquid containing antibodies (↑); the liquid is serum (p.180) taken from an animal producing the antibodies.

inoculation (*n*) the putting of a vaccine or an antiserum into the blood of a vertebrate (p.148) animal. **inoculate** (*v*).

allergy (*n*) a very strong reaction by the body of an animal to a particular antigen, e.g. some people have an allergy to pollen (p.211); it causes the nose to produce large quantities of mucus (p.190). **allergic** (*adj*).

immunity (*n*) the ability of a plant or animal to resist the attack of antigens (p.237), or pathogens (p.236). The animal defends itself by: (a) its skin not allowing pathogens to enter; (b) the acid in its stomach destroying pathogens; (c) phagocytes (p.183) digesting pathogens; (d) antibodies making antigens harmless. Immunity is needed against each particular disease or each particular antigen, so an animal can be immune to some diseases and not to others. **immune** (*adj*).

immunize (*v*) to give a person, or animal, an inoculation (↑) of vaccine or serum which provides artificial immunity (↑) against a particular disease. **immunization** (*n*).

抗體（名） 當抗原（第237頁）侵入脊椎動物（第148頁）體內組織時，該動物體內的淋巴和其他結構所產生的一種化學物質。抗體可與抗原化學結合，使抗原對身體無害。一種抗體通常只和特定的抗原結合。細菌（第145頁）的細胞壁產生抗原，抗體的形成使細菌不再為害；一些抗體使細菌更易受吞噬細胞（第183頁）攻擊。抗體由血液和淋巴（第182頁）輸送到動物的全身。

抗毒素（名） 一種能與毒素（第237頁）結合並使之變成無害的抗體（↑）。

疫苗（名） 一種含有毒性已減弱的病原體（第236頁）或已死病原體的液體。疫苗注射入脊椎動物（第148頁）體內，引致產生抗體（↑），即能和病原體的抗原（第237頁）相結合的抗體。

抗血清（名） 含有抗體（↑）的一種液體；它是從產生抗體的動物身上提取的血清（第180頁）。

接種（名） 將疫苗或抗血清注入脊椎動物（第148頁）的血內。（動詞為 inoculate）

過敏變態反應（名） 動物體對某種抗原產生很強烈的反應。例如有些人對花粉（第211頁）有過敏反應，花粉會使其鼻腔分泌大量黏液（第190頁）。（形容詞為 allergic）

免疫力（名） 動、植物抵抗抗原（第237頁）或病原體（第236頁）侵襲的能力。動物自衛的辦法是：(a) 靠皮膚阻止病原體侵入體內；(b) 以胃酸殺死病原體；(c) 用吞噬細胞（第183頁）吃掉病原體；(d) 藉抗體使抗原成為無害的物質。對於每種特定的病或每種特定的抗原需要有一定的免疫力，所以一種動物對某些病可以免疫，而對另一些病則沒有免疫力。（形容詞為 immune）

使免疫（動） 給人或動物接種（↑）疫苗或血清，使之對某種疾病有人工免疫力（↑）。（名詞為 immunization）

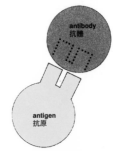

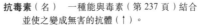

action of antibody and antigen
抗體和抗原的作用

acquired immunity this kind of immunity (↑) is obtained by a person when he suffers from a disease and gets well again. His blood contains antibodies (↑) against the pathogens (p.236) or toxins (p.237), which caused the disease, so that his body resists further attacks. The immunity is for a particular disease and the length of time for which he is protected varies with the disease, e.g. immunity against measles is for life, that against influenza is only for a short period.

後天免疫　人患病復原後獲得的一種免疫力（↑）。該人的血液裏含有抗體（↑），能對抗引致此種疾病的病原體（第236頁）和毒素（第237頁），其身體能抵抗受進一步的侵襲。後天免疫力是對某種疾病而言的，免疫時間的長短隨疾病而有所不同。例如麻疹是終生免疫性的，而流行性感冒則只在短期內免疫。

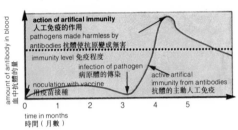

artificial immunity this kind of immunity (↑) is obtained by an inoculation (↑) and only lasts for a short time, e.g. between six months and six years, depending on the disease.

passive artificial immunity immunity obtained from an antiserum (↑). It is given to a person suffering from a disease to make him well.

active artificial immunity immunity (↑) obtained from a vaccine (↑). It is given to a healthy person, and prevents his being attacked by a particular disease by stimulating production of antibodies.

epidemic (n) a state of affairs in which a great many people are attacked by the same disease at the same time, so that many suffer from the disease. **epidemic** (adj).

endemic (adj) of diseases, always present in particular places or particular groups of people.

pandemic (adj) of diseases, spreading over the world.

sporadic (adj) of diseases, present from time to time in particular places, happening irregularly.

isolate (v) to keep a person separated from all other people so that he cannot infect (p.236) them.

人工免疫　由接種（↑）而獲得的免疫力（↑），免疫期短，例如在六個月到六年之間，視疾病而不同。

被動人工免疫　由於注射抗血清（↑）而獲得的免疫力（↑）。病人有了這種免疫力就能使疾病痊愈。

自動人工免疫　由於注射疫苗（↑）獲得的免疫力（↑）。健康人有了這種免疫力就能通過刺激產生抗體，從而預防某種疾病的侵襲。

流行病（名）　指很多人同時受到同一種疾病的侵襲，以致都患了同一種疾病的情況。（形容詞爲 epidemic）

地方性的（形）　指經常發生於一定地方或某些人羣中的疾病。

大流行的（形）　指世界範圍內傳播的疾病。

散發生的（形）　指不時存在於一定地方的，不定期發生的疾病。

隔離（動）　將一個病人與其他所有的人分開以免傳染（第236頁）給別人。

hygiene (*n*) the science of keeping healthy by cleanliness, proper diet, enough exercise, and not eating contaminated (↓) food or using contaminated water. **hygienic** (*adj*).

contaminate (*v*) to spread viruses (p.145), bacteria (p.145) or other pathogens (p.236) on food, water, clothes, or in the air, e.g. (a) houseflies, can contaminate food when they walk on it; (b) faeces from diseased persons can contaminate water.

pollute (*v*) to spread harmful or unpleasant substances in air, water or on the ground, e.g. (a) the smoke from fires pollutes the air; (b) animal faeces can pollute river water; (c) sea shores may be polluted by oil from ships. **pollution** (*n*).

antiseptic (*adj*) (1) describes chemical substances used on cuts and wounds to prevent pathogens (p.236) entering. (2) of conditions, preventing pathogens entering cuts and wounds. **antiseptic** (*n*).

aseptic (*adj*) of conditions, with no pathogens (p.236) present.

antibiotic (*n*) a substance which prevents bacteria reproducing; this allows the body's phagocytes (p.183) to destroy the bacteria. Antibiotics have no effect on most viruses (p.145), protozoa (p.143) or fungi (p.145). **antibiotic** (*adj*).

pasteurization (*n*) a process (p.129) of making liquids and food free from bacteria (p.145). The food or liquid is heated to a temperature at which the bacteria are killed, but the food or liquid is not spoilt.

anaemia (*n*) a condition in which the blood has too few red blood cells (p.179) or too little haemoglobin (p.179) in the red blood cells. The blood carries too little oxygen to the tissues, so a person has little energy. **anaemic** (*adj*).

leukaemia (*n*) a condition in which a person's blood contains too many white blood cells (p.180). The spleen (p.183) becomes very large and the person usually dies.

cancer (*n*) a growth of abnormal cells in epithelial (p.192) tissue. The growth increases with time and eventually may cause death. **cancerous** (*adj*).

衛生學（名） 講究清潔、注意適當的飲食、恰當的身體鍛鍊、不食受沾染（↓）的食物和不飲用髒水以保持身體健康的科學。（形容詞爲 hygienic）

沾染（動） 病毒（第 145 頁）、細菌（第 145 頁）或其他病原體（第 236 頁）散佈在食物、水、衣服被褥和空氣中。例如 (a) 蒼蠅在食物上爬過會沾染食物；(b) 病人排出的糞便會沾染水。

污染（動） 將有害的或氣味難聞的物質散佈在空氣中、水中或地上。例如 (a) 燒火發出的烟污染空氣；(b) 動物糞便會污染河水；(c) 船舶流出的油會污染海岸。（名詞爲 pollution）

抗菌的（形） (1) 描述用於傷口或創傷處以防止病原體（第 236 頁）侵入的化學物質；(2) 指能防止病原體侵入傷口或傷處的條件。（名詞爲 antiseptic）

無菌的（形） 指沒有病原體（第 236 頁）存在的環境。

抗生素（名） 防止細菌繁殖的物質；它使機體的吞噬細胞（第 183 頁）殺死細菌。抗菌素對大部分病毒（第 145 頁）、原生動物（第 143 頁）或真菌（第 145 頁）都無效。（形容詞爲 antibiotic）

巴氏消毒法（名） 除去液體和食物中細菌的方法（第 129 頁）。將液體或食物加熱到能殺死細菌（第 145 頁）的溫度。但食物和液體又不致受到破壞。

貧血病（名） 血液中紅血球（第 179 頁）太少或紅血球中的血紅蛋白（第 179 頁）太低的疾病。由於血液輸送到組織的氧太少，所以人的精力不足。（形容詞爲 anaemic）

白血病（名） 人的血液中含有過多白血球（第 180 頁）的疾病。病人脾臟（第 183 頁）腫大，通常都造成死亡。

癌（名） 上皮（第 192 頁）組織中異常細胞的增生。增生的速度隨時間而加快，最後導致死亡。（形容詞爲 cancerous）

pollution 污染

smoke 烟霧

factory 工廠

river 河流

effluent 廢水

atrophy (*n*) a decrease in size of an organ, or part of the body; usually caused by lack of use. **atrophy** (*v*).

infestation (*n*) the condition of being a host to parasites (p.234), other than pathogens (p.236), e.g. an infestation of fleas.

萎縮症（名） 器官或部分機體縮小，通常是由於不使用引起的。（動詞爲 atrophy）

侵擾（名） 成爲不是病原體（第236頁）的寄主而是寄生蟲（第234頁）的寄主的狀況。例如跳蚤的侵擾。

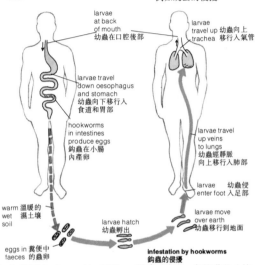

larvae at back of mouth 幼蟲在口腔後部

larvae travel up trachea 幼蟲向上移行入氣管

larvae travel down oesophagus and stomach 幼蟲向下移行入食道和胃部

hookworms in intestines produce eggs 鈎蟲在小腸內產卵

larvae travel up veins to lungs 幼蟲經靜脈向上移行入肺部

larvae enter foot 幼蟲侵入足部

warm wet soil 溫暖的濕土壤

larvae hatch 幼蟲孵出

larvae move over earth 幼蟲移行到地面

eggs in faeces 糞便中的蟲卵

infestation by hookworms 鈎蟲的侵擾

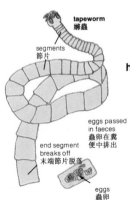

tapeworm 縧蟲

segments 節片

eggs passed in faeces 蟲卵在糞便中排出

end segment breaks off 末端節片脫落

eggs 蟲卵

tapeworm (*n*) a parasite (p.234) in the gut (p.162) of vertebrates (p.148). Its body is a chain of segments, each one complete in itself, *see diagram*. Each segment has both male and female gonads (p.210). The eggs of a tapeworm are passed out in the host's faeces.

hookworm (*n*) a parasite (p.234) in the gut (p.162) of man. Male and female hookworms fix themselves to the walls of the intestine and feed on the blood of their host. The male fertilizes (p.210) the female, and she produces eggs which are passed out in faeces. Larvae (p.151) hatch out of the eggs, grow, change into mobile (p.229) larvae, which enter the feet of human beings. Fully-grown hookworms in the gut are 2 to 3 cm long; they cause anaemia (↑) and general ill-health.

縧蟲（名） 脊椎動物（第148頁）腸道（第162頁）內的寄生蟲（第234頁）。蟲體是鏈狀節片，每一節片本身自成一體，（見圖）。每個節片兼具雄性腺（第210頁）和雌性腺。縧蟲卵隨寄主的糞便排出。

鈎蟲（名） 人的腸道（第162頁）內的寄生蟲（第234頁）。雌、雄鈎蟲本身固着在小腸壁上，以寄主的血液爲食。雄蟲鈎蟲授精（第210頁）給雌蟲，雌蟲產出的卵隨糞便排出。幼蟲（第151頁）從卵中孵出後，發育成爲能活動的（第229頁）幼蟲，它鑽進人足，侵入體內，在腸內發育成熟的鈎蟲體長2-3厘米；鈎蟲會引致貧血（↑）和一般性的健康不良。

INDEX 索引